Adapting to Climate Change: An International Perspective

Springer
New York
Berlin
Heidelberg
Barcelona
Budapest
Hong Kong
London
Milan
Paris
Santa Clara
Singapore
Tokyo

Joel B. Smith   Neeloo Bhatti   Gennady V. Menzhulin
Ron Benioff   Max Campos   Bubu Jallow
Frank Rijsberman   Mikhail I. Budyko   R.K. Dixon

EDITORS

Adapting to Climate Change: An International Perspective

Springer

Joel B. Smith
Hagler Bailly Consulting, Inc.
Boulder, CO 80306
USA

Neeloo Bhatti
Argonne National Laboratory
Argonne, IL 60439
USA

Gennady Menzhulin
State Hydrological Institute
St. Petersburg
Russian Federation

Ron Benioff
U.S. Country Studies Program
Washington, DC 20585
USA

Max Campos
Central American Project on Climate Change
San Jose
Costa Rica

Bubu Jallow
Department of Water Resources
Banjul
The Gambia

Frank Rijsberman
Resource Analysis
Delft
The Netherlands

Mikhail I. Budyko
State Hydrological Institute
St. Petersburg
Russian Federation

R.K. Dixon
U.S. Country Studies Program
Washington, DC 20585
USA

The information in this document has been subjected to technical peer review, but does not necessarily reflect the official views of any governmental or intergovernmental body.

CIP data available.

Printed on acid-free paper.

Softcover reprint of the hardcover 1st edition 1996

Acquisitions Editor: Robert C. Garber
Production managed by Laura Carlson; manufacturing supervised by Joe Quatela.
Camera-ready copy prepared by the editors.

9 8 7 6 5 4 3 2 1

ISBN-13:978-1-4613-8473-1 e-ISBN-13:978-1-4613-8471-7 SPIN 10522981
DOI: 10.1007/978-1-4613-8471-7

Introduction

Global climate change is one of the most important environmental issues facing the world today. The United Nations Framework Convention on Climate Change (FCCC) acknowledges the potential for global climate change to have major effects on the world economy. The work of the Intergovernmental Panel on Climate Change (IPCC) is focused on evaluating the scientific data on climate change and analyzing the potential responses to it.

One of the primary issues in the global climate change debate is how to adapt to any change that might occur. The process of identifying adaptation measures and evaluating their effectiveness is the focus of this book.

In dealing with climate change adaptation, the sequence of events in conducting these types of analyses can be generalized as follows:

- Develop scenarios for the possible range of climate change,
- Assess the vulnerability of various sectors of the national economy and infrastructure to climate change, and
- Identify and evaluate measures in each sector to adapt to the climate change

It is this third step that is the subject of this book. In presenting this material, Chapter 1 gives an overview of the concept of climate change adaptation and the general principles guiding the conduct of analyses in this area. Chapters 2-7 give the results of evaluating climate change adaptation options in the agriculture, water resources, coastal resources, forest and ecosystems, fisheries, and human settlements sectors. While these do not represent all sectors that will need to consider adaptive measures, they do reflect those which are considered to be among the most sensitive to global climate change.

It will be evident from the work presented here that the state-of-the-art of climate change adaptation assessment is evolving. The wide range of uncertainty in the magnitude of climate change itself makes the adaptation assessments even more difficult to carry out. Nevertheless, it is clear that this work can lay the foundation for countries to develop appropriate and cost-effective steps to adapt to climate change

On May 22-25, 1995, more than 100 scientists, engineers, and policy analysts representing 30 nations convened in St. Petersburg, Russian Federation, to discuss the methodologies for assessing adaptation to climate change and to share the results of country studies designed to determine the effectiveness of various approaches. The International Conference on Climate Change Adaptation Assessments, the first international conference to deal exclusively with issue, was chaired by S. Avdjushin (Russian Federation), R. Dixon (USA), and F. Rijsberman (Netherlands). It was cosponsored by the Russian Federal Service for Hydrometeorology and Environmental Monitoring, the U.S. Country Studies Program, and the Directorate General International Cooperation of the Netherlands Ministry of Foreign. The papers in this book represent the results of work by the participating countries in researching climate change adaptation. The conference statement, the "St. Petersburg Statement on Adaptation to Climate Change," is included here.

Conference Summary Statement

Approximately one hundred scientists, engineers, and policy analysts, representing about 30 nations, convened in St. Petersburg, Russia, from May 22–25, 1995, at the International Conference on Climate Change Adaptation. They discussed methodologies for assessing climate change adaptation and presented preliminary results of country studies dealing with adaptation to climate change.

The Conference agreed to the following definition of adaptation to climate change:

> *Adaptation to climate change includes all adjustments in behavior or economic structure that reduce the vulnerability of society to changes in the climate system.*

The discussions concerning response policies to climate change in the framework of the Intergovernmental Panel on Climate Change (IPCC), the Intergovernmental Negotiating Committee (INC), and the Conference of the Parties (COP) of the U.N. Framework Convention on Climate Change (UNFCCC) have emphasized stabilizing ambient greenhouse gas concentrations. The conference participants urge the Parties to the UNFCCC to consider more fully the development, evaluation, and costing of adaptation strategies, as well, to properly consider mitigation and adaptation options for all countries.

The conference participants reached the following conclusions in their deliberations related to adaptation to climate change in general, as well as adaptation for specific sectors of human activity. These conclusions do not necessarily reflect views of their governments and are offered to support preparation for adaptation as recognized in the UNFCCC (Art. 4.1(b)).

The IPCC Technical Guidelines for Assessing Climate Change Impacts and Adaptation were discussed at the conference and found to be a good framework for country study adaptation assessments. The draft IPCC working group II report now being circulated for comments, however, does not provide a particularly broad overview of adaptation options open to countries. Many more options, particularly related to anticipatory adaptation, are expected.

The experience that regions and peoples have had in adapting to their current climatic conditions for centuries can be a rich source of inspiration for developing adaptation response policies. This experience also shows that an increase in sustainability of economic development is likely to go hand in hand with increased capabilities to cope with adverse climate conditions and changes therein. Incorporating this formal (and informal) knowledge and experience into the adaptation assessment process will require a bottom-up approach that could benefit from the use of participatory planning procedures.

The preparation for climate change adaptation will require more quantitative analysis, including identification, analysis, costing, and evaluation of a broad spectrum of adaptation options related to ecosystems, human health, economy, and society, as well as support for effective communication of analysis results to stakeholder groups to increase their social acceptability. Feasible adaptation schemes must also address current anthropogenic stresses.

The UNFCCC recognizes the importance of adaptation (Art. 3.3.) and calls on Parties to prepare for adaptation as well as mitigation (Art. 4.1(b)). Funding through the Global Environment Facility (GEF), the interim financial mechanism of the UNFCCC, for developing countries that wish to address adaptation is, to date, limited to planning for adaptation (Stage I) as part of the national communications. Nevertheless, it is considered important that countries start to determine the cost of the options to adapt to climate change. The participants urge the COP of the UNFCCC to broaden the GEF mandate related to adaptation.

Recommendation

The participants in the International Conference on Climate Change Adaptation call on the Parties to the UNFCCC to pay more attention to the options to adapt to climate change, while continuing their efforts to stabilize greenhouse gas concentrations. The participants call on the IPCC to provide more guidance on the options available to countries to adapt to climate change and to assess their adaptations.

Acknowledgments

The participants wish to thank the sponsors of the meeting, the Russian Federal Service for Hydrometeorology and Environmental Monitoring, the U.S. Country Studies Program, and the Directorate General for International Cooperation of the Netherlands Ministry of Foreign Affairs, for their support in organizing this important meeting.

Contents

7 Human Settlements

1

OVERVIEW

Rapporteur's Statement

Joel B. Smith
Hagler Bailly Consulting, Inc.
P.O. Box 0
Boulder, CO 80306-1906, USA

Neeloo Bhatti
Argonne National Laboratory
9700 S. Cass Avenue, Bldg. 900/DIS
Argonne, IL 60439 USA

Both natural systems and human infrastructure must adapt to changes in climate in order to survive. The key questions to be answered in any analysis of climate change adaptation are:

- What adaptation will occur naturally?
- What intervention strategies designed to supplement autonomous adaptation should be considered?
- How can the best intervention strategies be evaluated?

This chapter provides an overview of the methodologies for analyzing adaptation strategies that have been applied around the world. Some of the important issues that have been identified from this experience are listed below:

Autonomous Adaptation. Some adaptation to climate change will occur without any special intervention by human institutions. Ecosystems will naturally adjust to changes in temperature, precipitation, water supply, and other climate-related factors. The rate of change of climate will determine how well these systems will adjust. A rapid rate of change may seriously inhibit the ability of some ecosystems to survive in their current regions.

In addition to the natural adaptation of ecosystems, human activities will adjust to climate change. For example, farmers will modify their planting dates, their crop selection, and their irrigation and fertilizing patterns as climate changes. Again, the rate of climate change will determine the success of these modifications.

When the autonomous adaptation is not effective enough to prevent serious damage to either ecosystems or to human infrastructure, additional intervention may be necessary.

Identification of Strategies. The process of identifying where additional adaptation intervention will be necessary will vary from country to country. Some sectors (e.g., agriculture, water resources, coastal resources) will be very sensitive in some countries and not sensitive in others. A systematic identification of alternative strategies is necessary to select the proper set of options for consideration.

Methodologies for Analyzing Adaptation Strategies. A number of different techniques have been employed for analyzing adaptation strategies. All have the common element of using a wide range of evaluation criteria. The criteria include population at risk, severity of impact, economic losses, ecosystem damage, and others. All methodologies address the need to measure the effectiveness of the adaptation strategies against these multiple criteria and to incorporate the information into a concise form useable by decisionmakers. The complexity of dealing with conflicting objectives and multiple criteria is recognized as one of the most difficult issues in conducting adaptation analyses.

Implementation of Adaptation Strategies. The implementation of appropriate adaptation strategies is the desired outcome of adaptation assessments. It is also the activity that involves the most risk, both financially and politically. Proper communication of adaptation assessment results to decisionmakers and the public is crucial in moving to successful implementation. The most acceptable adaptation strategies are likely to be those that have other economic benefits beyond what is to be gained by forestalling damage from climate change or have minimal costs.

Adaptation and Economic Development. Many adaptation strategies will have implications for a country's overall economic development. These implications may be positive or negative. There are advantages to a country including climate change adaptation into its overall development planning at an early stage to avoid the need for costly adjustments at a later date.

International Cooperation. Many of the adaptation strategies require cooperation across national boundaries (e.g., in river basins). There is a need for countries to collaborate on both the analysis of adaptation measures and on their eventual implementation.

An Overview of Adaptation to Climate Change

M. Toman and R. Bierbaum
Council of Economic Advisers, Office of Science and Technology Policy
Executive Office of the President
Old Executive Office Building, Room 328
Washington, DC 20500, USA

Abstract

Enhancing adaptation to climate change, including adjusting to altered biological, technical, economic, institutional, and regulatory conditions, is a crucial aspect of national and international efforts to better understand and respond to the risks posed by climate change. Socioeconomic, as well as ecological or technological, issues arise in judging the potential for adaptation with respect to different natural systems. In discussing attempts to better understand adaptation potential and to expand its scope internationally, six basic lessons emerge: (1) ecological and socioeconomic impacts are as important as changes in atmospheric chemistry; (2) beneficial climate change policies include opportunities for reducing risks from future climate change; (3) adaptation is a key complement to reduction of greenhouse gas accumulations; (4) adaptation potential varies significantly across natural and human systems and depends crucially on both financial capacity and social infrastructure; (5) a number of adaptation options can be pursued at relatively low cost, especially those that involve correcting existing economic inefficiencies; and (6) government has an important role in promoting effective research and development to increase adaptation options, in lowering domestic barriers for cost-effective adaptive measures, and in promoting low-cost international cooperation to address the global consequences of climate change.

Introduction

Considerable domestic and international attention has focused on possibilities for reducing the risks of climate change by limiting future greenhouse gas (GHG) emissions. However, adaptation is a crucial aspect of national and international efforts to better understand and respond to the risks posed by climate change. Adaptation to climate change includes any adjustment to altered conditions, such as biological, technical, economic, institutional, or regulatory.

Adaptation is especially important in view of the growth of atmospheric concentrations of GHGs (Intergovernmental Panel on Climate Change 1994). If current emission trends continue, atmospheric concentrations by the middle of the next century will be large enough to increase the average global temperature by 1.5-4.5°C. Even if global emissions were to remain at 1990 levels indefinitely, growth in GHG concentrations still would change average temperature, rainfall, and sea level. To stabilize atmospheric concentrations at less than twice their current levels requires a significant decrease in global emissions below current levels, and decades will be required for the world to achieve a new climate steady state. Because the potential ecological and socioeconomic impacts of adjusting to the new steady state are great, it is crucial to better understand opportunities for adaptation and to promote actively the dissemination and use of adaptation strategies.

The following sections examine (1) development of the concept of socioeconomic, as well as ecological or technological, adaptation, to climate change; (2) a brief survey of some of the most significant issues in judging the potential for adaptation with respect to different natural systems; and (3) ways to better understand adaptation potential and to expand its scope internationally.

Ecological and Human Adaptation

For years, scientists have focused on understanding the climatic consequences of the increasing atmospheric concentrations of GHGs. Such efforts are important in understanding and managing global change, but other issues also need to be examined. For human well-being and the design of policies to enhance that well-being, climate change is important primarily because it can significantly alter the characteristics of natural systems. Thus, climate change analysis is incomplete without assessing ecological consequences.

Humankind is concerned with the state of ecological systems because of the diverse benefits derived: extracted materials (e.g., food, fiber, wood products), transport and processing of waste products, stores of biodiversity, and nonconsumptive uses (e.g., recreation, intrinsic cultural and spiritual values). Ultimately, climate change is intimately connected to both socioeconomic and ecological issues.

Figure 1 illustrates these interconnections, while introducing the roles of adaptation and mitigation. The top four boxes illustrate the loop from GHG emissions to climate responses to natural system responses to socioeconomic consequences, which in turn affect emissions. As socioeconomic consequences are manifested, human behavior and institutions adapt. From a policy perspective, however, it is important to address how the anticipation of future climate

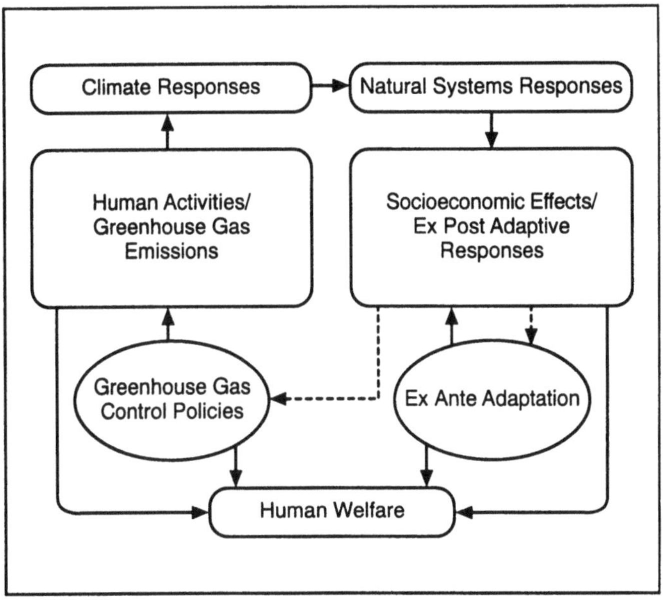

Figure 1. Climate Change Impacts, Adaptations, and Mitigation (adapted from Darmstadter and Toman 1993)

change consequences can set adaptive responses in motion today, actions that will help to reduce or absorb the effects of future changes in natural systems. This anticipatory effect is shown in Figure 1 by the dashed arrow. Similarly, the anticipation of climate change damages stimulates actions to control emissions of GHGs.

How do policymakers decide which adaptive measures to pursue and how intensively to undertake them? A broad conceptual answer to this question is to undertake those measures whose "advantages" are great enough to justify their costs. Advantages can encompass a variety of monetary, physical, and qualitative measures. In practice, applying this criterion becomes challenging because of (1) the large uncertainties surrounding potential climate change damages, (2) the slightly smaller uncertainties surrounding the potential performance of different adaptive measures, and (3) the economic and ethical questions that arise in balancing interests among different nations and between current and future generations. Nevertheless, adaptive measures ultimately must be chosen.

One of the important benefits of attempting to enhance adaptation (as well as reducing GHGs) is the decrease in uncertainty about potentially costly adverse effects in the future. Efforts to expand and share knowledge also are important steps toward improving decision making. In addition, many adaptive measures involve correcting existing economic inefficiencies that would probably inhibit adaptation to climate change (e.g., subsidized water use). For such measures, it is possible to compare tangible benefits to costs, as well as to attempt the more speculative investigation of adaptation benefits.

Other factors influence the assessment of costs and benefits of improved adaptation capacity. One of the key lessons emerging from studies of sustainable development is the need to confront environmental challenges at the appropriate geographic and temporal scale (e.g., Toman 1994). Ultimately, the services derived from natural systems depend on how well underlying ecological processes (e.g., water, nutrient, energy cycling) operate. These processes, in turn, operate at different spatial scales. Also, many smaller-scale processes tend to respond to stimuli more rapidly (so that temporal and spatial scales are related). Consequently, overcoming negative shocks to small-scale natural changes is generally easier and faster than when the impacts are wide ranging and long lived.

Changes in the composition of the atmosphere clearly constitute a global phenomenon, although the effects of climate change may be very different for various locations and natural systems. Such changes are long lived and, therefore, lead to problems of intergenerational equity as well as issues of fairness within the current generation. Thus, the potential effects of climate change must be approached with care.

Adaptation Potential and Challenges

The flexibility and robustness of natural systems are critical components in assessing the ultimate socioeconomic consequences of climate change. As discussed in more detail below, three basic principles underlie current knowledge about the potential for natural-systems adaptation:

- Highly managed systems (such as agriculture), given sufficient resources, are likely to be more adaptable (and at a lower cost) than less-managed ecosystems.
- Capacity for adaptation to a particular stress in any system greatly depends on (1) the level of understanding of ecosystem processes and options for preserving the flows of services provided by them, (2) the degree to which this knowledge is diffused among the many decision makers who are ultimately responsible for the functioning of natural systems and for the capacity of these systems to provide human benefits, and (3) the level of financial and human resources available to support adaptive actions and research to increase options.
- Adaptive potential is likely to be greater in countries where levels of capital, stores of human knowledge, and social institutions permit greater attention to adaptive efforts. Economic development that is sensitive to the performance of natural systems (one possible definition of sustainable development) is a powerful tool for promoting adaptation to climate change.

Table 1 elaborates these points and summarizes the adaptation potential of several systems, especially in developed countries. Two systems with the least human management — natural landscapes (such as wilderness areas) and marine ecosystems (including coastal wetlands) — have the most limited prospects for adaptation. Systems with greater human involvement — water resources, coastal structures, managed forests and grasslands, and agricultural systems — show more adaptation potential at some cost, depending on the underlying social capacity for adaptation. In fact, adaptation of agriculture may be even "probable," not just "possible," at least in societies with the necessary financial and technological means. A somewhat less optimistic view generally is appropriate where the means for adaptation are limited, as in many developing countries. (See Strzepek and Smith [1995] for further discussion.)

Table 1 also indicates that industry and human health are not particularly sensitive to climate change. A less optimistic view of the health risks may be appropriate (in terms of Table 1, placing health risks in the category of "adaptation possible at a cost" or even "limited adaptation"). For example, tropical diseases may spread north if temperature increases; greater heat-related mortality may also occur. Health risks may generally increase if fluctuations in climate conditions (like El Niño events) become more frequent.

Climate change also affects other human activities and values (Table 1). For example, climate change places stress on recreational capacities, thereby affecting tourism. Stress caused by human migration also may intensify.

Table 1. Sensitivity and Adaptability of Natural Systems and Human Activities/Values

Object of Concern	Low Sensitivity	Adaptation Possible at Some Cost	Limited Adaptation Potential
Agriculture		X	
Water resources		X	
Coastal settlements		X	
Managed forests and grasslands		X	
Natural landscapes			X
Marine ecosystems			X
Industry	X		
Health	X		
Recreation		X	
Migration		X	

Note: See text for discussion of caveats.
Source: U.S. National Academies of Science and Engineering/Institute of Medicine (1992).

A detailed synthesis of knowledge regarding adaptation potential to climate change provides further insight into the problems faced by agriculture, water resources, coastal settlements, forests, wetlands, and wilderness areas (Office of Technology Assessment 1993). The first three systems are more heavily managed and are expected to show a greater capacity for adaptation than the second three. The discussion below is a compilation of average tendencies; substantial regional variations are likely, and, consequently, there will be both winners and losers. Interdependencies across the categories (e.g., disruption of natural ecosystems could increase the risk of disease outbreaks) are not addressed.

Agriculture

Both temperature changes and changes in soil moisture from altered precipitation are likely to put pressure on agricultural activity on regional scales. These changes are likely to alter the range of a number of major food crops. The result need not be global strains on food supply, especially if the fertilization effect of higher carbon dioxide (CO_2) levels in the atmosphere is taken into account. Nevertheless, regional climatic shifts will alter the distribution and intensity of farming and require compensation to maintain agricultural productivity. Compensation includes continued research and dissemination of knowledge regarding crop strains suited to changed climate (e.g., more drought-tolerant wheat); removal of regulatory and institutional barriers to more efficient farm practices (e.g., reduction of distorted prices for crops, water, and fertilizer; regulations that encourage monoculture; and trade restrictions); and better contingency planning for natural disasters, which may increase in severity with a changed climate.

Water Resources

Even without the risk of climate change, effective management of water quantity and quality is a global concern. A drier climate in many areas will only increase the technical and environmental challenges of providing adequate supplies to growing populations (e.g., trade-offs between diversionary uses and such in-stream values as hydropower, navigation, and ecological health). More intensive use of existing water supplies, including groundwater, will increase the challenges to human and ecological health posed by water contamination. More efficient management of water supplies and demands, supported by economic incentives that allow water to be allocated to the highest valued uses and by better contingency planning for droughts and storms, is a critical first step. In the longer term, more efficient water management will affect the intensity of land use by limiting demands for scarce resources.

Coastal Settlements

Coastal areas obviously are vulnerable to inundation from a rising sea level in a warmer world, as well as to any increase in the incidence of violent storms. Human settlements can gain a measure of protection from flood-control works, although they are an expensive option. Ultimately, it may be more cost-effective in some areas to relocate inland; in other circumstances, relocation may be the most costly option. The costs of adaptation can be lessened by altering current land-use decisions through limits on indirect subsidies for coastal development, such as no-cost disaster insurance.

Natural Ecosystems: Forests, Wetlands, and Wilderness Areas

Even if coastal human settlements can adapt, coastal wetlands and other natural features of importance may not meet the challenge of rising sea level, given the uncertainties about ecosystem dynamics and the barriers imposed by human infrastructure (e.g., highways). Inland wetlands also face serious challenges; some wetlands could simply disappear in a drier climate. Policies that slow the ongoing loss of critical wetlands to development, reductions of other human-induced stresses (e.g., water pollution), and greater scientific understanding of wetland functions and how they can be restored are needed to address these problems.

Climate change is also likely to alter the natural range of many forest species. For managed forests (those cultivated for timber products), relocation is feasible in principle, but at a cost reflecting the acquisition or conversion of new lands, uncertain yields during the transition, and losses in productivity of existing stands stressed by the changing environment. The availability of seed banks and ongoing research into management practices in the face of a changing climate are high priorities. The transition for unmanaged forests — sources of biodiversity, recreation and spiritual values, and such important basic ecological functions as water management — would not be easy, because these ecosystems are more complex and less well understood and because of the problems in acquiring and maintaining land areas that allow natural progression with the climate change. Simply maintaining "island" forest sites may not be adequate, and the cost and uncertainty in maintaining corridors for migration of flora and fauna would be high. These problems also arise in preserving other wilderness areas.

Common themes in these assessments are the need for improved knowledge about ecosystem sensitivities, for technical innovation in promoting the physical adaptation of the system or coping with its reduced productivity, for dissemination of this information to those whose decisions affect natural systems, and for deployment of economic incentives to reduce inefficient use of "natural capital" as well as to stimulate innovation. Many of these options have substantial benefits in addition to their capacity for expanding adaptation potential. Such policies yield benefits in many different climatic outcomes, and the response measures can adapt to new information about climate change risks and potential responses.

These conditions for success in adaptation efforts underscore the dilemma facing many less affluent countries in developing their own adaptation strategies. Such countries may be disproportionately vulnerable to the effects of climate change because their natural systems play a relatively large role in economic activity. In countries where investment capital is scarce relative to meeting basic human needs, infrastructure for knowledge acquisition and dissemination is low, and economic and other social institutions are weak, expanding the capacity for adaptation is difficult. For example, agricultural systems may be less adaptable in agrarian societies; enhanced efficiency of water use in the face of reduced stream flows may be hard to establish; and protection from health risks will be difficult with limited public health resources and lack of air-conditioning.

The success of both overall economic development and targeted adaptive measures depends substantially on the efforts of the implementing countries. However, the advanced industrialized countries can and should support these efforts by sharing knowledge and providing access to analytical methods and adaptation technologies. The next section offers some suggestions for expanding such efforts.

Expanding Adaptation Potential

U.S. Adaptive Measures

The 1994 *Climate Action Report* submitted by the United States to the U.N. Framework Convention on Climate Change lists actions that the United States is undertaking to promote greater resilience of natural and socioeconomic systems facing the risk of climate change. Many of these steps are scientific in nature: research and development of new row crop and tree varieties, identification of wetlands at greatest risk, and improvement of the basic understanding of functions of unmanaged ecosystems. Other efforts are more operational in nature: coastal land-use restrictions, reform of disaster relief and insurance programs to effect more desirable land-use decisions, and programs to improve water management and allocation.

Many of these efforts are at an early stage, and a number of additional steps may become desirable as knowledge about the damages of climate change and the costs of alternative response options grows. Of particular importance are "low regret" strategies — those strategies that make sense for society even without climate change risks. These strategies could include reforms of agricultural programs and further progress in the management of water resources and wetlands. The fate of many of these efforts is unclear, especially in light of budget constraints.

Reorganization of Global Change Research

The U.S. Global Change Research Program (GCRP) was funded at $1.4 billion in fiscal year (FY) 1994 and $1.8 billion in FY 1995, but its future is now uncertain. The GCRP has undergone several evolutionary changes since its inception. Starting with a primary focus on atmospheric science, the GCRP has expanded to include impacts, mitigation, and adaptation. Overlying this expanded list are several cross-cutting issues related to ecosystem management and scale, environmental technology, and such science policy tools as risk analysis, socioeconomic issues, and data needs. Although debate continues in the United States regarding the allocation of resources among basic science, social science, and applied policy analysis, these changes in the GCRP seem beneficial.

International Cooperation

As described in the *Climate Action Report* (1994), the United States is engaged in bilateral activities to support GHG mitigation or sequestration in other countries. In addition to these efforts, the Country Studies Program supports sharing ideas as well as encouraging efforts by individual countries to review their vulnerabilities and adaptation potential. With activities involving more than 50 countries under way, the Country Studies Program is a good example of productive, low-cost international cooperation ($35 million appropriation).

The GCRP also supports U.S. participation in a number of international scientific research activities, including the World Climate Research Program, the International Geosphere-Biosphere Program, the Human Dimensions Program, and the Intergovernmental Panel on Climate Change. Participation in these programs can only serve to promote better understanding of adaptation. In addition, the United States supports the Climate Technology Initiative discussed at the Conference of Parties in Berlin. This program is intended to expand markets for cost-effective, climate-friendly technologies and access of developing countries to these markets. The program could support technical progress in adaptation as well as in the development and dissemination of cost-effective technologies that reduce emissions. Again, the fate of these international programs is somewhat uncertain in the current political climate. However, these programs seem to be avenues for reducing climate risks at affordable costs.

Conclusions

Some basic vulnerabilities of natural systems can be expected with a changing climate and some strategies will be needed to cope with these challenges at national and international levels. The following are some of the important challenges discussed here:

- Ecological and socioeconomic impacts are as important as changes in atmospheric chemistry.
- Beneficial climate change policies include opportunities for reducing risks from future climate change.
- Adaptation is a key complement to reducing GHG accumulations.
- Adaptation potential varies significantly across natural and human systems and depends crucially on both financial capacity and social infrastructure.
- A number of adaptation options could be pursued at relatively low cost, especially those that involve correcting existing economic inefficiencies that limit adaptation potential.
- Government has an important role in promoting effective research and development to increase adaptation options, in lowering domestic barriers for cost-effective adaptive measures, and in promoting low-cost international cooperation to address the global consequences of climate change.

The success of international adaptation efforts ultimately will be measured by their ability to expand the capacities of all participating countries to better understand and manage the effects of a changing climate relative to other pressing economic and environmental concerns. Future generations will benefit from how well countries have addressed the potential for climate change.

Acknowledgments

The views expressed here are the authors' alone and should not be attributed to the U.S. government. Helpful comments by Joel Darmstadter, Samuel Fankhauser, Jonathan Patz, and Joel Smith are gratefully acknowledged.

References

Climate Action Report, 1994, U.S. Government Printing Office, Washington, D.C., USA.

Darmstadter, J., and M.A. Toman, 1993, *Assessing Surprises and Nonlinearities in Greenhouse Warming*, Resources for the Future, Washington, D.C., USA.

Intergovernmental Panel on Climate Change (IPCC), 1994, *Climate Change 1994,* Cambridge University Press, Cambridge, United Kingdom.

Office of Technology Assessment, U.S. Congress, 1993, *Preparing for an Uncertain Climate,* OTA-O-567, U.S. Government Printing Office, Washington, D.C., USA (Oct.).

Strezepek, K.M., and J.B. Smith (eds.), 1995, *As Climate Changes,* Cambridge University Press, New York, N.Y., USA.

Toman, M.A., 1994, "Economics and 'Sustainability': Balancing Tradeoffs and Imperatives," *Land Economics* 70(4):399-413.

U.S. National Academies of Science and Engineering/Institute of Medicine, 1992, *Policy Implications of Greenhouse Warming: Mitigation, Adaptation, and the Science Base*, National Academy Press, Washington, D.C., USA.

Past Changes in Climate and Societal Adaptations

M.I. Budyko
State Hydrological Institute
23 Second Line
St. Petersburg, 199053, Russian Federation

Abstract

Humankind will probably face a grave food shortage in the first half of the next century. Anthropogenic climate warming may help to solve this problem. In the past, high carbon dioxide (CO_2) concentrations were associated with noticeably higher air temperature, an increase in precipitation, and more favorable conditions for most plants. Increased crop productivity due to changes in CO_2 concentration and agroclimatic conditions could provide enough food for a larger population and prevent an ecological catastrophe. By using climate data from the warm epochs of the Quaternary and the Pliocene, it is possible to evaluate climatic conditions for the near future and estimate probable increases in food production due to higher CO_2 concentrations.

Introduction

Almost all organic matter consumed by humankind and animals is produced by photosynthetic plants, which use carbon dioxide (CO_2), water, and minerals to absorb solar energy. Current atmospheric CO_2 concentrations are much lower than the optimum value for producing plants. The mean CO_2 concentration in the Cretaceous (67-133 million years ago) is estimated at 0.15-0.18%. During the Tertiary (2-67 million years ago), the mean CO_2 concentration gradually fell to a level close to that of modern times, about 0.03%. During the last two centuries, the CO_2 concentration has increased, slowly at first and then more rapidly, from 0.028% to 0.035%. This increase is due largely to human activity, primarily the combustion of fossil fuel.

Considerable progress in the paleosciences during the second half of the 20th Century has provided paleoclimatic information for most continents and oceans. Data for the continents are available for the Phanerozoic, while those for the oceans pertain to the Cenozoic and the Late Mesozoic.

The causes of climate change in the geological past were discussed in empirical studies by Russian scientists and in works on climate theory, including those of Manabe and Broccoli (1985) and Manabe and Bryan (1985). These studies revealed a cooling trend over the last 100 million years due to a gradual decrease in atmospheric CO_2 concentration. Fluctuations in atmospheric CO_2 concentration also intensified temperature changes in the glacial and interglacial epochs of the Pleistocene.

During the last 100 years, in both the Northern and Southern Hemispheres, the mean air temperature has tended to increase, especially in the 1930s and 1980s. Investigations concluded that the warming of the 1930s (mainly in the Northern Hemisphere) was due to greater transparency in the lower stratosphere, which contained a small amount of aerosol particles (Budyko and Izrael 1991). The warming of the 1980s was caused primarily by an increase in the concentrations of CO_2 and other gases, which enhanced the greenhouse effect. The first warming ended in the 1940s, while the second one is expected to continue for some time.

At the end of the 19th Century, Arrhenius theorized about possible warming as a result of an accumulation of CO_2 in the atmosphere due to the combustion of larger quantities of coal, oil, and other carbon fuels. However, this theory was not confirmed by empirical data and did not attract the attention of the scientific community at that time. In the early 1970s, however, a quantitative forecast of anthropogenic warming was published (Budyko 1972). This forecast proved to be sufficiently realistic.

Among the anthropogenic factors affecting global climate is direct heating of the atmosphere as a result of the transformation of energy into heat, in addition to that received from the sun. The atmosphere can also receive additional heat from a decreased surface albedo resulting from irrigation and construction of water reservoirs, roads, and buildings. The first factor has had a substantial local effect (temperature increases in large cities), and its effect on the global climate may become pronounced. The other factors influence the global climate only slightly.

The effect of human economic activity on atmospheric aerosols and its possible influence on future climate is not clear. The major anthropogenic factor in climate change is the enhancement of the greenhouse effect due to increased concentrations of CO_2 and trace gases, such as methane, nitric oxides, chlorofluorocarbons (freons), and tropospheric ozone. The effect of trace gases on the climate is increasing and may enhance global warming. This effect can be detected by different methods.

Budyko (1974) considered the possibility of global climate change by using the directed effect on atmospheric aerosol. It was also suggested that an unpremeditated atmospheric aerosol increase due to human activity could affect the climate. This supposition was discussed, but not generally accepted.

Adaptation policies should be assessed on the basis of reliable predictions of climate change, including data on its potential impacts. Assessing adaptive adjustments to climate change is a new, unevaluated area of science, especially on the global scale. The current task is to develop preliminary approaches to adaptive measures.

Paleoanalog Method: Global Scale

In the 1970s, Budyko (1977) discussed the possibility of detecting anthropogenic warming by using observed data on air temperature and direct radiation. At first, this theory was supported only by limited empirical material, but it was later substantiated in studies that used independent methods based on analyzing large amounts of observed data. However, some scientists believed that an insufficient understanding of the physical mechanism of climate change limited the possibility of reliably detecting global warming on the basis of observed data. This opinion was based on the use of assessments of climate sensitivity to changes in heat income obtained by a single method — climate model simulations that produced contradictory results. For example, according to reports from the U.S. Academy of Sciences (1982), a doubling of the CO_2 concentration would cause the mean global surface air temperature to increase by $\Delta Tc = 3.0 \pm 1.5°C$. This wide range of ΔTc values weakened the conclusion of these reports.

However, ΔTc could be determined accurately enough by empirical data. Estimates of this parameter obtained by paleoclimatic methods, taking into account changes in atmospheric chemical composition in the geological past, proved to be reliable (Budyko et al. 1987). In recent years, different empirical studies used a narrow range of $\Delta Tc = 3.0 \pm 0.5°C$. This range is three times smaller than that used in the U.S. Academy of Sciences reports. Both theoretical and empirical studies of paleoclimatic change show that mean temperature increase with global warming is proportional to the logarithm of relative increase in CO_2 content. Thus, the increase in mean air temperature (ΔT) due to an increase in CO_2 concentration can be determined by the following empirical relationship:

$$\Delta T = \frac{lg\,\alpha}{lg\,2}\,\beta, \tag{1}$$

where ΔT is expressed in degrees Celsius; α is the ratio of current and past CO_2 concentrations (with current warming, α should increase to take account of the influence of trace gases on the greenhouse effect); β characterizes the effect of the thermal inertia of the climate system on temperature changes (β is near zero for rapid and short-term CO_2 concentration fluctuations, close to 1 for slow and long-term temperature changes, and equal to 1 when this formula is used to compare stationary climate system states).

According to observed data from meteorological stations, the 10-year anomaly in mean air temperature in the Northern Hemisphere of the 1980s is higher than temperature anomalies for each decade since the late 19th Century. Because this climate change was forecast earlier, it can be concluded that the warming was due to anthropogenic, not natural, factors.

According to observed data, the CO_2 concentration was 0.335% in 1980 and 0.355% in 1990. Trace gases enhanced the greenhouse effect by about 50% over this period because of an increase in the CO_2 concentration. Therefore, Equation 1 shows that such an increase in greenhouse gas content with $\beta = 0.7$ should cause the mean air temperature to increase by 0.3°C from 1980 to 1990 (Budyko and Izrael 1991). An empirical estimate of mean temperature change over this decade is 0.25°C, in good agreement with the theoretical estimate. This comparison is the simplest substantiation of the theory of use of observed air temperature data to detect anthropogenic climate change.

The progress of global warming confirms the first realistic forecast of mean air temperature change (Budyko 1972). This forecast predicted a possible 2.0-2.5°C increase in mean air temperature by 2070, compared with the mid-20th Century, on the basis of an analysis of the effect of increases in CO_2 concentration on temperature. Later national and international reports on climate change presented new forecasts of climate change, which as a rule, did not contradict the first one.

The hypothesis that anthropogenic climate change can be forecast is based on the calculation of the mean global air temperature increase due to the greenhouse effect. Regional changes in air temperature and other meteorological parameters, corresponding to a certain level of warming, can be calculated by using mean temperature increases for past warm epochs. To assess regional climate changes due to global warming, General Circulation Models (GCMs), as well as the data on the climates of past warm epochs (paleoanalogs), can be used.

Analysis of centennial changes in mean surface air temperature, including seasonal and latitudinal features, shows that anthropogenic warming was largely disguised by natural temperature fluctuations before the 1970s. Observed data and model estimates both indicate that an enhanced greenhouse effect began to influence the centennial mean air temperature trend in the last quarter of the 20th Century. Thus, it can be concluded that a mean air temperature increase of 0.3°C from 1975 to 1985, calculated from averaged (for five-year periods) data on mean air temperature change in the Northern Hemisphere, was primarily due to anthropogenic factors.

Estimates of mean temperature increase for various time intervals from 1990 to 2050 are less reliable. The main factors contributing to the uncertainty in these estimates are inaccurate projections of greenhouse gas emissions into the atmosphere and not taking into account the influence of natural factors on temperature change, primarily atmospheric transparency fluctuations. Uncertainty caused by the first factor increases for estimates for more distant years. Uncertainty caused by the second factor decreases as the ratio between the value of expected anthropogenic temperature change and noise due to natural factors increases.

The accuracy of climate forecasts is highest for the relatively near term and decreases for the more distant future. On the basis of the available data, projections of greenhouse gas emissions for the late 20th Century and first decades of the 21st Century are assumed to be sufficiently reliable. The available estimates of climate change will likely be most reliable for 2000-2020. For 2020-2050, the accuracy of the projections decreases, and for the period after 2050, the projections are even more speculative. Thus, calculations concerning climatic conditions of the second half of the 21st Century should be viewed as possible scenarios of climate change rather than as forecasts.

Budyko and Izrael (1991) determined spatial-seasonal distributions of air temperature and precipitation changes by using paleogeographic data on the climates of three past warm epochs. According to these data, the mean air temperature increase in the Holocene Optimum (5,000-6,000 years ago), relative to the modern preindustrial epoch, approximately corresponds to that expected to occur by 2000; the mean temperature increase in the Last Interglacial (125,000 years ago) is close to that projected for 2025; and the Pliocene Optimum (3-4.5 million years ago) is similar to the warming expected to occur by 2025. (These climate change estimates pertain not to individual years, but to 5- or 10-year intervals, the middles of which are the years indicated.)

To determine the climate conditions of these warm epochs, several authors developed reconstructions that provide data on the distribution of temperature (winter and summer) and precipitation changes compared with those of the modern preindustrial epoch. The use of data on past warm climates, described in Budyko and Izrael (1991), seems to have solved the problem of predicting future climate conditions by an empirical method. However, two questions remain: do available paleoclimatic reconstructions accurately characterize the meteorological regime of the warm epochs, and are the selected epochs realistic analogs of future climate conditions?

Climate changes due to anthropogenic effects on the atmosphere in the late 20th Century and first half of the 21st Century must be similar to climate fluctuations in the past warm epochs if the analogs are to be used to predict the future climate. This condition is likely to be met if past warmings and current anthropogenic warming were both caused by the same reason (i.e., increased greenhouse gas concentration). If a past warming occurred for a different reason, using data on this warming as an analog of the future climate may need further study.

Both theoretical (e.g., Wetherald and Manabe 1975) and empirical (e.g., Budyko and Yefimova 1984) investigations revealed similarities between features of climate change due to different causes. However, differences were revealed in meteorological regime changes, particularly during the warmings of the 1930s and 1980s (Budyko 1988). These differences probably resulted because these warmings were caused by different factors (the first was due to an increase in the transparency of the lower atmospheric layers, and the second was caused by an increase in the greenhouse gas concentration). However, different causes of warming may not preclude the use of climate conditions during one warming as an analog for predicting climate during another because of the limited accuracy requirements for the analogs.

Paleoanalog Method: Regional Scale

An empirical approach for choosing satisfactory analogs of future regional climate change can be used to compare latitudinal distributions of relative values of mean zonal surface air temperature anomalies in the Northern Hemisphere for past warm epochs. These values are determined as departures of temperatures from their modern values, divided by the anomaly values for the hemisphere as a whole. As these distributions are a fundamental characteristic of global climate change, their similarities and differences are important in assessing the possibility of using data on past warm climates as analogs of expected warmings.

To use data on past climates as analogs of future climates, the dependence of climate parameter anomalies in different latitude zones on mean global temperature increase should be determined. If these anomalies are directly proportional to mean global temperature anomalies, the use of climate analogs is simplified because data on different warm epochs can be used for verification. If the dependence is nonlinear, the possibility of verification is limited.

Comparison of mean latitudinal temperature anomalies shows that temperature changes with different warming levels in the 0-60° N zone are in good agreement. A comparison of maps of winter and summer air temperature increases for the three warm epochs, compared with the climate of the 19th Century, shows that the distribution of temperature anomalies is similar. All three comparisons reveal a pronounced spatial variability in the temperature anomalies, which increase with increasing latitude in both summer and winter. They also increase in regions with a more continental climate.

The similarity of the temperature anomaly maps for the three warm epochs leads to the conclusion that at first approximation, regional changes in temperature anomalies are proportional to mean global temperature increases and, consequently, analogous. It can also be concluded that empirical methods of assessing spatial temperature distribution with global warming are sufficiently reliable. Thus, relative winter and summer anomalies can be averaged to obtain more accurate patterns of temperature distribution with global warming.

The effect of global warming on annual precipitation on continents is more complex. Global warming may cause absolute humidity to increase, which would, in turn, increase convective precipitation. Thus, moisture conditions on continents would become more homogeneous, which is characteristic of many past warm epochs.

Paleoclimatic maps of precipitation change in the Holocene Optimum, Last Interglacial, and Pliocene Optimum do not contradict this supposition. During the 20th Century, precipitation increased on continents in the middle and high latitudes of the Northern Hemisphere. In most regions of this zone, the mean global temperature increased by several tenths of degrees and caused annual precipitation to increase by about 10%. Extrapolation of this dependence leads to the conclusion that a 2°C warming would increase precipitation by a few dozen percent. This conclusion is in good agreement with empirical data on the precipitation regime in the warm epochs of the Pleistocene and Late Tertiary. However, the agreement between these values is not sufficient to assess the reliability of forecasts of regional climate change.

The accuracy of regional climate change forecasts based on prescribed atmospheric chemical composition is often assessed by comparing estimates of changes in certain climate parameters obtained by different methods. The importance of such comparisons can vary, depending on whether the theoretical or empirical climate models used in the calculations are sufficiently independent. Several investigations compared estimates of regional temperature and precipitation changes due to an enhanced greenhouse effect obtained by several theoretical climate models.

Budyko et al. (1990) compared estimates of regional climate change due to a doubled greenhouse gas concentration, obtained by five GCMs. Correlation coefficients for regional temperature anomalies, calculated by different models, ranged from +0.3 to 0.8 for winter and from +0.4 to 0.7 for summer. Correlation coefficients for precipitation changes ranged from -0.1 to 0.1 for both winter and summer.

More reassuring results were obtained by comparing estimates of the error of expected climate change forecasts, calculated by independent empirical methods. Budyko et al. (1990) and Budyko and Izrael (1991) presented the results of calculations that led to the conclusion that estimates of forthcoming regional climate change obtained by the paleoanalog method are rather high.

The accuracy of paleoclimatic materials can be determined by analyzing maps of temperature anomalies for past warm epochs. Regional distributions of relative temperature anomalies (i.e., differences between the regional temperatures of these epochs and those of the preindustrial epoch), divided by mean temperature anomalies for the Northern Hemisphere, are similar in all cases, for both winter and summer. Correlation coefficients for temperature anomalies for continental regions with a linear scale of the order of thousands of kilometers range from +0.5 to 0.9 when data on the past warm epochs are compared.

The agreement between temperature anomalies for different past epochs indicates similar reactions of the climate system to changes in heat income in each epoch. Thus, possible differences in the causes of climate change in various epochs do not prevent paleoclimatic data from being used as analogs of future climate.

The effects of errors should be considered in both projected changes in atmospheric chemical composition and short-term natural climate fluctuations. Such errors could affect the forecast mean temperature change presented in the early 1970s. However, up to 1995, such errors were insignificant. Additional information is needed to assess the reliability of forecasts of near future climate conditions.

Budyko and Groisman (1989, 1991) suggested a new empirical method for solving this problem. They compared climate change in the nontropical zone of the Northern Hemisphere from 1980 to 1985 with a forecast of climate change up to 2000 obtained by the analog method. This approach to testing the reliability of the analog method is justified because after 1975, the mean air temperature in the Northern Hemisphere increased considerably. Temperature changes averaged over 5- or 10-year intervals went beyond the range of temperature increases caused by natural factors. A series of high mean temperatures in the Northern and Southern Hemispheres occurred during the 1980s. Seven times in individual years over this period, temperatures reached record high values (i.e., higher than those ever detected). This fact supports the conclusion that for the near future, mean temperature change primarily depends on anthropogenic factors.

Budyko and Groisman (1989) also compared regional temperature and precipitation changes in the Northern Hemisphere, based on observed data for 1980-1985, with changes in these parameters forecast to 2000 calculated by paleoclimatic materials for the Holocene Optimum. They compared data for 11 regions in Europe, Asia, and North America. The analysis showed good agreement between winter temperature anomalies calculated by paleoclimatic data and those determined by observed data, for almost all the regions. Summer temperature anomalies were difficult to analyze, however, because their absolute values were too small. Good agreement was revealed when comparing annual precipitation anomalies for Europe and Asia, while the comparison was less satisfactory for North America.

Accuracy of Future Climate Predictions

Information available on the accuracy of future climate forecasts leads to several conclusions:

- The positive results obtained from comparing predictions of regional temperature and precipitation anomalies, on the basis of independent paleoanalogs of different warm epochs (and modern climate change), indicate the comparatively high reliability of empirical forecasts.
- Quantitative substantiation of the predictability of climate change is important. However, such predictability is limited: it often depends not on the accuracy of the theoretical or empirical climate models, but on the reliability of projections of future atmospheric chemical composition and information about other factors that affect climate.
- Taking into account these conclusions, researchers can now estimate future climate conditions reliably enough for practical purposes.

Adaptation to Climate Change

Global climate change will likely affect the future of humankind. The world population doubled in the last 35 years, reaching more than 5.5 billion. It is expected to double again just as rapidly. After that, the growth rate is expected to decrease, and population will reach a stable level of 10-12 billion. Many authors have discussed the problem of providing this population with all necessary resources (Rosenzweig and Parry 1994). Special attention has been given to agricultural productivity, which must be increased to about twice the current level over several decades.

No matter how strenuous efforts are to increase crop yields in developing countries, where population growth rate is especially high, the problem can be only partially solved. As a rule, these countries cannot store sufficient food to diminish the threat of famine during unfavorable weather conditions. At the same time, the amount of food consumed often does not meet necessary standards for life. Animal protein is of particular concern because it requires a higher level of agricultural production than does vegetable protein. Food production can be increased considerably in some developed countries; however, developing countries may find this food too costly.

There is no guarantee that the additional billions of people expected to be born in the near future can be provided with food. In the next century, maintaining living standards will be complicated by the need to produce more consumer goods that require additional agricultural resources. Restoring more favorable climate conditions, similar to those that existed over long periods with higher CO_2 concentration, may be possible.

Combustion of coal, oil, gas, and other carbon fuels unintentionally helped stop the depletion of CO_2, the primary resource for producing organic matter by autotrophic plants. It also increased primary productivity, which is the basis of existence for all heterotrophic organisms, including humankind.

The current CO_2 concentration increase of 25% of its preindustrial value has already resulted in a noticeable increase in total bioproductivity, particularly of crops. Data show that productivity may increase by as much as 5% solely because of the effect of increased CO_2. Thus, an increase in CO_2 concentration can provide food for about 250 million people. If the CO_2 concentration doubles compared with the preindustrial epoch and carbon fuel consumption is not restricted, crop productivity may increase sufficiently to provide food for an additional 1 billion people.

Quantitatively assessing the effect of climate change is more difficult. An additional global increase in crop yield due to increased precipitation and temperature may be comparable with a crop yield increase due to the direct effect of increased CO_2 concentration. If progress in agrotechnology leads to a total crop yield increase of 60% over the next 50 years, an increased CO_2 concentration would help to provide approximately 2 billion people with food in 2025-2050.

This estimate is only approximate. The profound effect of an increased CO_2 concentration on the productivity of plants should not be overestimated. This conclusion explains the interest in "returning to the epoch of more fertile biosphere," where increased plant productivity could provide food for additional billions of people.

A relatively small reduction in greenhouse gas emissions into the atmosphere would only insignificantly affect the expected temperature increase. To slow down global warming, greenhouse gas emissions must immediately be reduced by dozens of percent, which is likely to damage the world power industry and require expenditures many countries cannot afford. On the other hand, climate warming can have negative consequences on the local or even regional scale (droughts, floods, damage for coastal and island zones, etc.). Thus, climate change remains a complex problem.

Budyko et al. (1991) discussed the optimum strategy for development with anthropogenic climate change. They recommend studying in detail the possible consequences of measures to slow down climate warming.

References

Budyko, M.I., 1972, *Man's Influence on the Climate*, Gidrometeoizdat, Leningrad, USSR (former Soviet Union).

Budyko, M.I., 1974, *Climate Change*, Gidrometeoizdat, Leningrad, USSR (former Soviet Union).

Budyko, M.I., 1977, "Investigation of Modern Climate Change," *Meteorologiya i Gidrologiya* (11):42-51.

Budyko, M.I., 1988, "The Climate of the Late 20th Century," *Meteorologiya i Gidrologiya* (10):5-24.

Budyko, M.I., and P.Ya. Groisman, 1989, "The Warming of the 1980s," *Meteorologiya i Gidrologiya* (3):5-10.

Budyko, M.I., and P.Ya. Groisman, 1991, "Climate Change, Expected to Occur by the Year 2000," *Meteorologiya i Gidrologiya* (4): 84-94.

Budyko, M.I., and Yu.A. Izrael (eds.), 1991, *Anthropogenic Climatic Change*, Arizona University Press, Tucson, Ariz., USA.

Budyko, M.I., Yu.A. Izrael, M. MacCracken, and A. Hecht (eds.), 1990, *Prospects for Future Climate*, Lewis Publishing, Inc., New York, NY, USA.

Budyko, M.I., Yu.A. Izrael, and A.L. Yanshin, 1991, "Global Warming and Its Consequences," *Meteorologiya i Gidrologiya* (12):5-10.

Budyko, M.I., A.B. Ronov, and A.L. Yanshin, 1987, *The History of the Earth's Atmosphere*, Springer-Verlag, New York, N.Y., USA.

Budyko, M.I., and N.A. Yefimova, 1984, "Annual Trend of Meteorological Parameters as a Model of Climate Change," *Meteorologiya i Gidrologiya* (11):5-10.

Manabe, S., and A.J. Broccoli, 1985, "A Comparison of Climate Model Sensitivity with Data from the Last Glacial Maximum," *Journal of Atmospheric Science* 42:2,643-2,651.

Manabe, S., and K. Bryan, 1985, "CO_2-Induced Change in a Coupled Ocean-Atmosphere Model and Its Paleoclimatic Implications," *Journal of Geophysical Research* 90:11,689-11,707.

National Academy of Sciences, 1982, *Carbon Dioxide and Climate: A Second Assessment*, Washington, D.C., USA.

Rosenzweig, C., and M.L. Parry, 1994, "Potential Impact of Climate Change on World Food Supply," *Nature* 2(1):65-70.

Wetherald, R.T., and S. Manabe, 1975, "The Effect of Changing the Solar Constant on the Climate of a General Circulation Model," *Journal of Atmospheric Science* 32:2,044-2,059.

Assessing Climate Change Adaptations: The IPCC Guidelines

T.R. Carter
Agricultural Research Centre of Finland
Postal Address: Finnish Meteorological Institute
Box 503, FIN-00101 Helsinki, Finland

Abstract

The Intergovernmental Panel on Climate Change Technical Guidelines for Assessing Impacts of Climate Change Impacts and Adaptations represent an initial attempt to integrate methods for formulating adaptation into a framework for assessing the impacts of climate change. These Technical Guidelines distinguish between two adaptation responses: autonomous adjustments and adaptation strategies. Autonomous adjustments are natural or spontaneous adjustments that will probably occur in response to climate change. Three groups are distinguished: (1) in-built adjustments, (2) routine adjustments, and (3) tactical adjustments. Adaptation strategies are responses to climate change that require deliberate policy decisions, i.e., responses exogenous to a system. The Technical Guidelines identify seven main steps for the assessment of adaptation strategies. The methodologies outlined in each step embrace assessments conducted across a wide range of environments, economies, and societies subject to climate change.

Introduction

Background

In recent years, concerns have grown about greenhouse-gas-induced climate change and its possible implications. In response to these concerns, researchers have made considerable progress in formulating a general set of methods for assessing the impacts of climate change. In 1992, Working Group II of the Intergovernmental Panel on Climate Change (IPCC) published summary guidelines for assessing impacts of climate change as part of its supplement to the first assessment report (IPCC 1992). Carter et al. (1992) contains the full version of these preliminary guidelines.

However, Carter et al. (1992) omitted an important consideration: methods for assessing adaptation to climate change. In light of this omission and of the preliminary nature of the earlier report, the IPCC expanded the document into a set of Technical Guidelines (Carter et al. 1994).

The literature on adaptation to climatic variability is considerable and diverse. It generally focuses on methods of coping with extreme climatic events, such as droughts, floods, frosts, and storms. Methods cited range from highly formalized, technical procedures used in formulating management strategies for dealing with climatic risks in some regions and economic sectors, to more anecdotal, empirical studies of adaptation responses to climatic events in others. However, potential analysts do not have a general framework to guide them through adaptation assessments that embrace both climatic variability and climate change. The Technical Guidelines represent an initial attempt to integrate the treatment of adaptation to climate change within a formal framework of impact assessment.

A Framework for Impact and Adaptation Assessment

The Technical Guidelines outline a seven-step framework for conducting a climate impact and adaptation assessment (Figure 1). The steps are consecutive (open arrows), but the framework also allows for redefining and repeating some steps (bold arrows). Each step has a range of study methods. The first five steps are characteristic of many previous climate impact assessments. Further details of these steps are discussed in Carter et al. (1994, pp. 6-31). A brief description of these steps follows:

- *Define the problem.* Step 1 identifies specific goals of the assessment, the ecosystem, the economic sector and/or geographic area of interest (exposure unit), the time horizon of the study, its data needs, and the wider context of the work.
- *Select the method.* Step 2 is contingent upon the availability of resources, models, and data. Methods can range from qualitative and descriptive to quantitative and prognostic.
- *Test the method/sensitivity.* Step 3 is necessary before performing a full evaluation. This step commonly uses feasibility studies, acquisition and quality control of data required in the analysis, and validation and sensitivity analysis of predictive models.
- *Select the scenarios.* Step 4 projects future environmental and socioeconomic conditions that could occur, assuming no change in climate (the future baseline), and the corresponding conditions consistent with given projections of climate change. These "scenarios" provide the context for the impact assessment.

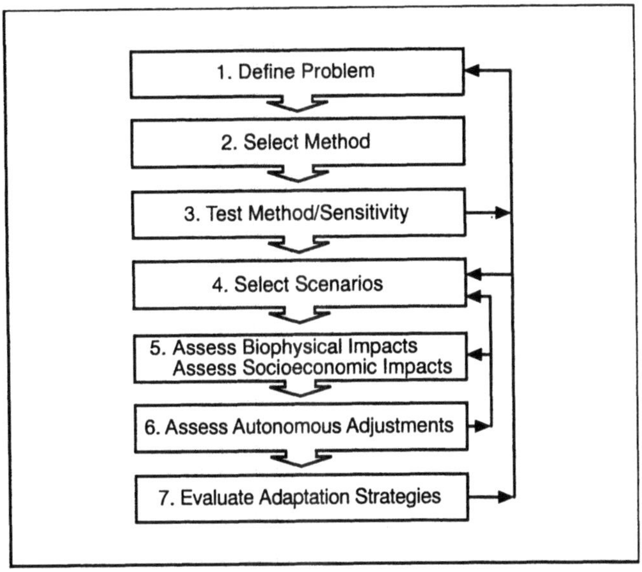

Figure 1. Seven Steps of Climate Impact and Adaptation Assessment (Source: Carter et al. 1994)

- *Assess the biophysical and socioeconomic impacts.* Step 5 compares the effects on an exposure unit of the scenario conditions with climate change with the effects of future baseline conditions without climate change.

Most earlier studies did not include steps 6 and 7, which address adaptation. The following section defines adaptation and distinguishes between autonomous adjustments and adaptation. The next section identifies three types of autonomous adjustment, along with examples of each. Later sections describe adaptation assessment methods, which are organized into a seven-step framework analogous to the general framework shown in Figure 1, and comment on the practical application of the Technical Guidelines in studies of climate change impacts and adaptation.

Adaptation to Climate Change

Impact assessments usually evaluate the effects of climate change on an exposure unit in the absence of any response that might modify these effects (cf., steps 1-5 in Figure 1). Two broad types of response are identified: mitigation and adaptation (Figure 2).

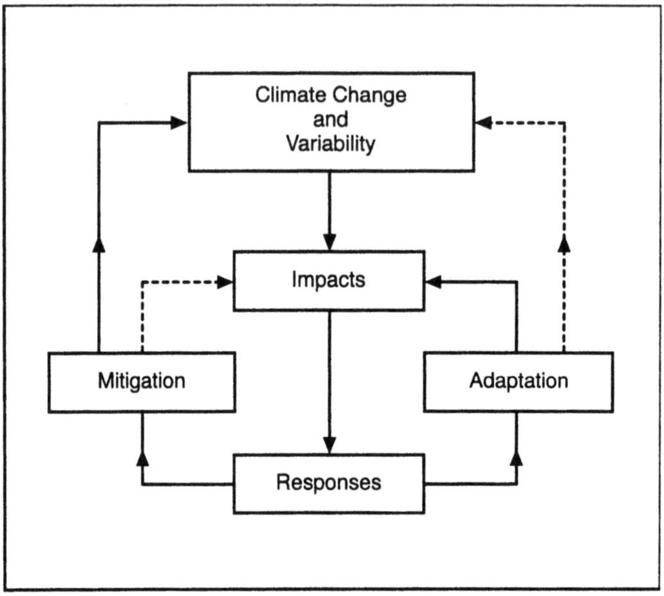

Figure 2. Pathways of Response: Mitigation and Adaptation (Solid lines indicate direct effects or feedbacks; dashed lines denote secondary or indirect effects. [Source: Smit 1993].)

Mitigation (or limitation) deals with the causes of climate change. It involves actions to prevent or retard the increase of atmospheric greenhouse gas concentrations. Mitigation was long regarded as the major option for "solving" the greenhouse gas problem. However, it is probable that realistic mitigation policies will not be able to prevent climate changes fully and that alternative adaptive measures will be needed.

Adaptation is concerned with responses to the effects of climate change. Adaptation is defined here as any adjustment — passive, reactive, or anticipatory — that can respond to anticipated or actual consequences associated with climate change. These consequences can be either adverse or positive.

Although the outcomes of mitigation strategies can have a bearing on impact assessments by modifying future climate, as well as socioeconomic and environmental conditions, mitigation is not formally included in these assessments. However, identifying and evaluating adaptation options should be an integral part of any realistic impact assessment.

A basic distinction can be drawn between automatic or in-built responses to climate change (termed autonomous adjustments) and responses that require deliberate policy decisions (referred to as adaptation strategies). The following section discusses these responses in detail.

Assessment of Autonomous Adjustments

Most ecological, economic, or social systems undergo some spontaneous or "autonomous" adjustments in response to changing climate. To account for these responses in the assessment process, it is helpful to distinguish among three groups of adjustment on the basis of their "degree of spontaneity": in-built, routine, and tactical adjustments.

In-built Adjustments

In-built adjustments are the unconscious or automatic reactions of an exposure unit to a climatic perturbation. Because they are usually associated with living organisms, these reactions are sometimes referred to as physiological adjustments. Straightforward examples include (1) perspiration by mammals exposed to high temperatures to induce evaporative cooling of the skin and (2) the closing of stomata in plants to reduce transpiration water loss under drought conditions. Such clear responses are usually predictable and easily modeled.

More difficult to assess are in-built adjustments to climate that occur over the long term (such as acclimation of tree species to changing conditions) or adjustments initiated only under particular combinations of conditions. For example, the onset of cold hardening in temperate plant species often depends on the coincidence of suitable temperature and photoperiod conditions, combined with a particular plant development stage. To understand and, hence, predict such adjustments, special monitoring or experimental programs might be required.

Routine Adjustments

Routine adjustments refer to everyday, conscious responses to variations in climate that are part of the routine functioning of a system. For example, farmers in the semiarid tropics frequently adjust the sowing time of their crops according to the onset of the rainy season. Similarly, in cool regions, it is often the disappearance of snow cover that determines the date of sowing. Such routine adjustments are often predictable and can be represented in models. Thus, the models perform internally adjustments that the system (i.e., the farmer) might do instinctively or routinely.

Tactical Adjustments

At a higher level than routine adjustments are responses that require a behavioral change, referred to as tactical adjustments. Tactical adjustments might be required, for example, following a sequence of anomalous climatic events that indicate a shift in climate. For example, Table 1 lists some alternative cropping practices for conserving soil moisture and nutrients, reducing run-off, and controlling soil erosion. These tactical adjustments — made as a result of several years of below-average rainfall in Canada — require the farmer to make a behavioral shift, but can still be accommodated internally within the system. Models can often be used to evaluate the effectiveness of such tactical adjustments.

However, it is not always easy to separate autonomous tactical adjustments directly related to climate change from adjustments made in response to other factors influenced by climate change. For example, a government response to repeated drought-related crop failures of maize in a semiarid region might be to

Table 1. Some Changes in Farming Practice Identified as Tactical Adjustments to a Drying Climate in Canada

Strategy	Tactical Adjustment Measures
Cropping alternatives	• Conventional bare fallow to conserve water
	• Stubble mulching to improve water uptake
	• Minimum tillage to reduce erosion
	• No-till farming
	• Plant furrows on contour to reduce run-off
	• Reduced field width to control soil erosion
	• Legume and sod-based rotations to conserve water
	• Wider row spacing to increase soil water per plant
	• Contour cropping to slope to conserve water
	• Deep furrow drilling to protect small grains
	• Maintain vegetative cover to reduce soil erosion
	• Avoid monocropping
	• Practice crop rotation
	• Alternative cropping patterns
Other strategies	• Chisel up soil clods to form small dams
	• Reduced crop population to match water supply
	• Lower planting densities
	• Vary fertilizer application optimize crop water-use efficiency
	• More efficient fertilizer and pest management to preserve water and soil
	• Use pre-emergence herbicides to control weeds
	• Take marginal land out of cultivation

Source: Smit (1993).

encourage the cultivation of drought-resistant sorghum. Various incentives (e.g., inexpensive seed grain, guaranteed prices) might be used. In this case, farmers would be responding not to low maize yields per se, but to the prospect of a subsidized sorghum crop — a tactical adjustment based on an externally imposed policy response to climate change. Such responses lie within the realm of adaptation rather than within autonomous adjustment and are the theme of the following section.

Evaluation of Adaptation Strategies

In addition to autonomous adjustments, a wide range of responses that would affect many systems or communities can be implemented exogenously by management or policy decisions at the regional or national level. These adjustments are adaptation strategies.

A single accepted set of procedures has not been established for formulating regional and national policies for adapting to climate change. This situation is probably due in part to the large differences in priorities between different regions. For example, researchers in low-lying Pacific islands are particularly concerned about adapting to anticipated sea-level rise, whereas scientists in Arctic regions are concerned about adapting to seasonal melting of permafrost. Furthermore, assessing adaptation options involves value judgments, which are subjective and can be controversial.

The framework adopted for developing adaptation strategies is designed to be applicable to different regions, sectors, and socioeconomic systems. The steps of analysis are loosely based on formal evaluation procedures used operationally to manage climatic risks in some developed countries (e.g., in water allocation and flood control). As with the general framework for impact assessment shown in Figure 1, this framework is also composed of seven steps: (1) define objectives, (2) specify climate impacts of importance, (3) identify adaptation options, (4) examine constraints, (5) quantify measures and formulate alternative strategies, (6) weight objectives and evaluate trade-offs, and (7) recommend adaptive measures.

Step 1: Define Objectives

General Goals

At the outset, it is important to know what goals and objectives guide the adaptation assessment. Often some general goals form the basis for an assessment, such as those defined by international institutions or agreed upon at conventions. Examples of such goals include (1) "the promotion of sustainable development" and (2) "the reduction of vulnerability." Because these goals are usually too broad and ambiguous to be of practical use in adaptation assessment, more specific objectives need to be defined.

Specific Objectives

In this context, specific objectives represent desired targets that can be evaluated by using specified criteria and constraints. They are usually based on public preferences, legislation, interpretation of the general goals, or a combination of these. Table 2 illustrates (1) some objectives that could be selected to achieve the two goals described above and (2) the evaluation criteria used to measure their success. Most are quantitative measures (e.g., income, employment); others (e.g., biodiversity) can be quantified, but not in economic terms. Evaluation criteria can also be subject to constraints, such as defining acceptable boundaries (e.g., minimum tax revenue, maximum probability of a flood).

Table 2. Example of a Multiple Criteria Evaluation Framework for Water Resource Management

Overall Goal	Specific Objective	Evaluation Criteria
Sustainable development	• Regional economic development	Income Employment
	• Environmental protection	Biodiversity Habitat areas Wetland types
	• Equity	Distribution of employment Minority opportunities
Reduction of vulnerability	• Minimize risk	Population at risk Frequency of event
	• Minimize economic loss	Personal losses Insured losses Public losses
	• Increase institutional response	Warning time Evacuation time

Source: Stakhiv (1994).

Step 2: Specify Climate Impacts of Importance

Impact Assessment

Step 2 of the analysis involves assessing possible impacts of climate change on the exposure unit being studied. Both damaging and beneficial climatic events need to be studied. Thus, in the former, appropriate adaptation options can limit damage, while in the latter, they can compensate for negative effects. For long-term climate change, impacts should be considered relative to impacts expected to occur in the absence of climate change.

In a full climate change impact assessment study, most of the analysis required for Step 2 would be completed beforehand (cf., Figure 1), but it is still important to interpret the results according to the specific objectives defined in the adaptation assessment. In addition, alternative methods for identifying critical climate impacts are available. A notable example is a vulnerability assessment.

Vulnerability Assessment

Vulnerability is the susceptibility of an exposure unit to harmful or adverse effects of climate. Thus, vulnerable systems, activities, or regions tend to be those most in need of planned adaptation.

Vulnerability assessment can be illustrated by referring to work on coastal zone vulnerability to sea-level rise. The IPCC has devised a Common Methodology to examine coastal nations' ability to cope with the consequences of global climate change, including accelerated sea-level rise (IPCC 1991). The methodology includes identifying the populations and resources at risk, investigating the costs and feasibility of possible responses to adverse impacts, and examining the institutional capabilities for implementing these responses. This methodology has been applied in nearly 50 regions worldwide, and a "vulnerability profile" has been created for each region. The profile portrays the potential impacts of a global sea-level rise of 1 meter by the year 2100, with and without protective measures (IPCC 1994).

Step 3: Identify Adaptation Options

Classification of the Options

One of the vital steps in the assessment process is identifying possible adaptive responses. Field surveys and interviews with appropriate experts can produce lists that consider the following:

- All practices currently or previously used;
- Possible alternative measures that have not been used;
- Newly created or invented measures;

- Frequency of particular actions, the circumstances, and the person involved;
- Effectiveness and cost of different actions;
- Possible impacts of technological change on adaptation strategies; and
- Actions that may lead to *maladaptation* to climate change.

To illustrate, Table 3 lists adaptation strategies, based on a literature review, compiled in a recent study of sectors likely to be affected by climate change in Canada. Other examples include traditional adjustment mechanisms for coping with interannual climatic variability in self-provisioning rural societies, which have been cataloged on the basis of field surveys (Jodha and Mascarenhas 1985;

Table 3. Adaptation Strategies for Sectors that Probably Would Be Affected by Climate Change in Canada and the Number of Measures Identified for Each (measures for changing farming practices are shown in Table 1)

Sector	Adaptation Strategy	No. of Measures
Agriculture	• Change land topography	11
	• Use artificial systems to improve water use/availability and protect against soil erosion	29
	• Change farming practices	21
	• Change timing of farm operations	2
	• Use different crops or varieties	7
	• Governmental and institutional policies and programs	16
	• Research into new technologies	10
The Arctic	• Protection from the natural environment	18
	• Protection of the natural environment	3
Coastal Areas	• Protection	14
	• Landward migration of wetlands	7
	• Water supply and sanitation	16
	• Elevation/floodproofing	9
	• Account for sea-level rise in land development policies	4
Ecosystems and Land Use	• Protection of existing habitats and species	5
	• Enlarge wildlife habitats to ensure species survival	9

Table 3. (Cont.)

Sector	Adaptation Strategy	No. of Measures
Energy Supply	• Alternative cooling systems and water conservation in thermal plants	7
	• Change the operations of hydroelectric plants	4
	• Augment hydroelectric production from other sources	3
	• Pricing mechanisms for hydropower	5
Fisheries	• Make fisheries sustainable with or without a changed climate	7
	• Develop alternatives to traditional uses of fishery resources	4
Forestry	• Protect existing forests	7
	• Introduce new species	4
	• Change or improve use of forests	9
Urban Infrastructure	• Revise standards and safety factors	10
Water Resources	• Alter water use	14
	• Water conservation and demand management	13
	• Improved water-use efficiency through management plans	13
	• Protect in-stream flows and water quality	5
	• Interbasin transfers	10

Source: Smit (1993).

Akong'a et al. 1988; Gadgil et al. 1988). Many responses involve policy changes, but autonomous adjustments are also included because they have often been identified in an earlier vulnerability assessment.

No Regrets Strategy

It is important to add that most of the adaptation measures currently used in resource management to contend with climatic variability are probably equally feasible, if not comparatively cost-effective, under a changed climate. Moreover, measures that should be adopted to account for climate change are also likely to be beneficial for other reasons. This has been termed the "no regrets" strategy.

Step 4: Examine Constraints

A number of constraints will probably be placed on adaptive measures that can actually be used in a particular region or jurisdiction (examples are given in parentheses):

- Legislative constraints (legal minimum levels of dissolved oxygen in rivers below which fish populations are threatened);
- Social constraints (prohibited access to resources threatened by climate change for cultural or religious reasons);
- Physical constraints (limits to a policy of controlled retreat of wetlands in response to sea-level rise, due to unsuitable geology, soils, or relief);
- Biological constraints (the maximum migration rate of a rare plant species in response to climate warming); and
- Economic constraints (the annual budget allocated to resource management in a region).

Such constraints can severely restrict the range of feasible choices; therefore, it is advisable to examine these constraints closely before evaluating adaptation options in detail.

Step 5: Quantify Measures and Formulate Alternative Strategies

Assessment of the Effectiveness of Management Measures

Once management measures have been identified, their performance needs to be assessed with respect to the stated objectives. In some cases, when appropriate data and tools are available, simulation models can be used (e.g., by simulating the effectiveness of different measures under a range of climatic scenarios). In other cases, published evidence, survey material, or expert judgment can be used. Both methods can yield either quantitative or qualitative results. In this way, it may be possible to obtain a relative, or in some cases, an absolute evaluation of the comparative effectiveness of different measures in fulfilling individual objectives. A formal and replicable method for conducting such an assessment is *multicriteria analysis*.

At this stage, costs and benefits probably play a prominent role in the evaluation. Moreover, the desired rate of adaptation probably has a strong influence on the economic viability of certain measures. The faster the need for adaptation, the more it is likely to cost.

Formulation of a Multiobjective Strategy

This type of analysis can be extended to assess the effectiveness of each measure across all the objectives, recognizing that some objectives may conflict with each other (e.g., regional economic development often conflicts with environmental protection). The assessment is a prelude to developing strategies. Each strategy involves a set of management measures that maximize the level of achievement of some objectives, without jeopardizing progress toward the remaining objectives.

Step 5 is illustrated schematically in Figure 3. The top half portrays a multicriteria analysis. Each row represents a different management measure (mm), and each column a different objective. The objectives satisfy a given goal, and the effectiveness of a management measure in achieving an objective is assessed according to evaluation criteria (cf., Table 2). The hatched areas show the highest ranked (most effective) measures for satisfying the objective in column 1, and the fully shaded areas show the highest ranked measures for satisfying the objective in column 2. The bottom half of Figure 3 shows how these management measures might be organized into alternative strategies (multiobjective strategy formulation). This example is simplistic; each strategy needs to consider the effectiveness of the measures across all objectives, rather than independently for specific objectives. Moreover, the strategy formulation also needs to account for the sequence (timing) and discounted cost of such measures.

Step 6: Weight Objectives and Evaluate Trade-offs

Steps 1-5 frequently yield conflicting results (i.e., adaptation strategies relating to specific objectives). Hence, Step 6 is a key step in the evaluation because it seeks to resolve the conflicts by weighting different objectives according to assigned preferences and then comparing the effectiveness of different strategies in meeting these revised objectives.

All the components of a strategy are compared with the same set of objectives, so that managers, policymakers, and the public can see the relative range of costs and benefits for each strategy, as well as the distribution of impacts among the sectors and population (equity). Only then can compromises or trade-offs be made among objectives and between management measures.

For example, four basic categories are used operationally to evaluate the effectiveness of different planning strategies in federal water projects in the United States (Stakhiv 1994):

- National economic development (based on measures such as gross domestic product, per capita income, or employment);
- Environmental quality (significant environmental resources and their ecological, cultural, and aesthetic attributes);
- Regional economic development (distribution of regional economic activity in terms of income and employment); and

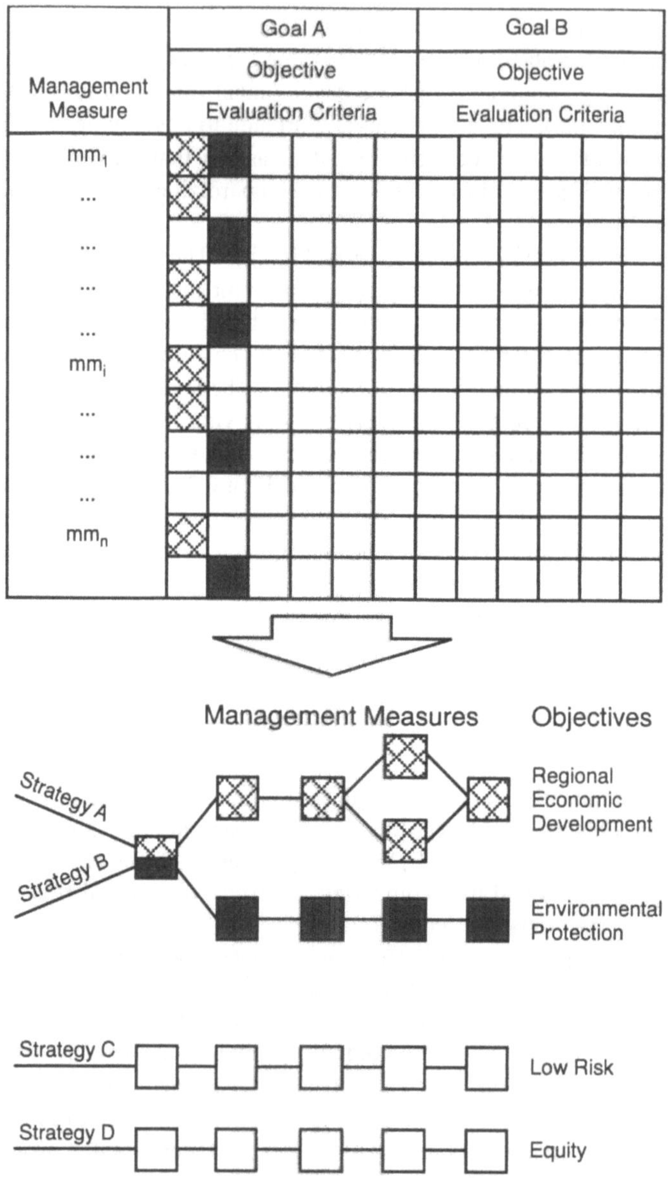

Figure 3. Some Procedures for Formulating Adaptation Strategies (Top: multicriteria analysis; bottom: multiobjective strategy formulation; for explanation, see text [Source: modified from Carter et al. 1994])

- Other social effects (including urban and community impacts; life, health, and safety factors; displacement; long-term productivity; energy requirements; and energy conservation).

All impacts and adaptive measures are evaluated according to these four categories. Selecting preferred strategies thus requires determining the trade-offs among the categories.

Step 7: Recommend Adaptive Measures

The results of the evaluation process should be compiled in a form that provides policy advisers and decision makers with information on the best available adaptation strategies. This information should indicate some of the assumptions and uncertainties involved in the evaluation, and the rationale used (e.g., decision rules, weightings, key constraints, institutional feasibility, national and international support, technical feasibility).

Conclusions

The Technical Guidelines do not provide a definitive approach to climate change impact and adaptation assessment — no such approach exists. Rather, they attempt to organize a set of methods into a convenient framework made up of the seven main steps of analysis. In practice, studies may adopt all or some of the steps. The sequence of steps may even be reversed (e.g., vulnerability assessments often consider critical impacts and adaptation strategies before evaluating impacts under given scenarios). Moreover, some of the methods are used repeatedly throughout an assessment, while others are not used at all.

The seven-step framework for adaptation assessment represents an initial attempt at formalizing a formidable and heterogeneous set of approaches for analyzing adaptation strategies. It can and should be revised in subsequent phases of the IPCC process, particularly to reflect the capabilities of developing countries, where resources are limited and many quantitative evaluation procedures and tools described in the present framework either are not available or cannot be readily applied.

To date, few studies have been performed that can be used as examples of this approach to climate change adaptation assessment. One interesting case is the Mackenzie Basin Impact Study in northwest Canada (Yin and Cohen 1994), which uses a number of the methods. The study has adopted a resource management framework for assessment, with four main components: goal setting, data acquisition and compilation, scenario development, and an integrated assessment model (Figure 4). The approach encourages various interest groups in the region to actively participate, so as to identify policy concerns relating to climate change across multiple sectors.

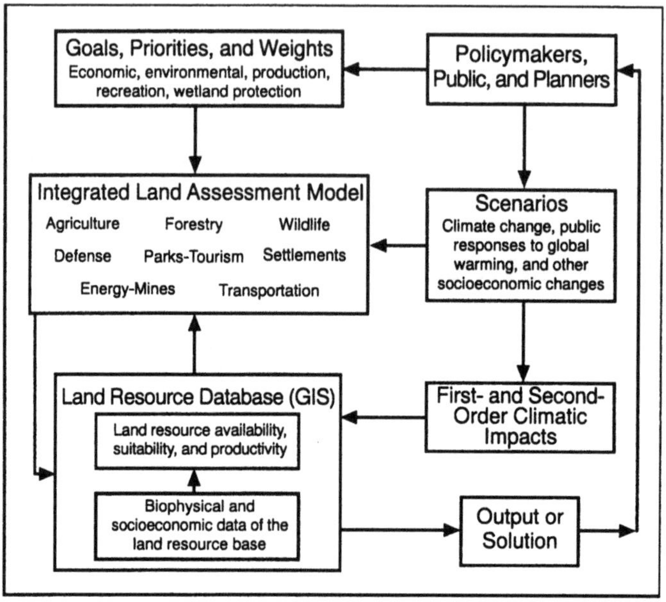

Figure 4. Integrated Land Assessment Framework Used in the Mackenzie River Basin Study (Source: Yin and Cohen 1994)

To improve their practical usefulness, the U.N. Environment Programme (UNEP) is currently preparing workbooks based on the Technical Guidelines, which attempt to "walk" the nonexpert through some typical impact and adaptation assessments that could be conducted at the national or regional levelfor particular ecosystems or economic sectors. When completed, the Technical Guidelines and UNEP workbooks will complement each other. The former is primarily for researchers, the latter for nonexperts. Both are subject to modification as methods of climate change impacts and adaptation assessment become more refined.

Acknowledgments

This study is largely based on the IPCC Technical Guidelines (Carter et al. 1994), which evolved into its present form thanks to contributions from an IPCC Expert Group of 53 individuals, as well as input from 26 reviewers. I would especially like to acknowledge the co-chairs of the Expert Group and my fellow lead authors, Dr. Shuzo Nishioka and Prof. Martin Parry, and lead author Dr. Hideo Harasawa. Any departures from the approved Technical Guidelines are my responsibility alone.

References

Akong'a, J., T.E. Downing, N.T. Konijn, D.N. Mungai, H.R. Muturi, and H.L. Potter, 1988, "The Effects of Climatic Variations on Agriculture in Central and Eastern Kenya," in M.L. Parry, T.R. Carter, and N.T. Konijn (eds.), *The Impact of Climatic Variations on Agriculture, Volume 2. Assessments in Semi-Arid Regions,* pp. 121-270, Kluwer Academic Publishers, Dordrecht, the Netherlands.

Carter, T.R., M.L. Parry, S. Nishioka, and H. Harasawa, 1992, *Preliminary Guidelines for Assessing Impacts of Climate Change*, Environmental Change Unit, University of Oxford, United Kingdom, and Center for Global Environmental Research, National Institute for Environmental Studies, Tsukuba, Japan.

Carter, T.R., M.L. Parry, H. Harasawa, and S. Nishioka, 1994, *IPCC Technical Guidelines for Assessing Climate Change Impacts and Adaptations*, Working Group II of the Intergovernmental Panel on Climate Change, University College London, United Kingdom, and Center for Global Environmental Research, Tsukuba, Japan.

Gadgil, S., A.K.S. Huda, N.S. Jodha, R.P. Singh, and S.M. Virmani, 1988, "The Effects of Climatic Variations on Agriculture in Dry Tropical Regions of India," in M.L. Parry, T.R. Carter, and N.T. Konijn (eds.), *The Impact of Climatic Variations on Agriculture. Volume 2. Assessments in Semi-Arid Regions,* pp. 495-578, Kluwer Academic Publishers, Dordrecht, the Netherlands.

IPCC, 1991, *Assessment of the Vulnerability of Coastal Areas to Sea Level Rise: A Common Methodology,* Rev. 1, Intergovernmental Panel on Climate Change, Response Strategies Working Group, Coastal Zone Management Subgroup, Ministry of Transport and Public Works, The Hague, the Netherlands.

IPCC, 1992, *Climate Change 1992: The Supplementary Report to the IPCC Impacts Assessment,* W.J. McG. Tegart, G.W. Sheldan, and D.C. Griffiths (eds.), report prepared for the Intergovernmental Panel on Climate Change Working Group II, Australian Government Publishing Service, Canberra, Australia.

IPCC, 1994, "Preparing to Meet the Coastal Challenges of the 21st Century," presented at the *World Coast Conference 1993*, Intergovernmental Panel on Climate Change, Nordwijk, the Netherlands.

Johda, N.S., and A.C. Mascarenhas, 1985, "Adjustment in Self-Provisioning Societies," in R.W. Kates, J.H. Ausubel, and M. Berberian (eds.), *Climate Impact Assessment: Studies of the Interaction of Climate and Society*, SCOPE 27, pp. 437-464, Wiley, Chichester, United Kingdom.

Smit, B. (ed.), 1993, *Adaptation to Climatic Variability and Change*, report of the Task Force on Climate Adaptation, The Canadian Climate Program, Occasional Paper No. 19, Department of Geography, University of Guelph, Ontario, Canada.

Stakhiv, E.Z., 1994, "Water Resources Planning of Evaluation Principles Applied to ICZM," in *Proceedings of the Preparatory Workshop on Integrated Coastal Zone Management and Responses to Climate Change, World Coast Conference 1993*, New Orleans, La., USA.

Yin, Y., and S.J. Cohen, 1994, "Identifying Regional Goals and Policy Concerns Associated with Global Climate Change," *Global Environmental Change* 4(3):246-260.

Adapting to Climate Variations in Developing Regions: A Planning Framework

Antonio R. Magalhães
Esquel Brazil Foundation
SQS 315 BL. A.-AP. 104
70384-010 Brasília, Brazil

Abstract

The experiences of different regions of the world in adapting to climate variability should be examined carefully before an adaptation policy is defined. Similarities in climatic phenomena under climate variability may indicate that current adaptation policies will be valid in the future. The drought-stricken region of northeastern Brazil has undertaken adaptation policies for more than a century. Its experience, particularly in assessing the causes of successes and failures of such policies, indicates that a comprehensive sustainable development strategy is needed to increase regional and societal capacity to face both present and future climate variability. A methodology for sustainable development planning — the Aridas Project — incorporates long-term sustainability and climate change adaptation into the regional development planning process.

Introduction

Climate Variability, Vulnerability, and Adaptation

Past and present climate variability has produced undesirable events. Droughts, floods, tornadoes, hurricanes, temperature peaks, and heat waves have caused material losses and human suffering. Although such events are predictable, many regions of the world still have difficulties in adapting to them.

The capacity of different regions to adapt to climate variability differs greatly. In general, developed regions are better equipped to resist climate variations than are developing regions. In Canada, for instance, the work of the Climate Branch "over many decades has enabled Canadians to adapt with great success to their climates" (Burton and Cohen 1993). However, even in developed countries extreme events still cause fear and destruction.

In developing countries, however, overall social, environmental, and economic vulnerability enhances the effects of droughts and other climatic events. Overpopulation (relative to current productivity, income, and natural resources), poverty, and land degradation translate into a poor capacity to face any kind of crisis. Poor people have no insurance against loss of income. Weak economic structures mean difficulties in maintaining jobs during an economic failure. Degraded marginal lands become totally unproductive when precipitation decreases. As a result, these regions have difficulty in facing climatic crises, although such crises are recurrent. Any extreme climactic event can become a social catastrophe when combined with the social-political characteristics of the region. For example, droughts and internecine wars in Ethiopia interact to increase the adverse effects of both.

Although developing regions are more vulnerable to climate changes than are developed countries, the degree of vulnerability varies in each specific region. Some parts of Asia and South America have reduced the effects of adverse climatic events significantly. Several adaptation strategies have been successful in these regions. These strategies may be useful in improving current adaptation policies and serve as analogs to possible future climate change adaptation strategies (Glantz 1988).

Nature of Climate Variability and Climate Change

One assumption underlies the use of past and present climate variability analogs in adapting to future climate change: future climate change will be perceived as a change in a pattern of already well-known climatic events (e.g., frequency or intensity). A certain region may face more frequent droughts or floods, or the droughts and floods may be more severe, or the rainfall may be more or less concentrated within certain time periods, but the elements of climate remain basically the same: "temperature, precipitation and storm run-off, drought, wind speed, snow loading, ice accumulation, blizzards, hail, tornadoes, frost, and severe cyclonic storms as well as many combinations of these" (Burton and Cohen 1993). In some places, however, the intensity of these climatic events may be greater than the region's capacity to cope with them.

Societies and governments have a long tradition in adapting to climate events. Many societies have introduced concern about the climate into their decision-making processes. Experiences in adapting to past and present climatic events will prove useful in future adaptation to climate change. These experiences provide information that enables countries to design and adopt more successful adaptation strategies.

The goal is not so much to learn what adaptive measures will be available (although this goal is important), but rather to learn about possible regional climate change scenarios and ways to make society aware of a future with a changed climate. Regional integrated assessments and scenarios, including climate variables, and environmental, social, economic and political scenarios, are increasingly important. Several locations have undertaken integrated assessments and scenarios (the MINK Project [Rosenberg and Crosson 1993], the Mackenzie Basin Impact Study Project [Cohen 1994], and the Aridas Project [Magalhães et al. 1994]). In general, however, the lack of credible regional climate change scenarios is a problem in preparing integrated regional studies.

This paper contributes to the debate on regional and local adaptation to climate change and offers a methodology for integrated studies of regional sustainable development that clearly considers future climate change scenarios. Past experience is valuable in discussing policies for adapting to future climate variations. The analysis and proposal are based on the author's experience with climate variability adaptation and sustainable development policies in drought-prone northeastern Brazil (NEB).

Adapting to Climate Variability: Northeastern Brazil

Northeastern Brazil is one of five main Brazilian regions. It comprises 1.5 million km^2, equivalent to 18% of the entire country, and is larger than France, Spain, and Portugal combined. Nine of the 27 Brazilian federal states or territories are in this region. The 1991 regional population was 42.5 million, or 29% of the Brazilian total.

Ecologically, the NEB is a diverse region, with four major ecosystems (Magalhães et al. 1988; Magalhães and Glantz 1992):

- The coastal forest zone, where sugar cane is grown and major cities are located (7% of area, 32% of regional population, average precipitation of 2,000 mm yr^{-1});
- The transition "agreste" zone (11% of area, 14% of population, average precipitation of 800-1,000 mm yr^{-1});
- The large semiarid interior (60% of area, 43% of population, average precipitation of 300-800 mm yr^{-1}); and
- The middle north transition to the Amazon (22% of area, 12% of population, high precipitation).

The semiarid NEB is drought prone and highly vulnerable to climate variations. It is an overpopulated region with a fragile ecosystem. Its socioeconomic vulnerability to droughts reflects the fact that the NEB is the least developed region of Brazil, although it was first settled in the 16th Century. The region has become more vulnerable to droughts, as inappropriate land use overstresses the natural land and water resources. High vulnerability implies low capacity to cope with crises. Hence, the impacts of droughts are very heavy (Table 1). In certain areas, small farmers may lose more than 90% of their crops during extreme drought. Cattle raising and irrigated crops are less vulnerable, but they also suffer heavy losses during extreme events, particularly back-to-back droughts.

Table 1. Aggregated Impacts of Droughts on the Economy of Northeastern Brazil, Sectoral Composition of Gross Domestic Product

Sector	Non-Drought (1978)	Drought (1983)	Drought (1993)
Agriculture (%)	18.2	7.6	8.3
Manufacturing (%)	23.8	23.6	29.6
Services (%)	58.0	68.8	62.1

Sources: Arraes and Castelar (1991), as quoted by Magalhães and Glantz (1992) and SUDENE (1994).

As more frequent and extreme droughts occur because of climate change, social problems increase (Table 2). Because a large portion of the population in developing regions is poor and illiterate, and have no savings to buffer a decline in income, they are much more vulnerable. In the NEB, the first consequence of a drought is mass unemployment. Farmers decide not to plant and, almost instantaneously, hundreds of thousands of people lose their jobs.

Strategies to augment the capacity of the population to cope with droughts will also increase its capacity to cope with other kinds of climate change. Strategies to adapt to climate variability have been adopted in the NEB for more than 100 years. The vulnerability is still high — and increasing, because of increasing population and land use — but there is a clear decline in real impacts, due to government and societal responses to drought. Two factors are at work:

- Resiliency is decreasing because of increased pressure on natural resources. When population and land use were both low (about 100 years ago), the environment was resilient to most droughts. Only severe climatic events would cause an environmental disruption with all its social consequences. Now, even small decreases in precipitation are perceived as droughts.
- Coping capacity has increased because of government and societal preparedness. Long- and short-term responses to droughts have been successful in improving the capacity of the population to face droughts. In the extreme drought of 1877, it is estimated that more than 500,000 people died of hunger and thirst. In modern droughts (1993, for instance), the population is still affected, but without the same dramatic and catastrophic impacts of the past.

Magalhães and Glantz (1992) drew the following conclusions from adaptation policies in the NEB:

Table 2. Social Impacts of Droughts in Northeastern Brazil: Directly Affected Rural Population

Drought Year	Number of Municipalities	Percentage of Municipalities Affected	Rural Population	Percentage of Rural Population Affected
1987-1988	1,287	79.9	15,632,173	91.8
1990	780	48.4	9,075,113	53.3
1991	775	48.1	8,849,557	52.0
1993	1,151	96.5	11,636,020	68.3

Source: Carvalho et al. (1994).

1. Drought policy . . . responds to societal and governmental perceptions about the drought phenomena.
2. Each new phase in the series of drought response strategies . . . represents a new dimension in the perceptions of droughts, which evolved from a simple concern about water shortages to a complex range of meteorological-economical-environmental interconnected variables.
3. The drought problem in a developing region . . . cannot be addressed in isolation from the issue of regional sustainable development. The only way that present and future populations may be made less vulnerable to droughts or other types of climatic variability, including those resulting from a possible climate change, is to ensure economic viability under an environmentally sustained basis, even for the lowest strata.

In assessing possible causes for past failure of government policies, Magalhães and Glantz (1992) point to lack of coordination between short- and long-term initiatives, lack of continuity, institutional instability, conflicting interests between different groups, and patronage in public drought policies. Recent policies, however, have been successful in removing some of these difficulties, particularly avoiding patronage through more open participation in the planning and decision-making processes.

Adapting to Climate Change: A Developing Region Perspective

In discussing adaptation to climate change, four questions need to be answered: (1) adapt to what, (2) what to adapt, (3) when to adapt, and (4) what conditions for adaptation?

To answer the first question, adaptation needs to be clarified. The answer depends on public perception or the way climate change will be perceived by the society and government. How will climate change be translated into climatic, environmental, social, economic, and political phenomena that affect people's lives on a daily basis, in the present as well as in the future? Developed areas have been successful in adapting to climate variability in terms of the "average climate" as well as extreme events (Burton and Cohen 1993). Some developing regions, such as the NEB, have also made progress (Magalhães and Glantz 1992). It is assumed that a climate change will alter patterns of climatic events, but the extent is unknown. Another unknown is the way in which governments and society will perceive such events. However, it is known that both public and government responses depend on how such events are perceived and diagnosed. Thus, more research is needed on climate change scenarios and impacts, with an emphasis on education. An effort should be made to identify relevant factors and to move both the public and the decision makers to adopt the necessary adaptive measures.

The second question refers to specific adaptation policies, both short and long term. Adaptation policies may include infrastructure building, environmental planning, increased education and awareness, security mechanisms, urban development, land-use planning, water resources management, and relief programs. The best adaptation policy will encompass a comprehensive strategy for sustainable development. As Ribot et al. (1993, 1995) state with regard to the semiarid regions of the globe: "There is an old solution to the problems these regions face — that is, development. But, this new development effort must occur within the constraints of ecological sustainability," in other words, sustainable development.

For more than a century, since the drought of 1877, Brazil has developed adaptation policies to face the impacts of climate variability in its semiarid region. Starting with simple relief actions, these policies have become more complex as the perception of drought becomes more comprehensive. Two main types of adaptive measures have been adopted: short and long term.

Short-term relief actions alleviate the heavy social consequences of droughts. This type of response consists of two main lines of action. The first is maintaining the income level of the rural unemployed population through huge Keynesian employment programs in the construction of public works. This policy has been very successful. During the 1983 drought, for instance, 3 million new employment opportunities were offered. The economy is thus maintained or even improved. The second line of action is distribution of water to the rural population or to cities where water resources have decreased. In 1983, about 5,000 trucks distributed water. Food security during dry periods has also been a concern. Imported food makes up for the loss of local agricultural production.

Long-term adaptive measures consist of ways to increase the permanent capacity of the region to face future droughts. With a broader perception of the social and economic nature of the droughts, adaptive measures are becoming more comprehensive. The main component is to create the capacity to store water for dry periods. The rivers in this region, except for the São Francisco and Parnaiba Rivers, are temporary (seasonal) rivers that flow only during the rainy season. Hundreds of dams were constructed, and billions of cubic meters of water were accumulated at strategic points in the NEB. Transportation infrastructure and irrigation projects are also important. In recent decades, rural development projects and irrigation projects have been adopted, along with efforts to diversify the regional economy to make it less dependent on rain-fed agriculture. The industrialization strategy for the NEB, adopted in the 1950s, has been relatively successful, but insufficient.

More recently, the NEB has benefited from climate research, particularly in regard to the El Niño phenomenon and the behavior of the Intertropical Convergence Zone and the Atlantic and Pacific sea surface temperatures. It is now possible to predict a drought in the northern part of the NEB three months in advance with 80% certainty. This information has been used to orient government agricultural policy, and the results have been positive. Estimates indicate that the use of climate prediction information has contributed significantly to increased agricultural production in drought years (Nobre 1994).

The third question refers to the timing of adaptation actions. Of course, a strategy of preventive adaptation is preferred. Such a policy depends, however, on the level of public awareness reached before a climate change occurs. Even if there is considerable public awareness, conflicts of interest will oblige policymakers to choose between climate change adaptation policies and other public policies (even in the case of "no regrets" policies). The choice becomes a political issue. The pressing social problems of developing countries, where people are struggling against hunger, starvation, and unemployment, will probably not allow much space for creating awareness of possible future changes in the climate. It is necessary to address the climate change issue in the context of other problems that affect the lives of the people in these regions. If nothing is done, adaptation to climate change must wait until the climate change is firmly established. Adaptation policies will be translated into emergency relief actions to alleviate the consequences of extreme events under the new climate change pattern.

The fourth question refers to regional conditions for adaptation. The adaptation response must come from the local community, whose participation in the planning and decision-making process is crucial. Imported solutions without democratic legitimization by the different groups of the affected community — or, at least, by those who hold decision-making power — will not work. Local communities and the people in general must participate in the decision-making process along with the government and the ruling elite. A thorough change of mentality and culture may be needed. If the climate change issue is dealt with in isolation from the more complex needs of local and regional socioeconomic development, adaptation policies will fail. Sustainable development is the answer to make the region, government, society as a whole, and every citizen become more resilient and less vulnerable. Human resources development may be the main component in a sustainable development strategy concerned with increasing regional resilience to climate hazards.

Climate Change Adaptations and Sustainable Development: A Planning Framework

Since 1992, a project has been underway in the NEB to develop a comprehensive methodology for sustainable development planning in semiarid regions — the Aridas Project (Magalhães et al. 1994). One objective was to develop a methodology to introduce climate change issues into development planning. The study was sponsored by the federal government and the nine regional state governments, with the participation of nongovernment organizations, universities, and government agencies. A general policy was proposed for the entire region. Some 50 sectoral studies were produced, with specific proposals seeking sustainability in such areas as education, health, water resources, land resources, biodiversity, energy, agriculture, manufacturing, urban development, land tenure systems, and vegetation. At this time (1995), several states of the NEB are preparing development plans based on the Aridas methodology.

Methodological Framework

The first step in the Aridas Project was to define "sustainable development" as development that is able to be durable. The definition of sustainability must go beyond the environmental dimension. It must also include the social, economic, and political dimensions. Development planning should integrate all these dimensions in both the short and long term.

The holistic nature of sustainable development requires interdisciplinary teams and broad public participation in the planning process. Teams were organized on the basis of the specific conditions of each region. For the NEB, the Aridas Project was organized into seven different interdisciplinary and interinstitutional working groups: natural resources and the environment (including climate), water resources, human resources, economy, rural and urban land use (including institutional analysis), development policies, and outreach.

Each team performed the same basic tasks in all studies. Beginning with a diagnosis of the current state of sustainability and its sensitivity to present climate variability, two future scenarios were prepared: a "business-as-usual" scenario and a "desired" sustainable development scenario. Future scenarios were tested against regional climate change scenarios. A sustainable development strategy was then derived for regional, state, and sectoral policies. Broad community participation validated the different analyses and proposals, enhancing public awareness and political support of the new strategies. Finally, development plans and programs were redefined to encompass the requirements of the Aridas methodology.

The nine steps of the process were as follows:

- Organize information and data for each area.
- Analyze present regional sustainability on the basis of sustainable development indicators for each area.
- Analyze present vulnerability to climate variability.
- Analyze past and present adaptation responses and lessons learned about the causes of success and failure.
- Prepare a business-as-usual scenario (years 2000, 2010, and 2020), based on a projection of present indicators.
- Prepare the desired sustainable development scenario for the future, based on broad societal participation in the several states of the NEB.
- Analyze future vulnerability to climate, with and without climate change (a scenario of regional climate change was needed to complete this step).
- Prepare a sustainable development strategy based on the previous tasks.
- Prepare state and sectoral sustainable development plans.

Each task should combine a scientific-technical contribution and a societal participation process.

Climate Change Dimension

For the first time, the Aridas methodology introduced possible climate change into the process of development planning. A climate change scenario was produced (Nobre 1994). The Aridas process itself has increased awareness of climate change in the planning and academic communities and in other communities involved in the project. At first, it was difficult to convince social researchers and policy analysts of the importance of addressing the issue of climate change because of the relative importance of pressing present-day survival problems compared with the uncertainty of future climate change.

Conclusions

The experience of the NEB constitutes a rich example of a climate variability adaptation policy. Many aspects have been successful. Government and public policies have helped to increase regional social and economic resilience to droughts, but have paradoxically increased land use. Population growth and unequal economic development have increased social and environmental vulnerability.

Two positive results are worth examining. One is the success of short-term strategies to alleviate the effects of recurrent droughts. Although the strategy needs improvement, it has met its objectives. The other positive result is a permanent increase in resistance capacity. Economic development policies and long-term infrastructure projects for water resources have helped to increase overall regional capacity to adapt to droughts. Lessons from the successes and failures of the NEB experience may be useful in assessing how adaptation policies will work and how to avoid mistakes in the policy-making process.

Several points emerge from the NEB experience in adapting to climate variability. During normal climate variability, the nature of climatic events will basically be the same. Hence, the nature of the adaptation policies may also be similar — although not necessarily the same because the intensity and frequency of the events may change and society also changes. The appropriate response must involve more public awareness, education, sustainable development, and cooperation, both regional and international. The ultimate objective should be to increase the capacity of each individual to face climate changes and other crises that may affect his or her life.

A methodology for sustainable development planning was proposed. The Aridas process incorporates a concern for adaptation to climate variability into regional and national development planning processes. In developing regions with pressing social and economic problems, this process may be the appropriate way to make people care not only for their present survival, but also for their own future and the future of the people of other parts of the world.

Acknowledgments

The author thanks the participants at the Climate Change and Adaptation Conference held in St. Petersburg, Russian Federation, on May 22-25, 1995. The author especially thanks M. Glantz and I. Burton for useful comments, as well as the organizers of the conference, in particular J. Smith for his continuous encouragement. Any errors are the responsibility of the author alone.

References

Arraes, R.A., and I. Castelar, 1991, "Efeitos da Seca nas Finanças Públicas do Ceará," in A.R. Magalhães and E. Bezerra Neto, *Impactos Socials e Econômicos de Variações Climáticas e Respostas Governamentals no Brasil*, Imprensa Oficial do Ceará, Fortaleza, Brazil.

Burton, I., and S.J. Cohen, 1993, "Adapting to Global Warming: Regional Options," in A.R. Magalhães and A.F. Bezerra (eds.), *Proceedings of the International Conference on the Impacts of Climatic Variations and Sustainable Development in Semi-Arid Regions, A Contribution to UNCED*, Vol. IV, pp. 871-886, Esquel Brazil Foundation, Brasília and Fortaleza, Brazil.

Carvalho, O., C.A.G. Euler, M.M.C.L. Matos, 1994, *Variabilidade Climática e Planejamento da Ação Governamental no Nordeste Semi-Árido — Ávaliação da Seca de 1993*, unpublished, SEPLAN-IICA, Brasília, Brazil.

Cohen, S.J. (ed.), 1994, "Mackenzie Basin Impact Study, Interim Report 2," *Proceedings of the 6th Biennial AES/DIAND Meeting on Northern Climate and Mid-Study Workshop of the Mackenzie Basin Impact Study*, Environment Canada, Ottowa, Canada.

Glantz, M.H., 1988, *Societal Responses to Regional Climate Change: Forecasting by Analogy*, Westview Press, Boulder, Colo., USA.

Magalhães, A.R., and M.H. Glantz, 1992, *Socioeconomic Impacts of Climate Variations and Policy Responses in Brazil*, U.N. Environmental Programme/ SEPLAN-CE/Esquel Foundation, Brasília, Brazil.

Magalhães, A.R., E. Bezerra Neto, and S.S. Panagides, 1994, *Projeto Aridas: Uma Estratégia de Desenvolvimento Sustentável para o Nordeste. Documento Básico*, unpublished, Secretaria de Planejamento da Presidência da República (SEPLAN/IICA), Brasília, Brazil.

Magalhães, A.R., H.C. Filho, F.L. Garagorry, J.C. Gasques, L.C.B. Molion, M. da S.A. Neto, C.A. Nobre, E.R. Porto, and O.E. Rebouías, 1988, "The Effects of Climate Variations on Agriculture in Northeast Brazil," in M.L Parry, T.R. Carter, and N.T. Konijn (eds.), *The Impact of Climatic Variations on Agriculture, Vol. 2: Assessments in Semi-Arid Regions*, pp. 273-380, Kluwer Academic Press, Dordrecht, the Netherlands.

Nobre, P., 1994, *Clima e Mudanças Climáticas no Nordeste*, preliminary version, Projeto Aridas, SEPLAN, Brasília, Brazil.

Ribot, J.C., A.R. Magalhães, and S.S. Panagides, 1995, *Climate Variability, Climate Change and Social Vulnerability in the Semi-Arid Tropics*, Cambridge University Press, Cambridge, United Kingdom.

Ribot, J.C., A. Najam, and G. Watson, 1993, "Climate Variation, Vulnerability and Sustainable Development in Semi-Arid Regions," in A.R. Magalhães and A.F. Bezerra (eds.), *Proceedings International Conference on the Impacts of Climatic Variations and Sustainable Development in Semi-Arid Regions, A Contribution to UNCED*, Vol. IV, pp. 73-184, Esquel Brazil Foundation, Brasília and Fortaleza, Brazil.

Rosenberg, N.J., and P.R. Crosson, 1993, "Understanding Regional Impacts of Climate Change and Climate Variability: Application to a Region in North America with Climates Ranging from Semi-Arid to Humid," in A.R. Magalhães and A.F. Bezerra (eds.), *Proceedings International Conference on the Impacts of Climatic Variations and Sustainable Development in Semi-Arid Regions, A Contribution to UNCED*, Vol. IV, pp. 481-518, Esquel Brazil Foundation, Brasília and Fortaleza, Brazil.

SUDENE, 1994, *Nordeste: Indicadores Econômicos e Sociais da Região Nordeste e Brasil (synthesis)*, Superintendency of the Development of the Northeast, Recife and Pernambuco, Brazil.

The Growth of Adaptation Capacity: Practice and Policy

Ian Burton
Atmospheric Environment Service
Environment Canada
and
Institute for Environmental Studies
University of Toronto
33 Willcocks Street
Toronto, Ontario, Canada M5S 3B3

Abstract

Countries can strengthen their capacity for adaptation to climate change and variability and develop appropriate public policy. Six reasons for developing and adopting adaptation strategies now are provided. Adaptation to extreme events (natural hazards) can reduce long-term vulnerability to climate change. A framework for identifying and classifying adaptations is presented and applied to the adaptation options described in Intergovernmental Panel on Climate Change reports. It is found that some major categories of adaptation, such as insurance, disaster assistance, and changes of land use and location, receive little or no attention.

Introduction

Efforts to prevent the impacts of climate change require the stabilization of levels of greenhouse gases in the atmosphere. Stabilization in turn requires substantial reductions in the emissions of greenhouse gases (especially carbon dioxide [CO_2]) from human activities. It is becoming increasingly clear that such reductions will be difficult to achieve. Beyond the first gains from improving energy efficiency, the costs rise sharply; however, cost is not the only factor. Other factors include access to proprietary technology and unresolved political debates (i.e., how much and how rapidly should emissions be reduced, how should nations share responsibility for such reductions).

In signing the U.N. Framework Convention on Climate Change (UNFCCC) in Rio de Janeiro in June 1992, representatives from developed countries (Annex I countries) committed to attempt to stabilize emissions at 1990 levels by the year 2000. Progress reports presented at the First Conference of the Parties in Berlin in April 1995 clearly show that many countries are failing to reach this modest target (U.S. Climate Action Network and Climate Network Europe 1995).

In weighing the need for action to limit the emissions of greenhouse gases, countries are finding that economic considerations are becoming increasingly important. Thus, governments would like to determine the impacts of global warming more precisely. Parties to the UNFCCC want to compare the costs of action vs. the costs of inaction. Clearly, the costs of inaction (i.e., let global warming proceed) can be reduced to some extent by taking adaptive measures. Hence, it is meaningful to assess the potential contribution of adaptive measures in reducing the impacts caused by climate change. For this reason, adaptation assessments are needed, especially assessments of the costs of adaptation and the amounts of reduction of adverse impacts.

The practical science of adaptation to *climate change* is new, whereas the practice of adaptation to *climate* is very old. Currently, few adaptation policies for climate change are in place. Research on adaptation is therefore urgently needed, and it is encouraging that researchers are beginning to examine these issues. However, people have always adapted to climate, and many policies and practices are already in place (i.e., building design standards, agricultural practices, laws dealing with water use and allocation). These policies generally treat climate as though it were constant over decades. The new challenge is to take the vast array of adaptations to climate and make them appropriate for climate change.

The capacity for adaptation varies considerably, and it can be changed in a conscious and planned fashion. Also, adaptive capacity can decrease as well as increase. In adaptation assessments, it is important therefore not to assume certain levels of adaptation and what they might achieve. It is also crucial to consider how adaptive capacity might be increased and prevented from decreasing. The study of adaptation is an empirical and practical science. It is possible to go into the field and observe adaptation at work, to recognize the consequences of the lack of adaptation, or worse, maladaptation.

The main purpose here is to suggest ways by which countries can strengthen their capacity for adaptation at the practical level and develop public policy to facilitate and support adaptation. However, before these objectives can be addressed, the widespread neglect (and even opposition) to the idea of adaptation must be recognized. From the outset, it is necessary to make a case for adaptation and show why it should form an integral part of a comprehensive climate policy.

Limitations of Adaptation

The debate on climate policy initially focused on limitation and adaptation issues. These issues were seen as complementary approaches, and the question was quantitative: how much of each would be needed to be effective. This social construction into a simple dichotomy of responses has proved unfortunate because under the political pressures for action, the complementarity between limitation and adaptation has changed. They are now too often perceived as substitutes rather than as mutually supportive. Fear has increased that a successful demonstration (or a mere argument about the potential for successful demonstration) of adaptive capacity may weaken the resolve on the part of governments and others to reduce greenhouse gas emissions. Evidence shows that society can readily adapt to climate and has been doing so successfully for some time (Ausubel 1991). However, adaptation to climate *change* is a more complicated and urgent matter.

An important assessment by a panel of the U.S. National Academy of Sciences (NAS [1992]) concluded that in most areas of climate impact, adaptation could be successfully achieved "at a cost." By implication, these costs were considered to be low or at least manageable within the United States (Table 1).

Others argued that the potential impacts of global warming were so catastrophic that greenhouse gas emissions should be reduced quickly and substantially. Thus, the Toronto Conference of 1988 recommended a 20% reduction in emissions by 2005 using 1988 as the base year (World Meteorological Organization 1989).

Scientific and economic debate about the impacts of climate change and their likely costs, and the costs of limitations and the potential for adaptation, quickly became polarized. It became increasingly difficult to keep policy and politics out of science. Complementary approaches — part of a mixed strategy — now became mutually antagonistic. Those carrying out research on adaptation were seen as proponents of adaptation and as supporters of the cause of those whose interests led them to oppose limitation actions. This group included the fossil fuel industry and the oil exporting countries and (by implication at least) those nations and societies with high per capita energy use.

Many scientists and scientific institutions carrying out research on limitations saw themselves as "saviors," trying to reduce greenhouse gas emissions to protect the global environment, to save small island states from the rising sea, and to protect the development interests of poorer countries with (as yet) low per capita energy use.

It became unpopular at an international level to advocate adaptation to climate change. While the UNFCCC recognizes adaptation, it is relegated to a

Table 1. Sensitivity and Adaptability of Human Activities and of Nature to Climate Change Alone[a]

Subject	Class
Farming	Sensitive; can be adapted at a cost
Managed forests and grasslands	Sensitive; can be adapted at a cost
Natural landscape	Sensitive; adaptation is questionable
Marine ecosystems	Sensitive; adaptation is questionable
Water resources	Sensitive; can be adapted at a cost
Industry and energy	Low sensitivity
Tourism and recreation	Sensitive; can be adapted at a cost
Settlements and coastal structures	Sensitive; can be adapted at a cost
Health	Low sensitivity
Human migration	Sensitive; can be adapted at a cost
Domestic tranquillity	Sensitive; can be adapted at a cost

[a] Climate is assumed to change gradually by 1-5°C when the planet reaches equilibrium with an equivalent of a doubling of CO_2.

Source: NAS (1992, p. 505).

minor role. The overwhelming emphasis is on limitations. For example, Article II of the UNFCCC states:

> The ultimate objective of this Convention and any related legal instruments that the Conference of the Parties may adopt is to achieve, in accordance with the relevant provisions of the Convention, stabilization of greenhouse gas concentrations in the atmosphere at a level that would prevent dangerous anthropogenic interference with the climate system. Such a level should be achieved within a time-frame sufficient to allow ecosystems to adapt naturally to climate change, to ensure that food production is not threatened and to enable economic development to proceed in a sustainable manner.

The UNFCCC aims to respond *actively* to achieve stabilization. Moreover, although adaptation is viewed positively, the Convention wants it to occur "naturally."

This neglect of and opposition to proactive adaptation has primarily been driven by the politicization of the climate debate at the international level. This factor has continued up to and including the First Conference of the Parties to the Convention held in Berlin in March-April 1995.

To some extent, a similar conflict has occurred at the national level, especially in larger countries with highly developed regional and sectoral interests. This conflict has prevented the formulation of comprehensive response strategies that give significant weight to adaptation.

A compelling case can be made for adaptation, not as an alternative or substitute for reducing greenhouse gas emissions, but as a necessary tool that can be used by the global community as countries collectively proceed with the difficult task of managing the potential impacts of climate change.

Adaptation Now

A number of reasons can be given for taking adaptive measures at this time. Six reasons are given in Table 2.

The first and most compelling argument that supports adaptation causes the most misgivings. Adaptation is necessary because limitation cannot totally succeed. It is too late to prevent some effects of climate change. Countries expect some climate changes in the atmosphere because greenhouse gas concentrations have increased and will continue to increase. Emissions from developing countries have been increasing and will continue to grow as their populations and economies expand and per capita energy use increases. Developed (Annex I) countries have not stabilized their emissions, and although they have agreed

Table 2. Six Reasons to Adapt to Climate Change Now

1. Climate change cannot be totally avoided. Present levels of development already have created commitment to some change.
2. Anticipatory and precautionary adaptation is more effective and less costly than forced, last-minute, emergency adaptation or retrofitting.
3. Climate change may be more rapid and more pronounced than current estimates suggest. Unexpected events are possible.
4. Immediate benefits can be gained from better adaptation to climate variability and extreme atmospheric events.
5. Immediate benefits can also be gained by removing maladaptive policies and practices.
6. Climate change brings opportunities as well as threats. Future benefits can result from climate change.

to the Berlin Mandate to negotiate the amount of reductions after 2000, it is clear that progress is likely to remain slow (U.S. Climate Action Network and Climate Network Europe 1995). Adaptation therefore has become a necessity, not a matter of policy choice.

A second reason to develop adaptation strategies and policies is that early anticipatory adaptation will prove less costly than forced adaptive measures if delays continue. For example, adaptation to anticipated sea-level rise presently involves avoiding construction of hotels and residences and making other fixed investments in zones likely to be inundated or subject to more severe, higher, and possibly more frequent storms. Protecting or abandoning such investments later is likely to prove more costly than restraining development now and redirecting or relocating to less vulnerable sites. Similarly, farming systems, including irrigation works and so on, will be more sustainable if they are developed to withstand future changes in climate (i.e., more frequent, severe, prolonged drought) than if their design criteria assume the climate will remain stable.

A third reason to consider adaptation seriously today is that climate may change more rapidly than General Circulation Models suggest. There is considerable uncertainty in the relationship of emissions to concentrations, and in the relationships of concentrations to climate change. The climate system is complex and nonlinear and is certain to have unknown threshold levels. At this time, it is impossible to predict when these levels will be crossed. However, experience with other atmospheric issues (e.g., ozone layer depletion, acid precipitation) indicates that surprises are possible and likely. This logic also supports stronger attempts to stabilize and reduce emissions.

These reasons may not be sufficient to convince governments faced with other imminent problems to address adaptation to climate change. It is tempting to delay action until the need for action is apparent. However, three additional reasons (if carefully presented) should be more appealing to governments, even those governments severely constrained because of poverty and lack of development.

Adaptation to climate involves more than a gradual rise in sea level and increase in mean temperature. As global warming advances, changes can affect the frequency and magnitude of extreme events. Adaptation to climate is driven more by climate extremes than by climate norms. Damage caused by extreme atmospheric hazards has risen sharply in recent years (Figure 1 and Table 3). Adaptation is needed now to reduce the vulnerability to present-day climate extremes. Society and the economy stand to gain immediate benefits. Moreover, steps to improve adaptation to present-day extreme events provide a testing ground for the stronger adaptive measures that may be subsequently required.

Another feature of present-day relationships between socioeconomic and climate systems is maladaptation. Many policies and practices in place increase vulnerability to atmospheric events. Examples include agricultural policies that

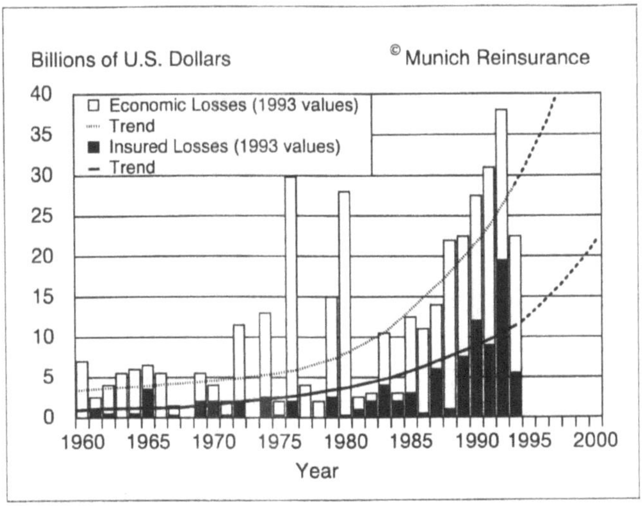

Figure 1. Economic and Insured Losses Due to Great Natural Disasters (1960-1993), with Trends Extrapolated to 2000 (Source: Munich Reinsurance 1994, cited in McCulloch and Etkin 1995; used with permission)

Table 3. Decadal Totals of Economic and Insured Losses (in billions of 1993 US$)

Losses	1960-1969	1970-1979	1980-1989	1984-1993	Last 10, 1960s
Economic	42.7	82.2	130.5	204.4	4.8
Insured	5.7	9.6	26.3	69.5	12.2

Source: Munich Reinsurance (1994); cited in McCulloch and Etkin (1995).

encourage risky farming decisions in regions prone to drought, or subsidized insurance and public relief and rehabilitation programs that encourage development on flood plains and in exposed coastal zones.

Paradoxically, the least welcome argument for adaptation is that climate change brings opportunities as well as threats. As global climate patterns change, some places will benefit from more favorable conditions. Most economies will have winners and losers within the same city or region (Burton 1992). While consensus is widespread that the net effects of climate change will be negative (i.e., the losses will outweigh the gains), it is nevertheless clear that both opportunities and potential gains will occur as well as adverse impacts. Adaptation plays a role in both.

Therefore, a proactive stand toward adaptation is justified; that is, develop an approach that will seek potential adaptive strategies, find ways to choose among them, and remove obstacles to their implementation. It is important to perform these activities without suggesting that less effort is required to reduce greenhouse gas emissions.

Adaptation Options

Identifying adaptation options is Step 3 in the Technical Guidelines developed under the Intergovernmental Panel on Climate Change (Carter et al. 1994) (Table 4).

On the assumption that policy- and decision makers have been persuaded to include adaptation as part of climate policy, how do countries develop suitable options or measures for climate adaptation? One approach is to survey the literature and collect examples and suggestions. A survey conducted in Canada by D. Herbert (Smit 1993) produced a list of 226 possible adaptation options described as an "inventory of adaptation strategies." Although the list covers a wide range of options, it is admittedly incomplete.

Another approach, developed some years ago under the auspices of the International Geographical Union, involved an international network of researchers who sought to better understand the process of adapting (or as they called it "adjusting") to atmospheric extremes, such as droughts, floods, tropical cycles, and other natural hazards (Burton et al. 1993). A series of 120 case studies was carried out in 23 countries (White 1974). In each case, field interviews were

Table 4. Development of an Adaptation Strategy

Define objectives
Specify important climatic impacts
Identify adaptation options
Examine constraints
Quantify measures/formulate alternative strategies
Weight objectives/evaluate trade-offs
Recommend adaptive measures

Source: Carter et al. (1994).

conducted with government officials and planners responsible for public safety and damage reduction. The field interviews resulted in lists of all adaptations in use, as well as those known to have been used in the past. The interviews also asked about potential new adaptations that might be available in the future. The empirical field approach resulted in a rich diversity of adaptive measures. These measures ranged from traditional agricultural practices of mixed cropping and intertilling to provide insurance against total crop failure in times of drought, to advanced engineering methods of water management, and government-sponsored insurance policies. An analysis of these data suggests that adaptations to extreme events can be grouped into six broad classes (Figure 2).

At one extreme (Choice 6), people do nothing and simply bear the loss. A conscious choice to take no anticipatory action may be appropriate if the consequences are likely to be trivial or if the probability of an event is extremely remote.

Where anticipated or experienced losses are higher, society often takes steps to share the losses (Choice 5). In traditional societies and in the past, informal arrangements were made to spread the losses over large numbers of people within extended family groups or within a village, tribe, or community-level institution. In modern states, private or publicly sponsored insurance can serve a similar purpose. Losses can also be shared by compensating the victims and providing emergency disaster relief and, subsequently, assisting in rehabilitation and recovery. The costs of this type of sharing come from general revenues in

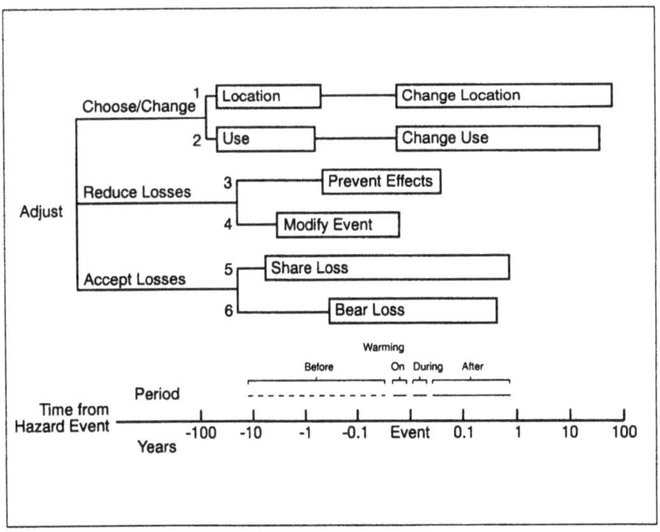

Figure 2. A Choice Tree of Adjustment to Extreme Events

the regions or countries affected. Also, on an international scale, disaster relief is now common (NAS 1994).

At the other extreme, decisions may be made to avoid the danger altogether by changing locations (Choice 1). For example, people have migrated from drought regions (i.e., the American Great Plains in the 1930s and Northeast Brazil intermittently). Mass migration usually occurs in response to major events, and this could happen in the case of climate change and sea-level rise. More routine locational choices include avoiding flood plain locations or removing people from flood plains after floods.

In Choice 2, people remain in the same place, but use their land for less vulnerable activities. For example, parks, golf courses, and parking garages may replace houses, hospitals, and schools on flood plains or exposed coastal sites. Farmers may plant crops less vulnerable to drought, hail, forest, or other atmospheric hazards.

Choices 3 and 4 include steps to prevent or reduce effects. Examples are (1) applying irrigation water to prevent drought, (2) modifying events by building a flood control dam, or (3) constructing a seawall to keep out storm surges. Examples of prevention include designing or operating structures to resist atmospheric extremes. Houses are commonly designed to withstand extreme winds, in some cases up to and including tornado velocity, and can be elevated on stilts to keep out flood water.

This typology has been applied more recently to the draft chapters of the forthcoming Intergovernmental Panel on Climate Change (IPCC) Working Group II report on Impacts and Adaptations to Climate Change (Smith et al., forthcoming.). This exercise suggests continuing to refine the typology by expanding the "prevent the effects" category into five kinds (Figure 3) and adding research and education as separate responses, making eight major categories.

The IPCC draft report suggests 105 adaptive measures (Carter et al. 1994). The distribution of responses is shown in Figure 3. Of particular note is that some categories of adaptation seem to be totally or largely neglected in the literature summarized by IPCC. These include bearing and sharing the losses, and avoiding impacts by changing land or resource use or changing location. The frequency of adaptations in the legislative, regulatory, financial, institutional, and administrative categories is markedly higher than in structural/technological adaptations or market-based options. The category of on-site operations also seems remarkably low.

These observations should be treated with caution. They may reflect the way in which categories have been defined and the lack of detailed description in the IPCC reports, which sometimes makes classification arbitrary. It seems likely that if the IPCC Working Group II had received support at the outset of the study with a typology of adaptation options, they would have reported more specific options. The possibility is open for a more thorough canvass of adaptation options in the third IPCC assessment.

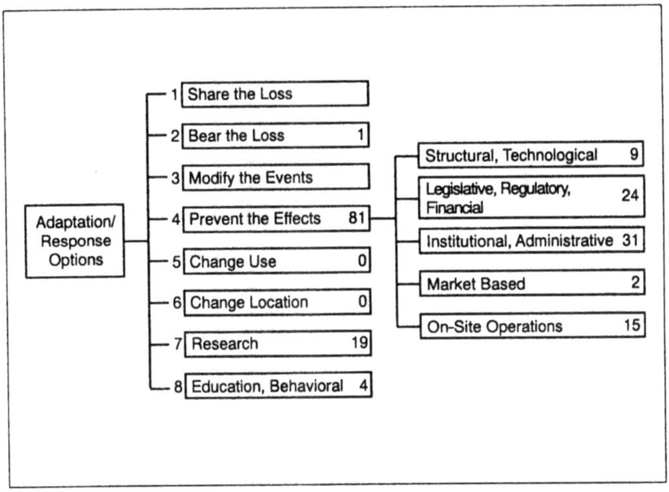

Figure 3. Classification and Frequency of Adaptation Options in IPCC Working Group II

Any such canvass of adaptation options in relation to climate change should distinguish between anticipatory and reactive options. As suggested in Figure 2, many actions can be taken before, as well as during and after, specific hazardous events. Climate change provides an opportunity to begin anticipatory adaptation now by improving response to present-day climate variability. In effect, the natural atmospheric hazards experienced today are a testing ground and a preparatory exercise for the more severe conditions that will develop as the climate regimes change.

Adapting to present-day climate variability also provides an opportunity to choose post-event adjustments that will not simply rehabilitate and restore pre-event conditions, but will reduce future vulnerability. For example, damaged buildings would not be reconstructed on flood plains or coastal locations in the same places and in the same way. Rather, they would be relocated or redesigned to be less vulnerable in the future.

Who Adapts?

An examination of the range of adaptation options shows that responsibility for adaptation is dispersed through all levels of government, the private sector, as well as individuals, families, and communities. To strengthen adaptive capacity requires developing a comprehensive and strategic view of adaptation in which relative roles and responsibilities are well defined. In particular, action at senior government levels is needed to facilitate and encourage adaptation at lower levels of government, among individuals, communities, and the private sector. Ill-considered government action can destroy adaptive capacity, which may help explain some current outstanding examples of maladaptation.

An examination of the national reports prepared for the UNFCCC's First Conference of the Parties suggests that no country has developed a comprehensive and strategic program for adaptation, and many countries have not started to consider such an idea. Given the current heavy focus on the means to reduce greenhouse gas emissions, it is not surprising that such adaptation programs have not been developed.

Without detracting from the limitation effort, scientists can make a case for adaptation today. The development of policies for adaptation should take into account the reasons for adaptation given in Table 2.

At the level of individual behavior and on-site operations, human ingenuity is likely to lead to much spontaneous adaptation, especially as the signals of climate change become more distinct. At the government level and, in particular, where major costs are to be incurred, adaptive capacity is likely to depend largely on available resources. Thus, the more highly developed and prosperous countries — whatever their experience in climate change — are more likely to be able to afford the costs of adaptation. The UNFCCC recognizes this circumstance in Article 4, Section 4: "The developed country Parties . . . shall also assist the developing country Parties that are particularly vulnerable to the adverse efforts of climate change in meeting costs of adaptation to those adverse effects" (U.N. Environmental Programme/World Meteorological Organization 1992).

A prerequisite for such assistance is the identification of adaptive options and the development of the capacity to adapt through both practice and policy initiatives. But the case for adaptation should not rest on the prospect of international assistance. Unlike the benefits of emission reductions, which are shared globally, the benefits of adaptive options are likely to fall more heavily in those countries and locations where the adaptation occurs.

References

Ausubel, J., 1991, "Why Climate Matters," *Nature* 350:649-652, April 25.

Burton, I., 1992, "Regions of Resilience: An Essay on Global Warming," in J. Schmandt and J. Clarkson (eds.), *The Regions and Global Warming: Impacts and Response Strategies*, pp. 257-274, Oxford University Press, New York, N.Y., USA.

Burton, I., R.W. Kates, and G.F. White, 1993, *The Environment as Hazard*, 2nd Ed., The Guilford Press, New York, N.Y., USA.

Carter, T.R., M.L. Parry, H. Harasawa, and S. Nishioka, 1994, *IPCC Technical Guidelines for Assessing Climate Change Impacts and Adaptations*, Working Group II of the Intergovernmental Panel on Climate Change, University College, London, United Kingdom, and Center for Global Environmental Studies, Tsukuba, Japan.

McCulloch, J., and D. Etkin, 1995, *Proceedings of a Workshop on Improving Responses to Atmospheric Extremes: The Role of Insurance and Compensation*, Toronto, Ontario, Canada, Oct. 3-4.

National Academy of Sciences (NAS), 1992, *Policy Implications of Greenhouse Warming: Mitigation, Adaptation, and the Science Base*, National Academy Press, Washington, D.C., USA.

NAS, 1994, *The National Report of Canada: Prepared for the IDNDR Midterm Review and the 1994 World Conference on Natural Disaster Reduction*, Yokahama, Japan, May 23-27, 1994, National Academy of Sciences, National Academy of Engineering, Ottowa, Canada.

Smit, B., (ed.), 1993, *Adaptation to Climatic Variability and Change*, report of the Task Force on Climate Adaptation, The Canadian Climate Program, Occasional Paper No. 19, Department of Geography, University of Guelph, Ontario, Canada.

U.N. Environmental Programme/World Meteorological Organization, 1992, *United Nations Framework Convention on Climate Change*, United Nations, Geneva, Switzerland.

U.S. Climate Action Network and Climate Network Europe, 1995, *Independent NGO Evaluations of National Plans for Climate Change Mitigation: OECD Countries.*

White, G.F., 1974, *Natural Hazards: Local, National, Global*, Oxford University Press, New York, N.Y., USA.

World Meteorological Organization, 1989, *Conference Proceedings from the Changing Atmosphere: Implications for Global Security*, Toronto, Canada, Publication 710, June 27-30, 1988, Geneva, Switzerland.

Using a Decision Matrix to Assess Climate Change Adaptation Options

Joel B. Smith
Hagler Bailly Consulting, Inc.
P.O. Box 0
Boulder, CO 80306-1906, USA

Abstract

A systematic method is needed to compare the effectiveness and costs of options for adapting to climate change. A decision matrix can be used to analyze the effectiveness of adaptation options in meeting specific policy goals under various climate change scenarios. Consistent quantitative measures of different goals should be used to enable comparisons to be made across multiple goals. The goals can be weighted based on their relative importance. The decision matrix can also be used to rank the relative cost-effectiveness of adaptation policies and practices, or if monetary measures are used, to examine whether benefits exceed costs and benefit-cost ratios.

Introduction

Many researchers and scientists have identified options and criteria for policymakers to consider in adapting to climate change. For example, the Office of Technology Assessment (OTA [1993]) and Smit (1993) each identified well over 100 adaptation options. Titus (1991) suggested considering such factors as economic efficiency, flexibility to perform well under a wide variety of climate change situations, urgency (need to implement the strategy now rather than waiting 10 or 20 years), cost, and institutional feasibility. Those adaptations addressing climate change events that may lead to irreversible or catastrophic consequences, long-term decisions that may be affected by climate change, and trends that may make adaptation more difficult in the future are the highest priority for being implemented now (Smith, in press).

These studies, however, do not suggest how to evaluate these adaptation options against each other or against current policies. For example, given the uncertainties about regional climate change, how is a policymaker to judge whether it is better to maintain current policies and practices on managing forests, or to adopt alternative policies and practices? This paper discusses the use of a decision matrix for evaluating and selecting options for adapting to climate change. The matrix helps users to identify policy goals and evaluate how well the current and alternative policies do in meeting the goals under current climate and climate change scenarios.

In most cases, policymakers will evaluate policies with multiple objectives. A method must be used to examine how well these objectives are satisfied by current and alternative policies. This method should enable one to make direct comparisons among the different policy objectives. The problem is that it may not be immediately obvious how to compare different objectives, such as maintaining water supplies for domestic use and meeting water quality standards. Meier and Munasinghe (1994) note that so-called "multi-attribute 'scoring' systems" have been gaining favor in the United States and elsewhere in recent years. For example, the Niagara Mohawk utility in New York state devised a scoring system to assess the multiple effects of sulfur dioxide and carbon dioxide emissions from power plants.

A decision matrix is simply a tool to organize information. Analysis of how well alternative policies perform must be done outside of the matrix, with the matrix serving as a way to display and analyze results. Assessment of biophysical or socioeconomic vulnerability of systems, such as forests, to climate change should be designed to provide results that can be entered into the decision matrix. To be sure, developing the information to be used in such a matrix can be difficult. If data or methods needed for conducting such an analysis are not available, it may not be possible to develop all the information necessary to fill in the decision matrix. It may be necessary to rely on expert judgment to supply results.

The use of a decision matrix is explained by examining a hypothetical example of water resource adaptation policies. As of this writing, such a matrix has not been used to examine climate change adaptation options; all numbers and results reported and discussed here are hypothetical. The purpose of this example is to explain how one can use the decision matrix, not to demonstrate that one water policy is superior to another.

Hypothetical Example

Assume that policies are being examined for operating a hypothetical reservoir management system for a river basin. Water in the river is used to meet consumptive and in-stream uses. Consumption consists of domestic uses and irrigation for agriculture. The in-stream uses consist of water quality and recreation. For simplicity, only these four uses are considered, although many others may exist for a reservoir system, such as hydropower production and flood control.

Assume that the current policy for managing the reservoir system is one in which priority is given to irrigation to receive water. The remaining water in the system is used to meet the needs of domestic users (including industrial), water quality, and recreation, in that order. In terms of the value of water per unit of use, however, domestic uses are greatest, followed by water quality, with irrigation and recreation having the same and least value. Thus, the government allocation is not consistent with marginal values of water.

Decision Matrix Organization

Table 1 displays the decision matrix format.[1] The columns list objectives and policies, climate, policy objectives, scores, and changes in costs of policies. The rows of values are for current and alternative policies. Within each policy are rows for the current climate and two climate change scenarios. The units in the decision matrix are a measure of success or, in this example, failure.

To compare how alternative policies perform in meeting objectives, quantitative measures of success or failure are crucial (Stakhiv 1993). Furthermore, such measures should be meaningful in that they represent the value associated with meeting an objective. The measure should be objective in that, to the extent possible, determining the quantitative value should not involve a subjective judgment.

Ideally, one numerical measure could be applied to all policy objectives. Evaluating costs and benefits of policies in dollars or other monetary units is an attractive option. Benefit-cost analysis enables one to compare all the benefits and costs of policies in a common metric. One can determine whether the costs of a policy are justified by its benefits, as well as whether one policy has greater net benefits (benefits minus costs) or a greater benefit-cost ratio (benefits divided by costs) than another policy (Gramlich 1981). Policies can be ranked based on their relative net benefits or benefit-cost ratio.

A difficulty of using benefit-cost analysis is that it may not always be possible to estimate monetary values for all benefits. In the water example, one would be able to estimate the dollar cost of implementing the current water allocation policy and alternatives, and estimating the dollar value of some uses of water is feasible. For example, one could estimate the benefit of irrigation water by estimating the change in value of output from a farm with and without the use of irrigation water (or by comparing farms that use irrigation water with

---

1 Note that the decision matrix enables one to "keep score" of how well a policy performs in meetings its objectives. Additional analysis may be needed to develop the scores that will be entered into each cell. For example, to examine the effect of different policies on the hypothetical reservoir, one may need a reservoir management model to estimate how water may be allocated. If one is examining climate change scenarios, a run-off model may be needed to estimate how climate change could affect water supplies.

Table 1. Water Allocation Policies Decision Matrix for Current Policy and Alternative Policies

Objectives and Policies	Climate	Policy Objectives: Consumptive Uses: Domestic[a] (No. of days <10 million kL/d)[b] ($w = 3$)[c]	Consumptive Uses: Irrigation (No. of days <30 million kL/d)[b] ($w = 1$)[c]	In-Stream Uses: Water Quality (No. of days <ppm DO)[d] ($w = 2$)[c]	In-Stream Uses: Recreation (No. of days <500 m elevation) ($w = 1$)[c]	Total Weighted Score	Average Total Weighted Score	Change in Costs ($\$10^6\ yr^{-1}$)
Current policy	Baseline	6	4	8	10	48	68	N/A[e]
	Scenario 1	15	8	20	40	133		
	Scenario 2	3	1	4	5	23		
Enhanced conservation	Baseline	4	1	6	8	33	40	+20
	Scenario 1	8	2	12	20	70		
	Scenario 2	2	0	3	4	16		
Market allocation	Baseline	3	5	8	10	40	55	+5
	Scenario 1	6	11	20	40	109		
	Scenario 2	0	2	4	5	15		

[a] Domestic, municipal, and industrial.
[b] kL/d = kiloliters per day.
[c] w = weight.
[d] DO = dissolved oxygen, measured in parts per million (ppm).
[e] N/A = not applicable.

those that do not). Estimating benefits of other uses of water, such as water quality, would be more challenging.

If it is not possible to estimate the dollar value of all objectives, another option is to use a similar quantitative, but nonmonetary, estimate of how well objectives are met. In this example, each policy objective, such as providing water for irrigation, has a standard, such as 30 million kL/d for irrigation during the irrigation season. The number of days the standard is not met is estimated, and the policy objective is to minimize this number of days of failure. The number of days is a realistic standard because it can easily be recorded by a reservoir system operator. The virtue of just counting the number of days that a standard is not met is, as seen below, that one can directly compare and combine this measure of effectiveness across the different policy objectives. It is not an ideal measure because it does not indicate the degree of damage when a standard is not met, nor the degree of benefit when a standard is met.

If it is not possible to use quantitative measures of ability to meet a policy objective, one could use an ordinal ranking system. For example, one could score how well an objective is fulfilled on a scale of 1 to 5. Scaling results can enable one to directly compare and even combine scores across categories, with weights, if appropriate. Because this approach is subjective, the scores may differ depending on who is involved in determining them.

Another approach is to use "high," "medium," and "low" ratings for how well objectives are met. Arranging these ratings in a matrix may clearly show differences in how well policies fulfill their objectives. For example, one policy could have mostly high ratings, while a second policy could have mostly medium ratings. The first policy would appear to meet the policy objectives more successfully. A problem with this type of approach is that if some policies score high on one set of objectives and low on others, while alternative policies score low on the first set of objectives and high on others (Table 2), it may be difficult to determine which policy is preferable.

Returning to the example, water resource managers are assumed to attempt to provide 10 million kL/d for domestic uses. Any day in which less than that amount is delivered would not satisfy domestic needs and is recorded as a failure. For water quality, the days on which a water quality standard for dissolved oxygen (DO) of 5 parts per million is not met are measured. If DO levels are below that standard, aquatic life could be harmed, and it is recorded

Table 2. Decision Matrix with High and Low Rankings

Policy	Objective 1	Objective 2	Objective 3	Objective 4
A	High	High	Low	Low
B	Low	Low	High	High

as one day of failure. Finally, recreational needs are assumed to be satisfied when reservoir levels are maintained at an elevation of 500 m above sea level. At that elevation, activities, such as boating and swimming, are possible.

To examine how well each policy performs in meeting all of the policy objectives, one could simply total the number of days of failure across them; however, this approach assumes that the value of meeting each standard is the same. Meeting some objectives may be more important to policymakers and the public than meeting other objectives. The objectives can be weighted depending on their relative importance. Weights are determined on the basis of the relative harm from failing to meet a daily standard. They should be determined by consulting with policymakers or by examining laws, regulations, or other materials that may indicate the relative importance of meeting objectives. (Note that if one is using monetary measures of benefits and costs, weighting is not necessary. One dollar of benefit is implicitly equal to another dollar of benefit.)

This example assumes that:

- Failing to meet the domestic objective results in three times as much harm as failing to meet the irrigation objective.
- Failing to meet the recreation objective has the same harm as failing to meet the irrigation objective.
- Failing to meet the water quality objective has twice the harm of failing to meet either the irrigation objective or the recreation objective.

A total weighted score for a policy under a particular climate scenario can be derived by simply multiplying the number of days of failure for each use by its weight and then summing across uses. A lower total weighted score implies that the reservoir system is meeting its objectives more successfully; a score of zero would mean that all objectives are satisfied.

The average total weighted score indicates how well a policy meets objectives across different climate scenarios. This score can be calculated by averaging the total weighted scores for a particular policy under different climate scenarios. These scores are displayed in the next to the last column in the decision matrix.

The final column in the decision matrix is change in costs. This column displays the difference in costs between alternative policies and the current policy.

In this example, the current policy and two alternative policies are examined; however, users need not limit themselves to two alternatives. A number of studies have identified adaptation policies. Among them are OTA (1993), Smit (1993), and Country Studies (1994). These may be used as sources of information on adaptation policy alternatives.

For each policy, how well the objectives are met under the baseline climate and under two climate change scenarios is examined. It is important to consider baseline climate for two reasons. The first, and most important, is that if one is examining policies that could be implemented now or soon, it is reasonable to analyze how the policies will perform in the short term and to expect the next decade or two to be quite similar to the baseline climate period. In addition, it is possible that climate will not change and baseline climate can be thought of as a scenario of future climate. To define baseline climate, one could use a recent 30-year record of climate observations, such as 1961-1990 or 1951-1980. A 30-year period is generally accepted defining a "climate" (Carter et al. 1994; Country Studies 1994; Greco et al. 1994).

The choice of climate change scenarios is not a simple matter because significant uncertainty exists about how regional climates will change (Houghton et al. 1992). There is uncertainty about such important factors as the magnitude of temperature increase; whether precipitation will increase or decrease; and how storms, droughts, floods, and other extreme events will change. One could construct dozens of scenarios to reflect these uncertainties. Unfortunately, it may not be practical to analyze the effectiveness of baseline and alternative policies by using dozens of scenarios. On the other hand, because these scenarios are being used to analyze policies, it is critical to ensure that qualitatively different types of plausible climate change scenarios are included. If only scenarios that result in drier conditions are used when it is possible for climate change to lead to wetter conditions, one could mistakenly conclude that adaptation policies that only address dry conditions, such as drought, are preferable over those that also address wetter conditions, such as flooding. By focusing only on dry conditions, vulnerability to flooding could be increased.

To simplify matters, at least two climate change scenarios should be used (in addition to the baseline scenario). These scenarios should reflect differences in the range of potential climate change in the region. For example, one scenario could have a relatively high degree of warming, and the other could have a relatively low amount of warming, or one could have a significant increase in precipitation, while the other could have a decrease in precipitation. If two scenarios are insufficient to capture qualitatively different and important climate change conditions, it may be necessary to use more than two scenarios.

These scenarios may not be equally likely. As discussed below, policymakers could assign probabilities to the scenarios based on scientific evidence or their beliefs about the likelihood of climate change.

The example decision matrix has two climate change scenarios. Scenario 1 is a hot and dry climate, and Scenario 2 is a hot and wet climate.

Current Policy

How well objectives are met for the current policy under the baseline climate is examined first, followed by an analysis of how well they are met under climate change.

Baseline Climate

Table 1 displays how well the reservoir system meets the objectives for the current policy under baseline climate. None of the objectives is fully met under the baseline climate. On average, the domestic needs are not met on six days per year, the irrigation needs are not met on four days per year, and so on. By using the weights, the total weighted score is 48. This number has no meaning by itself, but must be compared with total weighted scores for other policies to determine relative effectiveness.

Climate Change

The next step is to examine how well the current policy performs under the climate change scenarios. Table 1 displays how the current policy performs in meeting the standards under Scenarios 1 and 2. In Scenario 1, allocations for all uses are reduced. Because the current policy favors deliveries to irrigation over other uses, water for irrigation is cut back less than for other uses. This policy translates into a smaller increase in days failing to meet the irrigation objective than the increase in days failing to meet the other objectives. The number of days that the agriculture objective is not met doubles. The number of days that the domestic and water quality objectives are not met increases by two and one-half times, and the number of days on which the recreation objective is violated increases by four times. The total weighted score for the current policy under Scenario 1 is 133, almost three times higher than the total weighted score under the baseline climate. A drier climate increases the failure rate significantly over the baseline climate.

Under Scenario 2, more water is available for consumptive and in-stream uses, so the number of days of failure drops for all categories.[2] Because the current policy is to try to meet agriculture's needs before other uses, the excess water is used to benefit agriculture more than to benefit the other uses. The number of days of failing to meet the irrigation objective is reduced by 75%, and the number of failure days for the other uses is reduced by 50%. The total weighted score for Scenario 2 is 23, a reduction of slightly more than 50% from the baseline climate total score.

The score of 68 is a simple average of the total weighted scores from the baseline climate, Scenario 1, and Scenario 2. This score has little meaning by itself, but will be useful to compare the effectiveness of current policies with alternative policies. Climate scenarios can be weighted by the probability of occurrence. The score for each scenario could be multiplied by the probability of that scenario occurring. The scores for each scenario could be added to yield a combined score. Even if the probabilities are unknown, one could test the ro-

---

2 Increased run-off could lead to more flooding — a potential damage from climate change. In the example, flood control is not a policy objective.

bustness of results by assuming different probabilities and examine the effect on relative benefits across policy options. Probabilities can also be used to reflect one's beliefs about the likelihood of climate change happening. If one thinks that climate change is unlikely, one could attach a lower probability to the baseline climate and higher probabilities to the climate change scenarios. If one thinks climate change is very likely, one could attach a higher probability to the baseline climate. If one thinks a drier climate is more likely to happen than a wetter one, the probabilities for these climate scenarios could be adjusted accordingly. Because little information is available about the probability of these climate change scenarios occurring, a weighting scheme was not used in this example.

The current policy appears to be vulnerable to a drier climate; the days of failure increase significantly under Scenario 1. Thus, the analysis of alternative policies under baseline and changed climates is justified.

Alternative Policies

In this hypothetical example, two alternative policies are examined. The first is a policy of enhancing conservation efforts. Under the enhanced conservation policy, the government would subsidize farmers to use more efficient irrigation technologies, and these technologies would reduce agriculture's water needs by 20%. The second policy alternative allows the free market to allocate water use. This policy would enable users with the highest marginal value for water to pay water users with lower marginal values to use their water. The weighting scheme assumed in this example implies that the highest marginal value uses of water are domestic and then water quality, and that irrigation and recreation have the same lowest marginal value use.

Baseline Climate

The first step is to examine how these alternative policies perform under the baseline climate. If the alternatives better meet the policy objectives than the current policy, the alternatives have benefits independent of climate change (Country Studies 1994).

Enhanced conservation policy involves increasing the efficiency of irrigation water use by 20%, so agriculture needs less water per day to meet its needs. Rather than needing water at 30 million kL/d, agriculture would now need 24 million kL/d. With agriculture's needs reduced, more water is available in the system to meet all needs; however, agriculture still receives the highest priority for allocation of water. The number of days when agriculture fails to receive its need is reduced from four to one (Table 1). With excess water in the system, reservoir operators are able to reduce the number of days of failure for other uses, but by a lower percentage than the reduction for agriculture. The domestic failures are cut by one-third, water quality failures by one-fourth, and recreation failures by one-fifth. The total weighted score for enhanced conservation is 33, which is about 30% lower than the score for the current policy.

Turning to the market allocation policy under the baseline climate, higher-marginal-value domestic users bid water away from lower-marginal-value irrigation users, so the domestic number of days of failure drops to three, while the number of days on which the irrigation target is not met increases to five. Trading does not affect water allocated for in-stream uses, so their days of failure remain unchanged from the current policy. The total weighted score for market allocation is 33, which is less than the total weighted score of 40 for the current policy. This occurs because the three-day reduction for domestic counts as nine (three [days] multiplied by the weighting of three), while the one-day increase for agriculture counts only as one.

Because both alternative policies have total weighted scores lower than the current policy under the baseline climate, one can conclude that both policies have benefits independent of climate change.

Climate Change

The next step is to examine how the alternative policies perform under the two climate change scenarios. The enhanced conservation policy has fewer days of failure in all categories and a lower weighted score under Scenario 1 than the current policy does, because with lower demand for agriculture, more water is available to meet other needs. The enhanced conservation policy also has fewer days of failure and a lower total weighted score under the wet scenario (Scenario 2) than does the current policy. The average total weighted score for the enhanced conservation policy is significantly lower than that for the current policy, so the enhanced conservation policy appears to be more flexible (i.e., it performs better under a variety of climate change scenarios than does the current policy).

The story is similar, although less dramatic, for market allocation. In Scenario 1, water use shifts from agriculture to domestic (increasing agriculture's failure rate by one day frees 30 million kL, which lowers domestic failure by three days). Because water is traded only between agriculture and domestic sectors, the number of days of failure for in-stream uses does not change. The total weighted score for market allocation in Scenario 1 is 24 less than the total weighted score for the current policy because water has shifted to a higher-value use. The market allocation policy also has a lower total weighted score in Scenario 2 than does the current policy. The average total weighted score for market allocation is lower than the average total weighted score for the current policy. Like the enhanced conservation policy, the market allocation policy appears to be more flexible than the current policy of water management.

On the whole, both policy alternatives have fewer days of failure than the current policy under both the baseline climate and climate change scenarios.

Enhanced conservation does the best at meeting the policy objectives because its average weighted score of 40 is the lowest.

The decision matrix allows ranking of these policies in order of which policy best meets the policy objectives; the matrix also indicates the relative difference in meeting these objectives. As the cost column indicates, the conservation policy is the most expensive policy, with a cost increase of $20 million over the current policy, followed by market allocation with a cost increase of $5 million over the current policy. Since the benefits of a policy are not expressed in dollars in this example, the benefits and costs of each policy cannot be directly compared. So, net benefits or benefit-cost ratios cannot be used to rank the three policy options. By using the quantitative scores, however, it is possible to compare cost-effectiveness. For example, the enhanced conservation policy reduces the combined score by 28, at a cost of $20 million, or $714,286 per weighted day of failure. The market allocation reduces the score by 13, at a cost of $5 million, or $384,615 per weighted day of failure. The market allocation policy buys more reductions in weighted days of failure per dollar than the enhanced conservation policy. Thus, market allocation appears to be the more cost-effective investment.

Conclusions

The advantage of using the decision matrix approach is that the benefits and costs of adaptation policies can be presented to policymakers in a consistent, systematic manner. The matrix enables a clear identification of policy objectives and state standards, if appropriate. If nonquantitative measures for meeting policy objectives are used, the matrix can still be used to display how alternatives compare in meeting policy goals. If the degree to which policy objectives are met (benefits) can be quantified, the relative effectiveness of policy alternatives can be examined. Each policy can be quantitatively compared with objectives under baseline climate and climate change scenarios. If nonmonetary measures of benefits are used (such as days of failure), alternatives can be ranked on the basis of their relative cost-effectiveness. If monetary measures of benefits can be used, policymakers can directly compare benefits and costs to determine which policies have the greatest net benefits or which have the greatest benefit-cost ratio. Thus, the matrix provides a way to organize analysis of adaptation policies.

In addition, the matrix provides a forum for organizing scientific and technical research on climate change vulnerability and adaptation. By requiring the identification of policy objectives and alternatives as well as climate change scenarios, the matrix enables users to define results needed from vulnerability assessments. To be able to fill in the cells of the matrix, the vulnerability assessments must provide results on how current and alternative policies perform in meeting the defined objectives under baseline climate and climate change scenarios. Thus, the matrix can provide very useful guidance to scientists conducting vulnerability assessments on how those assessments should be structured to produce results useful for policy analysis.

References

Carter, T.R., et al., 1994, *IPCC Technical Guidelines for Assessing Climate Change Impacts and Adaptations*, Department of Geography, University College London, United Kingdom.

Gramlich, E.M., 1981, *Benefit-Cost Analysis of Government Programs*, Prentice-Hall, Inc., Englewood Cliffs, N.J., USA.

Greco, S., et al., 1994, *Climate Scenarios and Socioeconomic Projections for IPCC Working Group II Assessment*, Intergovernmental Panel on Climate Change Working Group II, Consortium for International Earth Science Information Network, Saginaw, Mich., USA.

Houghton, J.T., et al., 1992, *Climate Change 1992 — The Supplementary Report to the IPCC Scientific Assessment*, World Meteorological Organization/U.N. Environmental Programme, Intergovernmental Panel on Climate Change, Cambridge University Press, Cambridge, United Kingdom.

Meier, P., and M. Munasinghe, 1994, *Incorporating Environmental Concerns into Power Sector Decisionmaking*, World Bank Environment Paper No 6, The World Bank, Washington, D.C., USA.

Office of Technology Assessment, 1993, *Preparing for an Uncertain Climate*, Washington, D.C., USA.

Smit, B. (ed.), 1993, *Adaptation to Climatic Variability and Change: Report of the Task Force on Climate Adaptation*, University of Guelph, Guelph, Ontario, Canada.

Smith, J.B. (1996), "Setting Priorities for Adapting to Climate Change," *The Environmental Professional*, in press.

Stakhiv, E., 1993, "Examining Water Resource Impacts of Climate Change," paper presented at the U.N. Environmental Programme — Canada Workshop on Impacts and Adaptation to Climate Variability and Change, Toronto, Ontario, Canada.

Titus, J.G., 1991, "Strategies for Adapting to the Greenhouse Effect," *APA Journal* 311:311-323.

U.S. Country Studies Program, 1994, *Guidance for Assessing Vulnerability and Adaptation to Climate Change*, Version 1.0, Washington, D.C., USA.

The Potential Costs of Climate Change Adaptation

S. Fankhauser
Secretariat, Global Environment Facility
The World Bank
1818 M Street
Washington, DC 20433, USA

Abstract

The various costs imposed on countries as a result of global warming are considered by interpreting adaptation as part of a comprehensive climate change response strategy. These costs include those associated with damages, as well as adjustments in strategies now in place for adapting to the current climate. Imposed costs are defined as the least-cost combination of adaptation costs, residual damages, and indirect effects. A possible framework is suggested for calculating imposed costs and assessing individual adaptive measures.

Introduction

The principal objective of greenhouse gas (GHG) mitigation activities is to reduce the amount of global warming likely to occur. In contrast, climate change adaptations aim to reduce the adverse impacts caused by warming. Although the most well-known adaptive option is coastal protection, adaptation also includes other measures, such as adjusting agricultural and forest management, implementing early warning systems for extreme weather events, and establishing migration corridors for migrating species. Because vulnerability to climate change is likely to decrease as national income increases (Intergovernmental Panel on Climate Change [IPCC] 1995), economic development achieved in a sustainable manner could also be regarded as an adaptation policy (Schelling 1992).

Although adaptation is increasingly being recognized as a response to climate change, its value as a policy tool remains debatable. Many authors have called for adaptation to be the key response to climate change, arguing that society will be able to adapt more or less fully to changes in climate conditions (e.g., Ausubel 1991). However, other authors not only question the power of adaptation, but also fear that emphasizing adaptation would convey the message that climate change is inevitable (affordable mitigation measures have little effect) and/or that climate change is manageable (mitigation measures are unnecessary).

While climate change adaptation is no panacea for adverse impacts, damage-limitation measures are an important aspect of any pragmatic climate policy. Given the long atmospheric lifetime of GHGs and the inertia of both the climate and the socioeconomic system, a certain amount of climate change seems unavoidable and may already be happening (see the GHG emission scenarios in IPCC [1995]). Adaptation could therefore be a major part of the national climate change response strategy, especially in countries that are either highly vulnerable to climate change or already have low GHG emissions (where attempting a further reduction would be expensive). Even in these cases, adaptation should complement, rather than substitute for, GHG abatement measures.

Adaptation in the Climate Convention

The U.N. Framework Convention on Climate Change (UNFCCC) recognizes the importance of adaptation. Article 3.3 of the UNFCCC calls for Parties to take

> precautionary measures to anticipate, prevent or minimize the causes of climate change and mitigate its adverse effects. . . . [M]easures should . . . be comprehensive, cover all relevant sources, sinks, and reservoirs of greenhouse gases and *adaptation* . . . . [emphasis added]

General reference is made to adaptation in Articles 4.1(b), 4.1(e), and 4.1(f). Article 4.4 deals with financing adaptive measures.

The types of adaptive measures were also analyzed at the tenth session of the Intergovernmental Negotiating Committee (INC-10) of the UNFCCC. The decision of INC-10 (Decision 10/3) has been endorsed by the first session of the Conference of the Parties (COP-1). INC-10 identifies three stages in the adaptation process:

- *Stage I. Planning:* Emphasis is on impact studies to identify "particularly vulnerable countries or regions." Attention is focused on general assessments of "policy options for adaptation and appropriate capacity building."
- *Stage II. Preparation:* "Measures, including further capacity building, which may be taken to prepare for adaptation, as envisaged by Article 4.1(e)" are to be readied for those countries singled out as particularly vulnerable in Stage I.
- *Stage III. Initiation:* "Measures to facilitate adequate adaptation, including insurance, and other adaptive measures as envisaged by Articles 4.1(b) and 4.4" for particularly vulnerable countries are to be initiated.

Table 1 lists possible adaptation policies following this three-stage categorization. Stage I adaptation in developing countries is eligible for funding under the UNFCCC, provided these measures are undertaken as part of the formulation of their national communications. Stage II and III measures are not yet eligible for funding.

In this paper, a long-term analytical perspective encompassing the entire adaptation process has been adopted. A long-term approach is essential for defining short-term decisions on Stage I measures. The paper does not deal with such financing issues as the distribution of imposed costs between countries through cost recovery, insurance, the Global Environment Facility (GEF), or any other funding source. COP will provide guidance on such issues. The GEF, in its capacity as the interim financial mechanism of the convention, has been entrusted by COP with the task of meeting the agreed full costs of activities required by Article 12.1 of the convention. These activities include adaptation activities undertaken by developing countries in the context of formulating national communications related to implementation of the convention. Such activities could involve studies of the potential impacts of climate change, identification of adaptation options, and relevant capacity building.

Conceptual Issues

Climate change adaptation is one of several instruments available to deal with the problem of global warming. The value of adaptive measures must therefore be judged in the broader context of a comprehensive response strategy that includes all of the instruments available.

Effect of Adaptation and Mitigation on Climate Change

For analytical convenience, it is useful to divide the available instruments into two types: adaptive measures, which reduce the negative impacts of a given amount of warming, and mitigation measures, which reduce the amount of warming that will occur. In the global context, the challenge is to find the mix of adaptation and mitigation projects that would best serve an agreed-upon set of global objectives. A reasonable goal could be to minimize the overall human welfare loss due to global warming. The problem then would be to determine the level of mitigation m and adaptation a that could minimize the sum of the three cost elements — mitigation cost MC, adaptation cost AC, and residual damage cost D:

$$\text{minimize } MC(m) + AC(a) + D(m,a). \qquad (1)$$

Table 1. Possible Adaptation Policies

Sector	Stage I: Planning (short term)		Stage II: Preparation (medium term)	Stage III: Initiation (long term)
	General Capacity Building: Impact Studies; Identification of Vulnerable Areas	Identification and General Assessment of Policy Options	Further Capacity Building in Vulnerable Regions; Development of Appropriate Adaptation Plans	Formulation of Measures to Facilitate Adaptation in Vulnerable Areas; Feasibility Studies; Insurance
Agriculture	Regional climate change and climate change impact studies: • Expected yield changes • Likelihood of droughts Vulnerability assessment: • Food security/possibility of famines	Individual adaptation options • Change farming practice (fertilizer use, etc.) • Change timing of action • Use different crops • Keep different livestock Public-sector adaptation options • Improve irrigation systems • Change land topography • Provide insurance/disaster relief	Farmer training Research on heat/drought-resistant plants	Pilot projects in farming adaptation Formulate disaster relief plans: • Emergency plans for famines • Crop insurance
Forestry	Forestry damage studies: • Forest migration patterns • Existing stress/over-exploitation of resources	Protection of existing forests: • Fire prevention • Suppress impacts of diseases, droughts, etc. Introduction of new species Forest management options: • Change in cutting practice • Sustainable forest use	Training and education: • Fire prevention • Disease prevention • Sustainable forest management Conserve gene pools: • Install seed banks	Pilot programs Efficient forest management: • Create right incentives • Abolish subsidies

Table 1. (Cont.)

Sector	Stage I: Planning (short term)		Stage II: Preparation (medium term)	Stage III: Initiation (long term)
	General Capacity Building: Impact Studies; Identification of Vulnerable Areas	Identification and General Assessment of Policy Options	Further Capacity Building in Vulnerable Regions; Development of Appropriate Adaptation Plans	Formulation of Measures to Facilitate Adaptation in Vulnerable Areas; Feasibility Studies; Insurance
Health/air pollution	Study of direct health impacts: • Heat stress Study of indirect health impacts: • Migration of vector-borne diseases • Increased air pollution	• Improve health/sanitary standards • Precautionary policies (vaccination) Air pollution policy options: • Impose air-quality standards • Emissions taxes/permits	Training and information: • Training of medical staff • Information for vulnerable groups Research in improved prevention: • Vaccines	Facilitate behavioral adaptation by individuals: • Provide information • Create right incentives Encourage structural adaptation: • Issue planning/building guidelines • Recommend air quality levels
Coastal zones	Collection of regional climate change data: • Sea-level data (tide gauge stations) • Storm data Coastal vulnerability studies: • Based on IPCC's common methodology (IPCC 1994) • Vulnerability of human settlements • Vulnerability of ecosystems	Option "protect": • Hard protection (dikes, bulkheads) • Soft protection (beach nourishment) Option "retreat": • Restrict development (set back zones) • Resettle affected people Option "accommodate": • Adjust economic activities (convert farms to fish ponds) • Insurance	Prepare ground for Integrated Coastal Zone Management (IPCC 1994): • Institutional arrangements • Build technological capacity • Build human capacity (training) • Provide information to public	Implement ICZM: • Provide incentives (market-based instruments) • Design regulatory measures • Set standards (water quality) • Plan physical structures • Ongoing monitoring of coastal processes • Forecast storms

Table 1. Possible Adaptation Policies (Cont.)

Sector	Stage I: Planning (short term)		Stage II: Preparation (medium term)	Stage III: Initiation (long term)
	General Capacity Building: Impact Studies; Identification of Vulnerable Areas	Identification and General Assessment of Policy Options	Further Capacity Building in Vulnerable Regions; Development of Appropriate Adaptation Plans	Formulation of Measures to Facilitate Adaptation in Vulnerable Areas; Feasibility Studies; Insurance
Water	Regional climate change predictions: • Change in precipitation rates Impact on water supply: • Water quality • Water quantity Predictions of water demand: • Nonclimate-change-induced (e.g., population growth) • Climate-change-induced (e.g., increased irrigation)	Supply management options: • Invest in reservoirs and infrastructure • Optimize systems (e.g., interregional water transfers) • Recycle water for lower-quality use Demand-side management options: • Invest in water-saving technologies • Change water-use practices	Create institutions and train staff: • Create water supply agencies • Develop hydrological models R&D into desalination and water-recycling schemes Education/information for households	Pilot studies for supply measures Pilot studies for demand measures Efficient water management: • Develop drought management plans • Formulate water quality standards • Remove market distortions (subsidies)

Sources: Compiled from IPCC (1992, 1994) and Smit (1993), based on the stages set out by COP.

The values MC and AC depend positively on their arguments — the more extensive a policy measure (i.e., the higher the amount of m and a), the higher the cost of its implementation (the higher the values of MC and AC). In some cases, AC may also depend on the degree of warming and thus indirectly on the level of mitigation [i.e., $AC\ (a,m)$]. For example, providing irrigation water may become more expensive under global warming. This difficulty is ignored in this analysis. Residual damage cost is a negative function of both a and m. The higher the level of mitigation and/or adaptation, the lower the damage from climate change. (Strictly, damage is a function of GHG concentration, which in turn depends on mitigation.)

At the global level, there is a trade-off between damage costs and the costs of mitigation and adaptation. The lower the desired level of damage, the higher the costs of meeting that target. As an example, consider the decision to intensify the search for a heat-resistant cultivar. Research and development (R&D) expenditures will increase, which means that AC will rise. At the same time, the chances of finding a suitable variety increase, leading to a decrease in the expected D from global warming.

In a cost-benefit framework, the optimal adaptation and mitigation levels can be found by comparing, for each policy type, the marginal costs of further action with the marginal benefits of avoided damage. (In formal notation, an optimum is obtained if $MC' = -D_m$ [optimal mitigation] and $AC' = -D_a$ [optimal adaptation], where the prime and subscript indicate marginal values [partial derivatives], i.e., $-D_m = -\partial D/\partial m$ is the marginal benefit [avoided damage] from mitigation, MC' is the marginal mitigation cost, etc.) Further GHG mitigation projects are justified as long as the additional costs of implementation are smaller than the additional benefits from prevented warming. Similarly, further adaptation is warranted as long as additional costs are lower than additional benefits from reduced damage levels. To minimize the total costs of global warming, marginal costs must intersect with the respective marginal benefits for each type of action.

Thus, a successful international response to climate change will contain elements of both adaptation and mitigation. However, in this paper, adaptation decisions are analyzed independent of mitigation considerations; i.e., an exogenously given amount of climate change is assumed (which is implicitly based on the success of internationally agreed-upon mitigation). This assumption adequately reflects the situation of most countries, which are, with one or two exceptions, too small to influence the global climate individually (Adger and Fankhauser 1993). The aim of a nation is to find the degree of adaptation necessary to minimize the costs of a given amount of climate change. In notational terms, the country seeks to

$$\text{minimize } AC_i\,(a) + D_i\,(a \mid cc), \tag{2}$$

where AC_i and D_i are the adaptation and damage costs of country i, respectively. Global costs are the sum of individual country costs; i.e., $\sum AC_i = AC$ and $\sum D_i = D$. The remainder of this section addresses how to determine the overall national burden that a given amount of climate change will impose on an individual country.

Adjustment of Current Adaptation Levels

Adaptation activities to accommodate the current climate and cope with existing variations are a part of everyday life. They include such measures as the irrigation of farm land, disaster relief schemes, weather forecasts, space heating and cooling, design of infrastructure, and coastal protection. The question of optimal climate change adaptation includes not only determining new measures, but also modifying the mix and strength of existing measures and determining what adaptation standards are required.

The optimal level of adaptation is that which minimizes the combined costs of adaptation plus residual damage (including any indirect effects). A certain level of adaptation, considered optimal in the present climate, is implicit in the way societies perform certain functions. For example, coastal communities first determine the acceptable risk of flooding and then design protection structures accordingly. Households in northern latitudes decide the extent to which their homes should be insulated and the type of heating required.

Climate change will affect the cost-benefit ratio of current adaptation decisions, and therefore a different degree of adaptation is required. Adaptive efforts under climate change may be either greater or smaller than exist today.

The rationale for adjusting standards is illustrated in Figure 1, which uses the provision of irrigation water as an example. Analytically, the problem is to minimize $AC_i + D_i$, first under current and then under changed climate conditions. The horizontal axis measures the adaptation level (e.g., the number of hectoliters of water provided), while the vertical axis measures the associated costs and benefits. In the absence of global warming, the marginal costs and benefits from irrigation are such that a_0 hectoliters are required (the combined costs of adaptation and residual damage are lowest at this level).

A change in climate will affect the marginal schedules of both benefits and costs. Global warming could reduce run-off and water supply, so the costs of providing water may rise. Therefore, marginal adaptation costs (MAC) shift upward. Marginal adaptation benefits (MAB) could shift upward as well. The beneficial effect of water on crop yields could increase at the margin because increased irrigation may compensate for more and longer periods of drought. The net effect of these two shifts is unclear. The irrigation level desired (a_1) could be either higher or lower than in the baseline case.

Climate change may not always cause a shift in both MAC and MAB, and sometimes the two curves could shift in opposite directions. However, the general argument will hold for all adaptive measures.

Imposed Costs of Climate Change

The effects of climate change are still uncertain and will differ across regions. However, climate change will incur net costs for most countries (Fankhauser

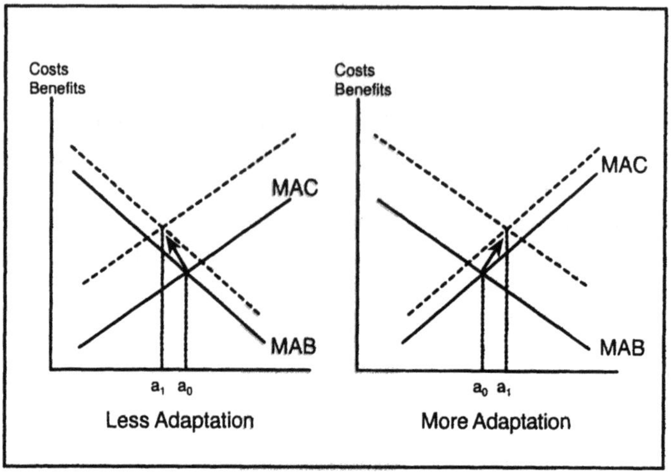

Figure 1. Impact of Climate Change on Current Adaptation Levels

1995a; IPCC 1995). Even if the economically optimal level of adaptation were undertaken, climate change would still reduce the aggregate social welfare of countries. This change in welfare can be viewed as the imposed cost of climate change. The principal sources of welfare loss will be the costs of damage and adaptation. There could, in addition, be several indirect effects. The different types of costs that will occur as a consequence of climate change are classified in Table 2, which provides a hypothetical example.

Table 2 presents two scenarios: one with and one without climate change. Column 2 contains the hypothetical reference scenario, describing the country situation in the absence of climate change impacts. Column 3, the climate change scenario, describes the country situation expected to occur, including climate change impacts. These climate change impacts are assumed to be known without ambiguity.

Two different adaptation strategies for an individual country (country subscripts are omitted for clarity) are considered. Baseline adaptation, corresponding to adaptation level a_0 in Figure 1, is the strategy that would be best to follow in the absence of climate change. Extended adaptation, corresponding to level a_1 in Figure 1 (assuming $a_1 > a_0$), is the optimal strategy for countries to follow in the climate change scenario. Thus, the occurrence of climate change leads to a shift from the top-left quadrant of Table 2 to the bottom-right. This shift imposes costs, mainly in the form of higher adaptation costs AC and climate change damage CD. Moving from baseline adaptation (indicated by subscript B) to extended adaptation (indicated by subscript E) incurs extra adaptation costs of $AC_E - AC_B = 80$. The extra damages exclusively, due to anthropogenic climate change by definition, only occur in the climate change scenario. In the top-left quadrant, $CD_B = 0$; in the bottom-right quadrant, there are residual damages of $CD_E = 40$.

Table 2. Imposed Costs of Climate Change: A Classification[a]

Type of Adaptation	Reference Scenario		Climate Change Scenario	
Baseline	Adaptation costs, AC_B	90	Adaptation costs, AC_B	90
	Climate change damage, CD_B	0	Climate change damage, CD_B	130
	Ordinary climate damage, OD_B	50	Ordinary climate damage, OD_B	50
	Other relevant costs, OC_B	60	Other relevant costs, OC_B	60
	Total reference costs	200	Total costs	330
Extended	Adaptation costs, AC_E	170	Adaptation costs, AC_E	170
	Climate change damage, CD_E	0	Climate change damage, CD_E	40
	Ordinary climate damage, OD_E	30	Ordinary climate damage, OD_E	30
	Other relevant costs, OC_E	50	Other relevant costs, OC_E	50
	Total costs	250	Total costs (minimum climate change costs)	290

[a] Numbers are illustrative and do not reflect relative magnitudes of the different cost components.

In addition, cost implications unrelated to climate change must be taken into account. Adaptive measures can sometimes serve multiple purposes. Extending adaptation would reduce the need for other investments that would have occurred in the baseline case. Hurricane shelters, for example, can be designed to serve other functions in the season when no hurricanes are expected (BUET/BIDS 1993). To account for such costs, Table 2 includes "other relevant costs" OC; extended adaptation is assumed to allow the country to reduce other expenditures in aggregate from $OC_B = 60$ to $OC_E = 50$.

The final cost category, ordinary climate damages OD, encompasses the adverse effects associated with the current climate, i.e., all climate-related costs that would occur even without climate change. Baseline adaptation avoids some, but not all, of these damages. In Table 2, residual damages of $OD_B = 50$ remain. Extended adaptation could further reduce ordinary climate damage to a value $OD_E = 30$.

In the absence of climate change, extended adaptation (a move from the top-left to the bottom-left quadrant) is not economic. The additional costs ($AC_E - AC_B = 80$) outweigh the benefits of ordinary climate damage avoided ($OD_B - OD_E = 20$) and other avoided costs ($OC_B - OC_E = 10$).

The imposed costs of climate change (ICC), defined as the difference in overall costs (adaptation plus damage plus indirect) between the climate change and the reference scenarios, take into account economically optimal adaptation. In the absence of climate change, costs of $AC_B + OD_B + OC_B = 200$ would have occurred (top-left quadrant of Table 2). The minimum climate change costs (minimum because adaptation keeps costs as low as possible) would be $AC_E + CD_E + OD_E + OC_E = 290$ (bottom-right quadrant). The ICC are the difference between these two elements, or ICC = 90.

The ICC do not simply include direct damages (a move from the top-left to the top-right quadrant, at costs CD_B). Climate change also requires a (costly) adjustment in adaptation strategies, which helps to reduce damage costs. As Table 2 shows, the ICC consist of four elements. Two of these are direct costs: (1) imposed adaptation costs ($AC_E - AC_B$) and (2) residual climate change damage under extended adaptation (CD_E). The other two are indirect, potentially offsetting costs: (1) the reduction in ordinary (nonanthropogenic) climate damage due to extended adaptation ($OD_E - OD_B$) and (2) other avoided costs ($OC_E - OC_B$). Only a few empirical studies have attempted to estimate the imposed costs of climate change, and they usually abstract from indirect effects (Fankhauser 1995a; IPCC 1995).

The concept of imposed costs differs from that of incremental costs, which is central to mitigation and other global change projects (GEF 1995). Incremental costs arise from decisions to implement measures that are not in a country's self-interest, but that benefit the global environment. Their magnitude has been agreed upon through negotiation (King 1993; Ahuja 1994). Imposed costs, in contrast, are a theoretical, nonnegotiated concept. Imposed costs are placed upon countries because of climate change and the need to regionally adjust socioeconomic activities and structures. Incremental costs and imposed costs are distinguished for analytical reasons only. The choice of terminology does not attempt to prejudge the outcome of the political debate on the funding of adaptation projects and imposed costs.

Assessment of Adaptation Projects

In practical terms, a country's adjustment to climate change will require a vast program of diverse adaptive measures. This section examines the implications of the analysis of imposed costs presented above for the assessment of adaptive measures made necessary by anthropogenic climate change.

Adaptation Benefits and Imposed Costs

Adaptation projects differ from other global climate change projects in one crucial respect — the absence of global environmental benefits. Most global change projects encourage actions that benefit the entire world, not solely a particular country. However, adaptation projects generally only benefit local and possibly regional areas. For example, the main beneficiaries of an irrigation project are the local farmers and households in the communities involved, whereas the benefit to the global environment will usually be insignificant. Exceptions occur when adaptation measures yield global side benefits, for example, in the form of preserved biodiversity, protected carbon sinks, or demonstrations with wide replicability (Fankhauser 1995b).

The assessment of an adaptation project thus involves a comparison of project costs with the domestic adaptation benefits. In a successful adaptation project, the benefits secured must be greater than the imposed costs. Other (e.g., equity) considerations may complement this efficiency criterion.

The different types of adaptation costs and benefits can be analyzed by using Table 2. Consider a comprehensive program to expand adaptation from the baseline to the extended level (a shift from the top-right to the bottom-right quadrant). Climate change is assumed as given, and Table 2 assumes impacts to be known with certainty. The question is to determine which measures are economic under the climate change scenario. Table 3 reproduces the relevant information from Table 2.

The key adaptation benefit is avoided climate change damages, i.e., a lower vulnerability of a region to climate change. For example, introducing a hurricane forecasting system yields regional benefits in the form of reduced damages from climate-change-induced hurricanes. In Table 3, extended adaptation avoids climate change damages of $CD_B - CD_E = 90$. This value represents the climate-change-related benefits of the extended adaptation program. In addition, non-climate-change-related effects may occur. The hurricane forecasting system, for example, not only predicts storms induced by climate change, but also predicts other extreme weather events. Thus, ordinary climate damage is reduced ($OD_B - OD_E = 20$).

Furthermore, climate change adaptation projects may complement or substitute for other baseline projects. For example, hurricane shelters can be designed for use in education, health care, and community activities, or for other functions when storms are not expected (BUET/BIDS 1993). Adaptation then reduces investment costs compared to the investment program in the baseline case. In Table 3, avoided costs are $OC_B - OC_E = 10$.

The extended adaptation program thus yields total benefits of 120, of which 90 are related to anthropogenic climate change. Total benefits compare favorably with imposed adaptation costs of $AC_E - AC_B = 80$. The program in the example is therefore economic. If the extended adaptation program is also compatible with national goals, satisfies equity considerations, and meets other

Table 3. Costs and Benefits of Baseline and Alternative Projects[a]: Hurricane Forecasting System

Adaptation/ Cost-Benefits	Adaptation Costs	Climate Change Damage	Ordinary Climate Damage	Other Relevant Costs
Baseline	$AC_B = 90$	$CD_B = 130$	$OD_B = 50$	$OC_B = 60$
Alternative	$AC_E = 170$	$CD_E = 40$	$OD_E = 30$	$OC_E = 50$
Project costs	$AC_E - AC_B = 80$	–	–	–
Project benefits[b]	–	$CD_B - CD_E = 90$	$OD_B - OD_E = 20$	$OC_B - OC_E = 10$

[a] Numbers are illustrative and do not reflect relative magnitudes of the different cost components.
[b] Project benefits are damages avoided.

relevant criteria, implementing the program would be in the country's interest.

In this example, extended adaptation is worthwhile on the basis of avoided climate change damage alone, since $(AC_E - AC_B) < (CD_B - CD_E)$. In general terms, an adaptation project is economic if the extra adaptation costs are less than the sum of *all* benefits, whether or not these are related to climate change. Projects therefore need to be evaluated according to their overall merit, including nonclimate change considerations.

As a second hypothetical example of the costs and benefits of adaptive measures, consider a project to replace flood-protection structures along a 500-m-long coastline. The old structures were 1 m high. Since the lifetime of the new structures is 100 years, precautions are taken to compensate for the effects of global warming. Therefore, the height is increased by 50 cm, and the length is extended by 10% to cover a previously unprotected area. What are the costs and benefits of this replacement project?

In the absence of climate change, the old structure would have been replaced with an equivalent one. The baseline project is therefore a structure 500 m long and 1 m high. The costs of this project (AC_B) must be compared with the costs of protecting 550 m of coastline with 1.5-m-high structures (AC_E). Suppose the costs of the baseline project are $AC_B = 700$, while the alternative, extended adaptation, costs are $AC_E = 1{,}200$. Imposed adaptation costs are then $AC_E - AC_B = 700$, as shown in Table 4. These imposed costs cover the expenses of building both higher and longer structures.

One way of measuring the benefit (avoided damage) of this project could be to compare the number of people at risk from flooding with and without extended adaptation (defined as the number of people living in an affected area times the probability of a flood). Suppose that 140,000 people were at risk under the baseline project (total figure over 100 years), and that this number is 60,000 more than would have been at risk without anthropogenic climate change. Further assume that the alternative project reduces this number to 40,000. The adaptation benefit of the project is therefore equal to fewer people at risk (140,000 – 40,000 = 100,000). It is assumed that these are the only benefits.

How much of this benefit is related to climate change? The answer depends on the split assumed between climate-change-induced and other types of floods. Suppose that the additional measures protect against only one type of flood, which has a 1 in 30 probability of occurrence. Also assume that global warming increases the frequency of floods from once every 30 years to once every 15 years (average over 100 years). The implication then is that every second storm is induced by climate change, which means that benefits are split evenly between climate-change-related benefits and ordinary damages avoided. Table 4 summarizes the calculations.

Financing Policy

The rules governing financing policy, which determine how a project's costs should be split between the parties concerned, must complement the economic

Table 4. Costs and Benefits of Baseline and Alternative Projects[a]: Flood Protection Structures

Adaptation/ Cost-Benefits	Adaptation Costs ($)	Climate Change Damage (10^3 people at risk)	Ordinary Climate Damage (10^3 people at risk)	Other Relevant Costs ($)
Baseline	$AC_B = 700$	$CD_B = 60$	$OD_B = 80$	$OC_B = 0$
Alternative	$AC_E = 1{,}200$	$CD_E = 10$	$OD_E = 30$	$OC_E = 0$
Project costs	$AC_E - AC_B = 500$	–	–	–
Project benefits[b]	–	$CD_B - CD_E = 50$	$OD_B - OD_E = 50$	$OC_B - OC_E = 0$

[a] Numbers are illustrative and do not reflect relative magnitudes of the different cost components.
[b] Project benefits are damages avoided, i.e., a lower number of people at risk with the alternative project.

decision rule described above. One question that might arise with respect to funds provided under the UNFCCC is whether grants should cover the total adaptation costs imposed by a project ($AC_E - AC_B$) or only adaptation costs less-nonclimate change benefits, i.e., adaptation costs minus the net value of ordinary climate damage avoided ($OD_B - OD_E$) and other avoided costs ($OC_B - OC_E$).

One way of dealing with this question is to treat nonclimate change benefits in the same way as concurrent domestic benefits are treated in evaluating mitigation projects. The following procedure to deal with concurrent domestic benefits has been proposed in the context of incremental cost estimation (GEF 1995). The methodology can be applied directly to the treatment of nonclimate-change-related adaptation benefits.

If the costs and benefits of a project can be separated, the easiest solution is to define a hypothetical "subproject" that only provides nonclimate change benefits. Consider again the example of hurricane damages. Part of a strategy to adapt to this effect of climate change could be to establish an emergency force that will assist the area if a disaster occurs. Many of the costs of such a project are variable and rise with the number of disasters. Although it is not possible to know whether an individual hurricane is induced by anthropogenic climate change, information may be available on the fraction of hurricanes that could be induced by climate change. A split between the global and nonglobal warming costs of the relief project is then theoretically possible.

(The question of how global warming might affect the severity and frequency of hurricanes is controversial [for a brief discussion, see IPCC 1995].

According to a frequently quoted estimate, a doubling in atmospheric carbon dioxide concentration [$2XCO_2$] could increase the frequency of hurricanes by up to 50% [Emanuel 1987]; under $2XCO_2$, a region currently hit by two hurricanes per year would experience three. In other words, about one-third of all hurricanes in a $2XCO_2$ world would be induced by anthropogenic climate change.)

Alternatively, it may be possible to interpret nonclimate change benefits as avoided costs that would have been incurred in the baseline case. For example, if cyclone shelters are used as schools during noncyclone periods, the resulting nonclimate change benefits could be estimated as the avoided construction costs for schools that would have occurred in the baseline case (the costs of providing the necessary equipment constitute separable costs).

If neither of these two methods can be applied, it could be argued that the benefit in question is not a national priority, since it is not worth providing in the baseline case. Assigning the benefit a value of zero may be the most pragmatic solution (GEF 1995).

An interesting case emerges if adaptation projects yield global (as opposed to local) side benefits. Integrated Coastal Zone Management, for example, will often not only serve adaptation but also help to protect international waters and prevent the loss of biodiversity — both focal areas of GEF interest. Significant global benefits must be accounted for. Projects could then become eligible for GEF financing for their adaptation benefit, their global benefit, or a combination of both.

The feasibility of the above approach is yet to be tested. Splitting total benefits into climate change and nonclimate change categories in any credible way may prove difficult. Any calculation will be complicated by the immense uncertainty about climate change impacts (although the examples given here provide some ideas of how to tackle the problem). Difficulties could also emerge with respect to win-win projects, i.e., cases where nonclimate change benefits are sufficiently high to make an adaptation project worthwhile independent of climate change considerations.

Distributional Impacts and Equity

The primarily domestic character of adaptation benefits has strong distributional implications for the possible funding of such measures under the UNFCCC. In contrast, the location of mitigation projects is often of secondary importance because benefits are predominantly global. For example, abatement of a tonne of carbon in India yields the same climate change benefit as a tonne abated in Brazil. But in the case of adaptation, location is crucial — an irrigation project in Africa benefits a different group of people than would coastal protection on a small island state. The geographic distribution of adaptation projects funded under the UNFCCC may become an important issue.

Summary and Conclusions

The objective of climate change adaptation is to reduce the adverse impacts of global warming. Adaptation includes all adjustments that help to reduce the vulnerability of society to climate change. It can take the form of both behavioral changes and technological adjustments and can involve tactical as well as strategic adjustments. Most adaptation will probably occur on an individual level, for example, adjustments in farm management, behavioral adjustments by households, or migration away from the most-affected areas. Nevertheless, publicly administered adaptation efforts may also take place, including measures to remove existing obstacles and facilitate individual adaptation through information, education, etc. A framework for the analysis of such adaptive measures has been set forth.

Climate change imposes a variety of costs on society. Two main types of imposed costs can be distinguished: (1) imposed adaptation costs and (2) residual climate change damages not avoided through economically efficient adaptation. In addition, two, potentially offsetting, indirect costs may occur: (1) reduction in ordinary (nonanthropogenic) climate damage due to extended adaptation and (2) other avoided costs. Discussions of financing policy focus on how these costs should be shared between a country and the international community.

For individual countries, the economically efficient level of adaptation to a given change in climate is the level that keeps the sum of costs — adaptation plus residual damages plus any indirect effects — as low as possible. Determining this level is difficult. Although an approximate picture of climate change vulnerability is emerging, the exact effects of climate change and, by implication, the benefits from adaptation are still uncertain. The appropriate level of adaptation will therefore also remain unclear, even though the direction of adaptation adjustments will often be known. Decisions regarding suitable adaptive measures have to be made on the basis of expected (probability-weighted) costs and benefits. Resolving uncertainty requires detailed vulnerability assessments. COP-1 has proposed such studies as a first step in a three-stage adaptation strategy.

The assessment of adaptive measures requires a comparison of project costs and benefits. Adaptive measures differ from other global change projects because their benefits occur locally rather than globally. The key benefit of climate change adaptation occurs in the form of avoided climate change damages. In addition, nonclimate-related benefits may occur in the form of avoided damages associated with current weather patterns and other avoided costs.

The concept of imposed costs is different from that of incremental costs. Incremental costs arise from the decision to implement measures that are not in a country's self-interest but which have global environmental benefits. Imposed costs, in contrast, are placed upon countries because of the need to adapt to climate change.

The conceptual discussion of imposed costs and the assessment of adaptation projects is still at an early stage. Further analysis is required, especially with respect to practical issues. Case studies are needed to test whether it is useful and practicable to make the various distinctions suggested in this paper.

Acknowledgments

This analysis is an abbreviated version of Fankhauser (1995b). The views expressed are not necessarily those of the Global Environment Facility or its associated agencies. The author is grateful to Dilip Ahuja, Ian Burton, Ken King, David Pearce, and two anonymous referees for their comments on earlier drafts.

References

Adger, N., and S. Fankhauser, 1993, "Economic Analysis of the Greenhouse Effect: Optimal Abatement Level and Strategies for Mitigation," *International Journal on Environment and Pollution* 3(1-3):104-119.

Ahuja, D., 1994, *The Incremental Cost of Climate Change Mitigation Projects*, Working Paper No. 9, Global Environment Facility, Washington, D.C., USA.

Ausubel, J.H., 1991, "Does Climate Still Matter?" *Nature* 350:649-652.

BUET/BIDS, 1993, "Multipurpose Cyclone Shelter Programme," Bangladesh University of Engineering and Technology (BUET)/Bangladesh Institute of Development Studies (BIDS), final report to U.N. Development Program, The World Bank, and the Government of Bangladesh, Dhaka, Bangladesh.

Emanuel, K.A., 1987, "The Dependence of Hurricane Intensity on Climate," *Nature* 326:483-485.

Fankhauser, S., 1995a, *Valuing Climate Change. The Economics of the Greenhouse*, Earthscan, London, United Kingdom.

Fankhauser, S., 1995b, *The Costs of Adapting to Climate Change*, Working Paper No. 13, Global Environment Facility, Washington, D.C., USA (in press).

Global Environment Facility, 1995, "Incremental Costs and Financing Modalities," Report to the GEF Council, Document GEF/C.2/6, Rev. 2, Washington, D.C., USA.

Intergovernmental Panel on Climate Change (IPCC), 1992, *Global Climate Change and the Rising Challenge of the Sea. Supporting Document for the IPCC Update 1992*, World Meteorological Organization and U.N. Environmental Programme, Geneva, Switzerland.

IPCC, 1994, *Preparing to Meet the Coastal Challenges of the 21st Century*, World Coast Conference 1993, Intergovernmental Panel on Climate Change, World Meteorological Organization and U.N. Environmental Programme, Geneva, Switzerland.

IPCC, 1995, "The Second Assessment Report of Working Group III," Intergovernmental Panel on Climate Change, Geneva, Switzerland.

King, K., 1993, *The Incremental Costs of Global Environmental Benefits*, Working Paper No. 5, Global Environment Facility, Washington, D.C., USA.

Schelling, T., 1992, "Some Economics of Global Warming," *American Economic Review* 82(1):1-14.

Smit, B. (ed.), 1993, "Adaptation to Climatic Variability and Change. Report of the Task Force on Climate Adaptation," The Canadian Climate Program, Downsview, Ontario, Canada.

Malawi: How Climate Change Adaptation Options Fit within the UNFCCC National Communication and National Development Plans

Jean Theu
Ministry of Economic Planning and Development
Health and Environment Section
Lilongwe, Malawi

Geoffrey Chavula
The Polytechnic University of Malawi
Department of Civil Engineering P/B 303
Blantyre, Malawi

Christine Elias
World Resources Institute
1709 New York Avenue, NW
Washington, DC 20006, USA

Abstract

The relationship between climate change adaptation policy and economic development and environmental management programs in Malawi is discussed. Malawi's current Statement of Development Policies stresses household food security and sustainable economic development. These objectives are identical to the ultimate goal of the U.N. Framework Convention on Climate Change. The Ministry of Economic Planning and Development will coordinate Malawi's adaptation policy analysis. As the coordinator of Malawi's development program, the Ministry is best placed to ensure that climate change considerations are integrated across sector programs. The Ministry will also coordinate a review of planned projects to determine how they could be modified to take future climate change into account. A team of sector experts, who are also participating in the vulnerability assessment, will analyze the climate change adaptation policy.

Malawi and the U.N. Framework Convention on Climate Change

The ultimate objective of the U.N. Framework Convention on Climate Change (UNFCCC [1992]) is as follows:

> . . . to achieve, in accordance with the relevant provisions of the Convention, stabilization of greenhouse gas concentrations in the atmosphere at a level that would prevent dangerous anthropogenic interference with the climate system. Such a level should be achieved within a time-frame sufficient to allow ecosystems to adapt naturally to climate change, *to ensure that food production is not threatened, and to enable economic development to proceed in a sustainable manner* [italics added].

The last two goals (italicized) are also central to Malawi's Statement of Development Policies (Ministry of Economic Planning and Development [EP&D] 1987). Thus, the Convention can help Malawi meet both national and global objectives.

As a signatory to the Convention, Malawi is committed to preparing a national communication that includes (1) an inventory of greenhouse gas emissions, (2) an assessment of the country's vulnerability to climate change, (3) priority measures for adapting to these vulnerabilities, and (4) priority mitigation measures.

With a grant from the U.S. Country Studies Program (CSP), Malawi initiated a greenhouse gas inventory and a vulnerability and adaptation assessment. This assessment will focus on four sectors identified as potentially most vulnerable to climate change: agriculture, water resources, forests, and wildlife/parks. For each sector, policy and technical responses to the effects of climate change will be assessed, as well as steps that could be taken to build the local capacity needed to identify, evaluate, and implement these adaptive measures. Malawi is just beginning the inventory and vulnerability and adaptation assessment.

Malawi's Vulnerability to Climate Change

Climate Variability

Climate change models identify the Southern African region, including Malawi, as highly likely to be vulnerable to the anticipated effects of global greenhouse gas emissions, even though the entire region contributes less than 0.1% of these emissions. (Data on Malawi's emissions are not yet available.) Malawi has already proved to be particularly vulnerable to extreme climate variability, especially hot/dry conditions leading to droughts and hot/wet conditions leading to floods. For example, the 1948 and 1991-1992 droughts, caused by the El Niño-Southern Oscillation phenomenon, had serious repercussions on agricultural production and resulted in rampant famine. Similarly, the flash floods of Zomba and Phalombe in 1946 and 1991, respectively, caused great loss of life and severe damage to property. These floods were caused by high-intensity rainfall due to the intertropical convergency zone and tropical cyclones. The intensity and magnitude of these extremes may increase under global climate change. It is therefore important for Malawi to emphasize assessment of climate change impacts and adaptation options, while meeting its commitment to inventory greenhouse gas emissions and identify mitigation measures.

Adaptation to Current Climate Variability

Malawi has been forced to develop various measures to cope with the negative impacts of climate variability. Some measures have been promoted through government policy; others have emerged at the household level in response to short-term food emergencies. The following examples from the agricultural and water resources sectors illustrate the diversity of current adaptive responses.

Malawi's agricultural development priorities are to ensure household food security and a stable flow of foreign exchange. Recurring droughts have forced the government to reevaluate strategies for achieving these objectives. One option is to import emergency food aid during drought years to relieve short-term food deficits, but this measure is expensive and does not address long-term insecurities inherent in persistent droughts. Through its agricultural extension service, the government is advising farmers to grow drought-resistant food crops, such as cassava, millet, and sorghum. However, maize is the preferred staple food, and convincing farmers to accept alternative crops as staple foods will require extensive public awareness campaigns and appropriate pricing policies. Farmers are also encouraged to plant fast-maturing maize varieties (Ministry of Disaster Preparedness, Relief, and Rehabilitation [DPR&R] 1994), but these varieties depend on regular and frequent rainfall once the seeds germinate. Erratic rainfall patterns in Malawi make dependence on these new varieties uncertain. Irrigation is an alternative, but not necessarily cost-effective for subsistence farmers. Another adaptive strategy is to encourage farmers to maintain their livestock herds at manageable levels to reduce demand for scarce forage and water. This strategy has had mixed results because many farmers in Malawi keep livestock for prestige and as a sign of wealth; scarcity of forage and water during droughts is not a primary consideration. Finally, extension agents are encouraging farmers to maintain sufficient food reserves by keeping, rather than selling, any short-term food crop surpluses. However, this adaptive measure contradicts the farmers' day-to-day requirement for cash for basic needs.

Moreover, farmers themselves are attempting to diversify their sources of income so that they can buy maize when they cannot grow enough to feed their households. However, income-generating opportunities are extremely limited in rural areas. Alternative sources of income often depend on Malawi's natural resources (e.g., forests and fish stocks), which are degrading as demand for them increases. A systematic assessment of these and other possible adaptive strategies (e.g., constructing water retention and soil conservation structures, resettling people from marginally productive lands, or improving communal livestock grazing pastures) is now necessary and will be an important component of the climate change adaptation analysis.

Malawi's water resources are also vulnerable to climate variability. The system of rivers, lakes, and groundwater reservoirs, all of which are replenished by precipitation primarily from the intertropical convergence zone, supports agriculture, livestock production, fisheries, and wildlife (Chavula 1995). Adaptation options are being considered to ease the serious impacts of drought on the domestic water supply. At present, more than 60% of Malawi's largely rural population obtains water from hand-dug wells and boreholes. As the water table recedes with recurrent years of drought and many hand-dug wells run dry, demand increases for deeper boreholes as an alternative source of domestic water. However, the hand pumps capping these boreholes frequently break down, and the shortage of spare parts and transportation to urban centers makes repairs costly and time-consuming. In response, the government has introduced village-level operation maintenance programs in rural areas, complemented by installing hand pumps that are easier to maintain. In the few areas where such projects have been implemented, rural communities have managed to keep their water sources operational more consistently. Nevertheless, their long-term sustainability has yet to be verified.

Malawi's hydropower is also affected by declining rainfall levels, as water flow in the country's rivers becomes more unreliable and droughts become more frequent. To supplement reductions in hydropower potential, Malawi plans to purchase hydropower from other countries in the region (e.g., Mozambique, from the Cabora Bassa Dam). Importing hydropower from neighboring countries is a costly alternative, given Malawi's limited foreign exchange reserves. In addition, the sustainability of such schemes depends entirely on the political stability of the region. Exploring the potential of solar energy as a source of electric power is one alternative. However, investment and maintenance costs for that option may be extremely high.

To date, these and other adaptive measures have not been systematically analyzed, either in terms of lessons learned or of their appropriateness for current climatic, economic, and political conditions. They will be the starting point for the climate change adaptation analysis. Further analysis is needed on the relative costs and benefits of these adaptation options. Nevertheless, these concrete experiences can be built upon as Malawi plans for future climate change. As the study progresses, more appropriate alternative measures may emerge that could improve Malawi's readiness for potential impacts of future climate change.

Malawi's Approach to Adaptation Analysis

Agriculture is the backbone of economic growth and development in Malawi, and national development priorities emphasize strategies to safeguard household food security and enhance sustainable use of environmental resources (water, wildlife, forestry, fisheries, etc.). Within the context of this development framework, four sectors were selected for the vulnerability and adaptation (V&A) assessment: agriculture, forests, water resources, and wildlife/parks. These sectors were selected by applying the criteria outlined in the National Environmental Action Plan (Ministry of Research and Environmental Affairs [R&EA] 1994) for prioritizing environmental issues. These criteria are as follows:

- Socioeconomic importance of the issue as a threat to sustainable development,
- Extent to which people are affected,
- Effects on vulnerable groups,
- Spatial extent of degradation or depletion, and
- Extent to which degradation or depletion is irreversible (R&EA 1994).

The goal of the adaptation assessment is to determine the implications of climate change for these focal sectors and to evaluate adaptation options in light of current national development plans and policies. For example, agricultural productivity has been seriously hampered in years of drought. If the magnitude and intensity of droughts could increase as a result of climate change, the policies, strategic plans, and programs designed to minimize the agriculture sector's vulnerability to drought may need to be reinforced. The adaptation analysis and any resulting policy changes will be conducted within Malawi's current national planning process.

National Development Plans

The responsibility for developing and analyzing the impact of macroeconomic policies and strategies rests with the EP&D. The Ministry coordinates preparation of long-term plans, such as the National Development Policy, and short- to medium-term strategies that support longer-term goals (EP&D 1987).

Within this broad policy framework, Malawi also has national plans to address specific priority development issues. The three most important examples are (1) the National Environmental Action Plan, prepared by the R&EA (1994), which is the Malawi government's strategy for environmental management; (2) the National Disaster Action Plan, prepared by the DPR&R (1994), which addresses disaster mitigation and management (including drought management and mitigation of hydrological hazards); and (3) the Policy Framework for Poverty Alleviation, prepared by the EP&D (1995), which is designed to raise the standard of living of the rural poor by increasing their access to income through small-scale business and employment opportunities.

The EP&D must review and approve any projects supporting these plans and requiring national resources for implementation and incorporation in the national budget. Projects requiring capital investments are also screened by the Ministry for inclusion in the Public Sector Investment Program (PSIP). A three-year rolling program, the PSIP consists of individual capital investment projects identified in consultation with sector and line ministries and evaluated against available resources, absorptive capacity, and national priorities. In support of the Policy Framework for Poverty Alleviation, special consideration is given to programs aimed at implementing community-based microprojects (i.e., projects identified by the members of the communities themselves to address specific community problems).

The Environmental Support Program (ESP) is currently being formulated to implement the National Environmental Action Plan. With funding from The World Bank, the ESP will become part of the PSIP and include investments designed to arrest or curb environmental degradation and contribute to sustainable natural resources management. The EP&D is closely involved in this process and must eventually approve the program. Six task forces are working to establish selection criteria and propose projects to be included in the final ESP. The findings and recommendations of the ESP's Climate and Air Pollution Task Force will be an important resource for the adaptation policy analysis (their work is now in progress, but conclusions are not yet available). The goal of this task force is to ensure that climate considerations are incorporated into the ESP and the PSIP by integrating the Malawi Country Studies Team's findings into the planning and investment decision process.

Adaptation Analysis Team and Strategy

The overall coordinator of environmental management in Malawi is the R&EA. In this capacity, the Ministry is coordinating the country study of greenhouse gas emissions and the V&A assessment. Four study teams have been organized to conduct the vulnerability assessment for each priority sector. Because the V&A assessment is just beginning, Malawi is integrating the adaptation analysis into the early stages of the vulnerability assessment. The adaptation analysis team will consist of one representative from each sector team. On the basis of the Country Studies Team's findings, the Ministry will submit a recommended adaptation strategy and investment plan to the EP&D.

Given the EP&D's central role in developing macroeconomic policy and in weighing national investment priorities, it is essential that this organization be involved from the outset in the adaptation assessment. To ensure this involvement, a representative of the Ministry has joined the Country Studies Team and will lead the adaptation policy analysis. This approach will ensure that climate change is addressed at the highest level of government planning and that recommended policy changes and investments receive full consideration. It is expected that the adaptation analysis team will be named and the studies initiated by August 1995.

The adaptation analysis will consist of the following elements:

- Assessment of current strategies to adapt to today's climate variability. Criteria for the assessment will include costs, long-term sustainability, social and political implications, market incentives/disincentives, negative impacts, and implications for other sectors.
- Identification of prudent anticipatory adaptive measures that could be incorporated into Malawi's PSIP, in particular, cost-effective near-term investments to avoid expensive or irreversible impacts in the future.
- Comparison of the benefits and costs of these adaptation options with other national priorities.
- Identification of possible modifications to projects planned for implementing the National Environmental Action Plan and the National Disaster Action Plan to ensure that climate change considerations are included.

It is envisaged that the adaptation policy analysis will include cost estimates for implementing the identified adaptive measures, where possible. Cost estimates will be vital in determining which measures merit inclusion in Malawi's national public investment program. This is the first priority. In addition, preliminary estimates will be made of the incremental implementation costs that may eventually be eligible for funding under Articles 4.3 and 4.4 of the Convention. To date, the Parties to the Convention have agreed that the Convention's interim funding mechanism, the Global Environment Facility, should fund studies of impacts and adaptation options (Stage I, planning). However, funding decisions about the adaptive measures themselves (Stage II, preparation, and Stage III, initiation) remain open until more information is available on the need for such measures (UNFCCC 1995). Cost estimates for Malawi's adaptive measures will begin to provide this type of information and will be valuable to the Parties in meeting their commitments "to assist developing country Parties to meet the agreed full incremental costs of implementing [the] measures" (UNFCCC 1995, Article 4.3).

Technical Assistance Requirements

The sector study teams for the V&A assessment include technical experts from government, academia, and nongovernmental organizations. External technical assistants from the CSP will provide additional support for the vulnerability assessment. The adaptation policy analysis is designed to integrate findings from the sector assessments into a cross-sector adaptation plan with clear priorities. In this light, an essential element of the assessment will be an economic analysis of the various options. The World Resources Institute (WRI) and the Biodiversity Support Project ([BSP], a consortium of the World Wildlife Fund, The Nature Conservancy, and WRI funded by the U.S. Agency for International Development in Washington, D.C.) are assisting the Malawi government in designing and conducting the adaptation policy analysis. WRI and BSP will provide technical assistance to support the analyses, as well as a small grant to Malawi's adaptation analysis team to finance in-country field work. In addition, Malawi will seek capacity- and institution-building assistance from the CSP, especially in costing individual adaptation options and in conducting cross-sector economic analyses of the various options.

Conclusions

From the preceding discussion, it is clear that the UNFCCC is consistent with Malawi's development strategy. Furthermore, the Convention has become a catalyst for initiating a systematic assessment of policy and investment options that will help the country adapt to the effects of current climate variability and better prepare for future climate change. Through a holistic and cross-sector approach to the analysis, adaptations to climate change will be considered as an integral part of a broader strategy of environmentally sustainable development.

References

Chavula, G.M.S., 1995, "The Vulnerability and Adaptation of Malawi's Water Resources and Agricultural Production Sectors to Climate Change," paper presented at the International Conference on Climate Change Adaptation Assessments, sponsored by the U.S. Country Studies Program, St. Petersburg, Russian Federation, May 22-25, 1995.

Ministry of Disaster Preparedness, Relief, and Rehabilitation (DPR&R),1994, *National Disaster Action Plan – Draft Report,* Lilongwe, Malawi, Jan.

Ministry of Economic Planning and Development (EP&D), 1987, *Statement of Development Policies 1987-1996,* Government Printer, Zomba, Malawi.

Ministry of Economic Planning and Development (EP&D), 1995, *Policy Framework for Poverty Alleviation Programme*, Lilongwe, Malawi.

Ministry of Research and Environmental Affairs (R&EA), 1994, *National Environmental Action Plan*, Lilongwe, Malawi.

U.N. Framework Convention on Climate Change (UNFCCC), 1992, New York, N.Y., USA.

U.N. Framework Convention on Climate Change (UNFCCC), 1995, *Report of the Conference of the Parties, Part 2: Action Taken by the Conference of the Parties, First Session*, Berlin, Germany, FCCC/CP/1995/7/Add. 1, March 28-April 7, 1995, New York, N.Y., USA.

2

AGRICULTURE

Rapporteur's Statement

T.R. Carter
Agricultural Research Center of Finland
Postal Address: Finnish Meteorological Institute
Box 503, FIN-001, Helsinki, Finland

Most studies described in this chapter address the vulnerability of a nation's agricultural system to climatic variations and changes. Most countries have identified some adaptive measures, but far fewer have actually evaluated these measures, combined them into strategies to meet given goals and objectives, or examined trade-offs among different objectives. The evaluation of adaptation options, which presumably is part of the decision-making process in the agricultural sector in most countries, clearly require more attention.

A number of important issues have been identified from work already carried out. These issues are discussed in the following paragraphs.

Suitability of Climate Change Scenarios Used for Agriculture Analysis. Many questions remain unanswered about the applicability of General Circulation Model (GCM)–based scenarios in some regions where recent trends are not in accordance with GCM predictions. Extrapolation of recent trends (e.g., of drying in sub-Saharan Africa), at least over the short term, might be another legitimate scenario. However, it is also important to distinguish between short-term variability and long-term anthropogenically induced change.

Limitations of Adaptation. Several physical constraints on adaptation have been identified:

- The response of crops to increasing fertilizer application follows a "law of diminishing returns."
- Similarly, the positive effect of carbon dioxide (CO_2) on C_3 crops begins to saturate above about 700 parts per million.
- Genetic limitations affect the potential for breeding new crop varieties to suit future climatic conditions (e.g., little scope exists to improve drought resistance in maize).

Improvements in Efficiency. Many adaptive measures in this category are valid under present-day climatic variability ("no-regret" strategies). Many concern water use and conservation, including long-distance transport of water for irrigation and improving the efficiency of water use in vegetation (e.g., in Egypt, plants are overirrigated).

Breeding. To guide breeders, some crop model simulations consider optimum characteristics of crops under changed climatic conditions (i.e., phenology, sowing). Researchers are examining the physiological adaptation of crops to stress events (both to the event itself and to recovery after the event). Breeding research needs to focus on genetic traits that exhibit resilience under climate change.

Soil Management. Proper soil management measures are critical to ensure sustainable production. In regions such as the Russian Federation, this aspect (especially maintaining nutrient levels) could be crucial for maintaining production potential during the next century (even without climate change).

Integration of Adaptive Measures. The need for an integrated approach for assessing adaptation options is pressing. All regions in which water is the major constraint on agriculture have competed for water. It impossible to assess the options for improving water availability for agriculture without considering the other sectors where water use is important (e.g., urban needs, industry, recreation). In the same context, it is imperative in evaluating impacts into the next century that analyses properly account for the future water demands of these competing sectors (regardless of climate change) due to population increase and changing economic activity.

An integrated approach to agriculture and water resources involves the following issues:

- Governments should allow water to be treated as an economic commodity to be bought and sold on the open market for its highest valued uses.
- Governments should allow farmers freedom in choosing cropping patterns that reflect economic and water resource conditions at the time of planting.
- In the face of reduced water resources, farmers should maximize the production of high-value crops — those providing the highest return per unit of water, whether rain-fed or watered by irrigation. This issue implies establishing liberal trading policies to sell crops with local advantage and import of more water-intensive crops.
- Governments should examine possible dietary changes by the population, moving from high-water-use crops, such as rice and sugar cane.
- In the face of changing climate, management of water, land, and crops should be integrated.

Agricultural Adaptation to Climate Change in Egypt

M.H. El-Shaer
Cairo University, Agriculture Faculty, Agronomy Department
17 Teiba St., El-Mohandeseen/Dokki
Giza, Egypt

H.M. Eid
Ministry of Agriculture, Soils and Water Resources Institute
17 Teiba St., El-Mohandeseen/Dokki
Giza, Egypt

C. Rosenzweig
Goddard Institute for Space Studies and Columbia University
2880 Broadway
New York, NY 10025, USA

A. Iglesias
Instituto Nacional de Investigaciones Agrarias, CIT-INIA
Carretera de la Coruna KN7
EC 804 Madrid, Spain

D. Hillel
University of Massachusetts
Amherst, MA 01003, USA

Abstract

Egypt may be particularly vulnerable to climate change because of its dependence on the Nile River as its primary water source, its large traditional agricultural base, and its long coastline, which is undergoing intensifying development and erosion. A simulation study characterized the potential impact of climate change on two reference crops; dynamic crop simulation models were combined with climate change scenarios derived from three equilibrium General Circulation Models. Under the future climate, yields and water-use efficiency (WUE) were lower than those under current climate conditions, even when the beneficial effects of carbon dioxide were taken into account. On-farm adaptation techniques that would result in no additional cost to the agricultural system did not improve the crop WUE or compensate for yield losses due to the warmer climate. Economic adjustments, such as improving the overall WUE of the agricultural system, soil drainage and conservation, land management, and crop alternatives, are essential. If appropriate measures are taken, the negative effects of climate change in agricultural production and other major resource sectors (water and land) may be lessened.

Introduction

Agricultural Regions

Agricultural land use in Egypt is determined by climate and water availability. The climate is very arid, with little or no precipitation (less than 250 mm/yr on the average). The Nile River extends about 1,700 km from south to north. The width of the Nile Valley ranges from 3-17 km in Upper (South) Egypt to 400 km in Lower (North) Egypt. The land available for agriculture is limited mainly to the Nile Delta and Valley and, to a smaller extent, to the Mediterranean coastal fringe and some desert depressions. The Nile Valley and inhabited lands account only for 3.3% of the national territory; the remainder is desert. The Eastern Desert and Sinai Peninsula account for about 25% of the land area, and the Western Desert accounts for about 70%.

The country's total area is relatively large (about 1.1 million km^2), spanning 9° of latitude (22.0-31.5°N) and presenting a north-south gradient of average temperatures and precipitations. This latitudinal gradient and the influences of the Mediterranean Sea, the Nile, and the desert bring about climatic differences among regions. Temperatures increase from north to south. The only region with appreciable rainfall is the Northern Coastal area, with 100-200 mm annually. Precipitation is 40-60 mm/yr within the delta and nonexistent in the desert and Upper Egypt Valley lands.

Egyptian agriculture is based entirely on irrigation and hence depends on a tenuous balance between the supply of water (the Nile and, to a lesser degree, groundwater) and the demand for it by crops. That balance is mainly dictated by the climate, inasmuch as climate determines both the supply of water from the Nile and the evapotranspirational demand for water imposed by the atmosphere. The water balance is also affected by the pattern of water use (i.e., the specific crops grown and the mode of irrigation), as well as by soil conditions and water quality (both of which appear to be deteriorating).

A combination of climate, water availability, and soils defines different agricultural regions in Egypt (Figure 1 and Table 1). Except in the North Coastal area, where short-cycle winter cereals can be rain-fed or grown with little supplemental irrigation, irrigation is required for crop growth in Egypt (El-Shaer 1994). Evapotranspiration values are much greater to the south, especially toward Aswan and the western desert depressions, which have the highest demand for irrigation water.

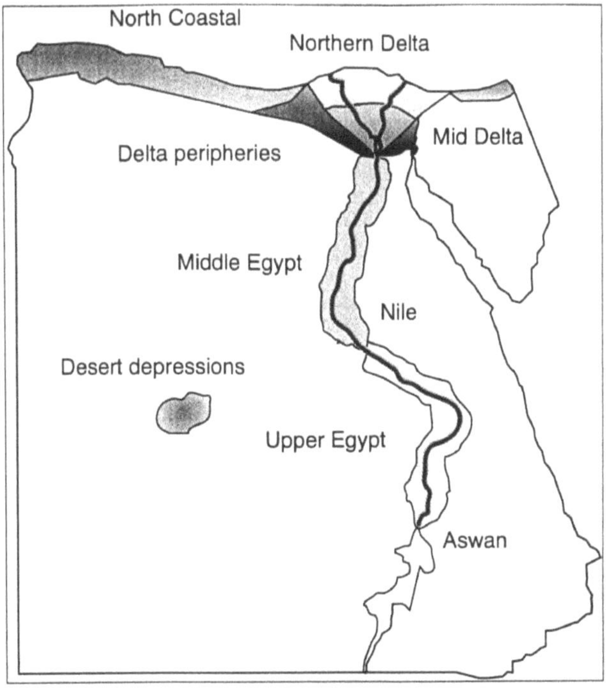

Figure 1. Agricultural Regions in Egypt

The Nile Delta and Valley are the main growing areas of Egypt. Because of the aridity of these regions and the poor management of surface irrigation and drainage systems, severe water table and salinity problems exist in many of the soils, especially near the Mediterranean coast and lakes. A well-established network of canals and drains covers the entire valley for surface irrigation, and open or tile drainage pumping stations in the northern catchments carry away drainage water or allow it to be reused. Crops can be produced in the Northern Delta only with efficient drainage. The mid-Delta is the richest crop production area. In general, the soils do not have major problems, and the water table is satisfactorily maintained below 1.5 m. The main constraints to crop production in the mid-Delta relate to intense agricultural practices and the pressures of many small holders on land use. Middle Egypt occupies the northern half of the Nile Valley south of Cairo, and Upper Egypt extends southward toward Aswan. Although the water table is relatively deep in these regions, soil salinity and sodicity, caused by poor management, are major concerns in some areas.

In contrast to the main agricultural regions, the delta peripheries and desert depressions have relatively advanced agricultural technology, and urban landholders and private companies have made large capital investments there. The

Table 1. Description of Primary Agricultural Regions in Egypt

Region	System	Problems
Northern Coastal[a]	Only rain-fed area, occasional supplemental irrigation Winter crops	Crust formation, nutrients, drought, wind
Northern Delta[b,c]	Multicrop, all seasons	Salinity, drainage, high water table
Mid-Delta[b,c]	Multicrop, all seasons	Intensive use of inputs and land
Delta Peripheries[a,b]	Multicrop, all seasons	High costs, low-quality water, soils, crust formation
Middle Egypt[b]	Multicrop, all seasons	Drainage, insufficient water, some salinity
Upper Egypt[b]	Multicrop, all seasons	Heat
Desert Depressions[a]	Mainly winter crops	Heat, wind, drought, limited underground water

[a] "New Lands" refer to desert areas outside the Nile Valley, to which water must be conveyed over some distance from the Nile or supplied from deep wells. The New Lands are west of the delta (Nubaria), east of the delta (Salhia and along the western side of the Suez Canal), in the northern Sinai, and in the New Valleys of the Western Desert. The soils here are generally sandy and calcareous, not nearly as naturally fertile as the alluvial soils. However, they are often readily drainable, though also more prone to salinity.

[b] "Old Lands" refer to lands along the Nile Valley and in the delta that are irrigated directly from the Nile or from groundwater fed by the Nile. These lands have been under long-term irrigation. The predominant soils are alluvial silt clay loams.

[c] Degraded (waterlogged and salinized) Old Lands that need to be rehabilitated are not included in the New Lands category. As these are reclaimed through drainage and leaching, they are referred to as "New-Old Lands."

major constraints in these areas arise from low fertility, low quality of water (often recycled from irrigation), and relatively high costs of production and marketing. Desert depressions are oases in the western desert where fossil water accumulates. Severe dry weather in the summer minimizes the acreage devoted to summer crops, but these areas are excellent producers of winter cereals and legumes. Programs are being developed for soil and water conservation, reuse of water for irrigation, drainage, wind breaks, sand dune stabilization, sustaining plant cover, and range land management. Further development of agriculture in the desert depressions is possible, in theory, because wells are available for irrigation. Unfortunately, poor drainage could cause lowland deterioration and saturation, water-table rise, and salinity.

Major Agricultural Systems

Egypt's warm mean annual temperature, high solar radiation, fertile soils, and abundant water supplies from the Nile have created a rich agricultural system that has been in place for approximately 5,000 years. The Old Lands (Figure 1 and Table 1) of the Nile Delta and Valley have been farmed continuously throughout this period. In the last 20 years, the Egyptian government has promoted expansion of agriculture into the New Lands (located in desert regions) and reclamation of the New-Old Lands (long-used areas now salinized or waterlogged) (Rosenzweig and Hillel 1994).

Agriculture in Egypt can be described as intensive (Rosenzweig and Hillel 1994): crops grow all year around, as long as water is available. Crops have been selected according to soil and water limitations in each region to produce the best yields and qualities. Major crops in Egypt include wheat (a staple food crop), maize (primarily for animal feed), clover, cotton, rice, sugar cane, fava beans, and soybeans.

In the Nile Delta and along the banks of the river, agriculture is characterized by complex year-long cropping patterns carried out by traditional farmers on small units of land with complicated land-tenure relationships. Two-thirds of the landowners in Egypt own less than 5 feddans (1 feddan equals 0.42 ha) (Central Agency for Public Mobilization and Statistics [CAPMAS] 1994). Agriculture in the Old Lands is managed so intensively that it may be called "gardening" rather than "farming."

Each year, 12 million acres (about 4.9 million ha) of crops are collected from 6 million acres of agricultural land. Twenty-eight major seasonal crops can be identified in the current cropping system (Table 2). This intensive use of the land implies many agricultural problems (nutrient depletion, salinization, waterlogging, etc.). Winter crops (46% of total crop land) are sown in October-December to May; summer crops (41% of total cropland) are sown in April-June to October; and Nili (or Kharif) crops (6% of total cropland) are sown in July-August to November. Perennial crops, such as sugar cane and alfalfa, are sown

Table 2. Area of Major Crops in Egypt by Season (in 1,000 feddans[a]) [b]

Winter Season		Summer and Nili Season	
Crop	Area	Crop	Area
Wheat	2,092	Cotton	840
Barley	248	Corn	1,962
Flax	29	Sorghum	355
Onions	32	Rice	1,216
Garlic	14	Sugar cane	267
Sugar beets	38	Potatoes	217
Beans	425	Peanuts	31
Chickpeas	14	Sesame	54
Fenugreek	11	Soybeans	52
Lentils	15	Vegetables	554
Lupin	7	Others	345
Clover	2,542		
Vegetables	350		
Others	35		
Total	5,852		5,898

[a] 1 feddan = 0.42 ha.
[b] Orchards total 896,000 feddans.
Source: Ministry of Agriculture (1993).

in either spring (March) or autumn (October). Cotton is a relatively long-duration summer crop and is planted in March-April. Vegetables are planted all year long; spring and autumn plantings are added to summer and winter plantings. Orchards (7% of total crop land) are transplanted in February. In the New Lands, crops are planted in fields larger than those in the Old Lands. Intensive management of modern irrigation is needed to sustain these crops at high productive levels.

Intercropping (interplanting of a major crop with a secondary crop) is another form of intraseasonal intensification (i.e., maize or sorghum with soybeans, cowpeas, or beans; sunflowers or sesame with peanuts; cotton with onions). Minimum tillage practices are sometimes followed in heavy clay soils for paddy rice fields in basin irrigated areas. Berseem clover or fava beans are often sown after rice is harvested, without any tillage, to improve compaction and infiltration and avoid clod formation by plowing.

Major Causes of Concern for Egyptian Agriculture

Any attempt to assess the future of Egyptian agriculture must consider the complex interactions among the factors that determine land use, choice of cropping systems, and socioeconomic characteristics and limitations (Rosenzweig and Hillel).

Population and Urban Growth

Inexorable population growth (increasing at the rate of 2.3% annually) and urban encroachment (estimated at 25,000–50,000 ha annually) will determine whether the Egyptian agricultural system can be sustained. The national production of many crops does not meet the current demand, and each year additional amounts have to be imported (i.e., up to 60% of total consumption of wheat). Because most of Egypt is a desert with limited areas for agriculture, the pressure to find new ways to increase agricultural productivity is great.

Loss of Agricultural Land and Deterioration of Crop Yields

Without changes in current crop patterns and water use, more land will be lost to waterlogging and salinization, as well as to urbanization. In Egypt, field water application efficiency values (the fraction of water applied that is actually used, or transpired, by the crop) are typically much less than 50%, and less than 30% in many cases. Such low values imply that water applied in the field exceeds the irrigation requirement of the crop by more than one-half (and often two-thirds). Excess irrigation reduces crop yields because it impedes aeration, leaches nutrients, and induces water-table rise, salinization, and the need for expensive drainage. Irrigation water quality will deteriorate, resulting in decreased agricultural productivity. Crop yields will diminish, in spite of the expected improvements in varieties, fertilization, and pest control. Especially vulnerable to progressive degradation of land and water resources are the ill-drained areas of the lower Nile Delta, which are already subject to land subsidence, water-table rise, and saltwater intrusion.

Climate Change

The future of Egyptian agriculture is thus hard to project, even if it is assumed that current climate conditions will continue. The task is more difficult because of the possibility of significant warming resulting from the enhanced greenhouse effect (Houghton et al. 1990, 1992; Tegart et al. 1990). The expected impact of climate change on the water supply (i.e., the flow of the Nile) is uncertain (Strzepek et al. 1994). Egypt appears to be particularly vulnerable to climate change because of its dependence on the Nile as its primary water source, its large traditional agricultural base, and its long coastline, which is undergoing both intensifying development and erosion (Rosenzweig and Hillel 1994; Rosenzweig 1995).

Equally serious is the potential effect of sea-level rise resulting from thermal expansion of seawater and melting of land-based glaciers. Even a slight rise in sea level will exacerbate the already active process of coastal erosion along the shores of the delta (currently 50 m annually of the head of the Rosetta branch of the Nile at Rashid), a process that accelerated after the Aswan High Dam was built. For a 1-m sea-level rise, 12-15% of the existing agricultural land in the delta may be lost (Nicholls and Leatherman 1994). Sea-level rise will also accelerate the intrusion of saltwater into surface bodies of water (lagoons and lakes in the Northern Delta) as well as into the underlying coastal aquifer (Sestini 1992; El-Raey 1995). The rise in the base level of drainage will further increase the tendency toward waterlogging and salinization in low-lying lands, and significant areas will become unsuitable for agriculture. At the very least, the costs of drainage will increase.

Characterization of Crop Changes with Climate Warming: A Simulation Study

A task for climate change studies is to estimate the potential impacts on yield, water use, and growing season of the main crops grown in each region. Models are available for most of the major crops grown in Egypt. Dynamic-process crop simulation models are recommended so that changes in agronomic processes, such as water stress, crop phenology, and crop failure, can be studied in detail. Crop models are also useful for testing potential adaptations to climate change, such as changes in planting dates and shifts in cultivars or crops (Rosenzweig and Parry 1994).

Simulating climate change effects on agricultural production in Egypt requires a coordinated effort in which data, computer software, and expertise from various disciplines and institutions are integrated. The first step is to calibrate and validate the models with local agronomic experimental data for a set of sites representative of major Egyptian agricultural regions (Eid 1994). Next, simulations with observed climate provide a baseline. Crop model simulations are then run with a suite of climate change scenarios. Finally, farm-level adaptations are tested to characterize possible adjustments to climate change.

This section describes a simulation study to characterize the potential effects of climate change on yields, irrigation demand, and phenology on representative crops in Egypt, and potential adaptations to such impacts. Wheat and maize were selected for the study because they represent different growing seasons, water needs, and physiological responses to an increase in atmospheric carbon dioxide (CO_2). Crops in four main agricultural regions were simulated: the Northern Delta region is represented by Sakha (about 60% of the wheat and 75% of the maize produced in Egypt are grown here); the mid-Delta region is represented by Giza; the North Coastal region is represented by Matrooh; and Upper Egypt is represented by Aswan (Figures 1 and 2).

Baseline Climate and Climate Change Scenarios

Daily maximum and minimum temperatures, precipitation, and solar radiation for Sakha (1975-1989) and Giza (1960-1989) were obtained from the Ministry of Agriculture in Cairo. Daily weather variables for Matrooh and Aswan were simulated from monthly means obtained from the Instituto Nacional de Meteorologia, Spain, by using the weather simulator in the Decision Support System for Agrotechnology Transfer, Version 3 (DSSAT 3) (Tsuji et al. 1995). Because of the limited data for these two sites, their results should be interpreted only as preliminary.

Climate change scenarios for each site were created by combining the daily climate data for each site (Rosenzweig and Iglesias 1994) with the output of three equilibrium General Circulation Models (GCMs): the Geophysical Fluid Dynamics Laboratory (GFDL), Goddard Institute for Space Studies (GISS), and United Kingdom Meteorological Office (UKMO) models. The Intergovernmental Panel on Climate Change (IPCC) Technical Guidelines endorse this approach (Carter et al. 1994). The three equilibrium GCMs used to create the climate change scenarios are at the high end of the IPCC range (4°C); nevertheless, they were chosen for consistency with other climate change impact

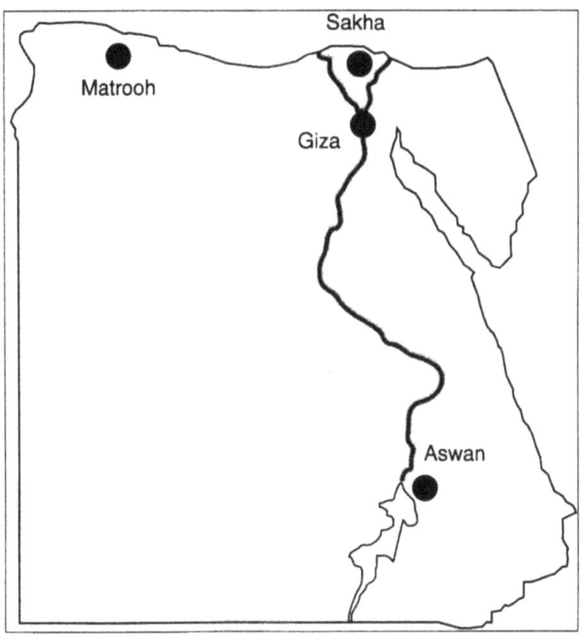

Figure 2. Location of the Crop Simulation Sites

studies in Egypt (i.e., effects on sea-level rise and water resources) for an integrated assessment of the potential effects of climate change. These GCMs have also been widely used in global impact studies (Rosenzweig and Parry 1994); therefore, their use in this study permits comparison with studies for other regions.

According to the scenarios (Eid 1994), Egypt will experience a significant rise in mean temperature in the middle of the coming century. Projected precipitation changes are more uncertain, and solar radiation changes are projected to be small. The different GCMs do not agree on the precipitation changes (about 50% increase with the GISS scenario and about 15% decrease with the GFDL and UKMO scenarios). Current low levels of precipitation in Egypt minimize the differences in the absolute values of projected future precipitations. For example, a 20% change in precipitation in the delta region translates into an absolute variation of ±8 mm/yr. The GISS scenario predicts the largest net precipitation change in the North Coastal region (an increase of 60 mm/yr).

Crop Models

Crop yields and demand for irrigation water were estimated with the CERES models in DSSAT 3. The DSSAT 3 crop models can simulate changes in crop photosynthesis and WUE due to changes in atmospheric CO_2 (Peart et al. 1989).

Crop Management Variables

Typical soils at Giza and Sakha are described elsewhere (Abdel Wahed 1983; El-Shaer 1994; Eid 1994). The soil description in the crop models includes texture, albedo, and specific water-related characteristics; however, it does not include salinity and other characteristics that can limit crop production. Generic soils (medium silty clay and medium sand) were selected to represent the soils at the study sites (Eid 1994; Tsuji et al. 1995). Crop varieties and planting dates were selected to represent the common practices for each region (Eid 1994). The crops were simulated in irrigated conditions in the delta and valley sites, and wheat was simulated in rain-fed conditions in the Northern Coastal area. Irrigation was simulated with the automatic irrigation option in the model (assuming 100% efficiency of the system, management depth of 1 m; irrigation triggered when plant-available water declines to 50% of capacity). Flood irrigation is not included in the crop models; therefore, only relative changes in irrigation are considered. Nitrogen was considered nonlimiting at all sites. Table 3 lists the crop management variables used in the study.

Table 3. Selected Sites and Crop Management Variables

Region	Site	Soil	Cultivar	Planting Date	Density (plant m^{-2})
North Coastal	Matrooh	Medium sand	Wheat: MEXPIK65	Dec. 15	200
Northern Delta	Sakha	Medium silt Clay	Wheat:	Dec. 15	300
			Maize: PIO514	June 1	5.7
Mid-Delta	Giza	Medium silt Clay	Wheat: MEXIPAK65	Dec. 1	300
			Maize: PIO514	June 1	5.7
Upper Egypt	Aswan	Medium silt Clay	Wheat: MEXIPAK65	Dec. 1	300
			Maize: PIO514	June 1	5.7

Crop Model Validation

The CERES models for wheat and maize were previously validated with local agronomic experimental data for Sakha and Giza (Eid 1994).

Characterization of Crop Changes

Simulations with the climate change scenarios resulted in lower grain yields for both crops at all sites, except for wheat at the northern site (Table 4). All GCM simulations included the direct beneficial effects of increased CO_2 on crop yield and evapotranspiration (see, for example, Acock 1990). The temperature-induced reductions in crop yield are due to a shortening of the grain-filling period (Table 4). The very high temperatures projected by the GCMs also affect other physiological processes not included in the crop models (i.e., pollinization failure); therefore, these yield changes may be underestimated. Simulated yield increases at the Northern Coastal site (Matrooh) reflect the precipitation increases projected by the GISS and GFDL scenarios during the crop season, and the increase in WUE of the rain-fed crop simulated with the direct beneficial effects of CO_2. Although the relative increases are substantial, the base yields at this site are very low (about 1.5 t/ha).

The agronomic WUE (yield/total crop evapotranspiration) decreased with temperature increase at all sites, except for rain-fed wheat at the Northern Coastal site (Table 4). Daily crop evapotranspiration increased in the warmer

Table 4. Simulated Changes in Crop Yield, Season Length, and Water-Use Efficiency (WUE)

Site	Scenario	Yield (%)	Season Length (days)	WUE (%)
		Wheat		
Matrooh	GISS	+80	-22	+100
	GFDL	+92	-20	+110
	UKMO	-21	-27	-14
Sakha	GISS	-23	-16	-14
	GFDL	-18	-16	-10
	UKMO	-17	-17	-12
Giza	GISS	-35	-7	+1
	GFDL	-31	-6	+3
	UKMO	-73	-5	-60
Aswan	GISS	-45	-10	-6
	GFDL	-40	-9	-13
	UKMO	CF[a]	CF	CF
		Maize		
Sakha	GISS	-19	-11	-3
	GFDL	-22	-11	-14
	UKMO	-18	-12	-17
Giza	GISS	-25	-22	-11
	GFDL	-17	-21	-14
	UKMO	-66	-25	-68
Aswan	GISS	-60	-25	-56
	GFDL	-58	-25	-52
	UKMO	CF	CF	CF

[a] Crop failure.

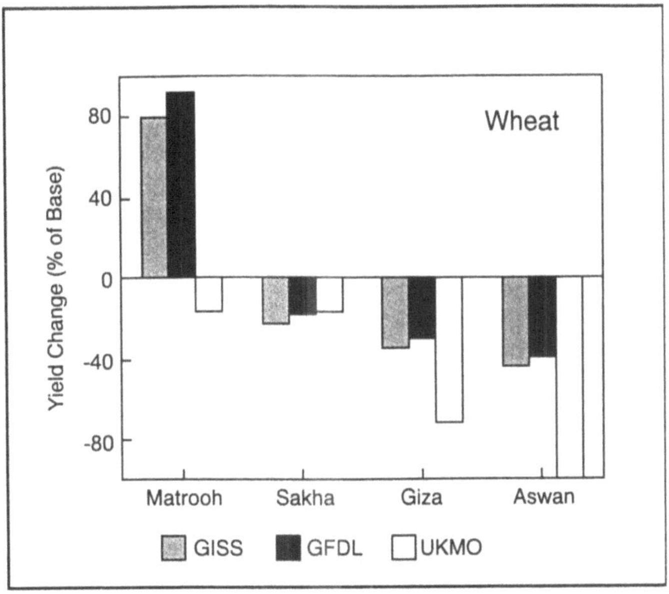

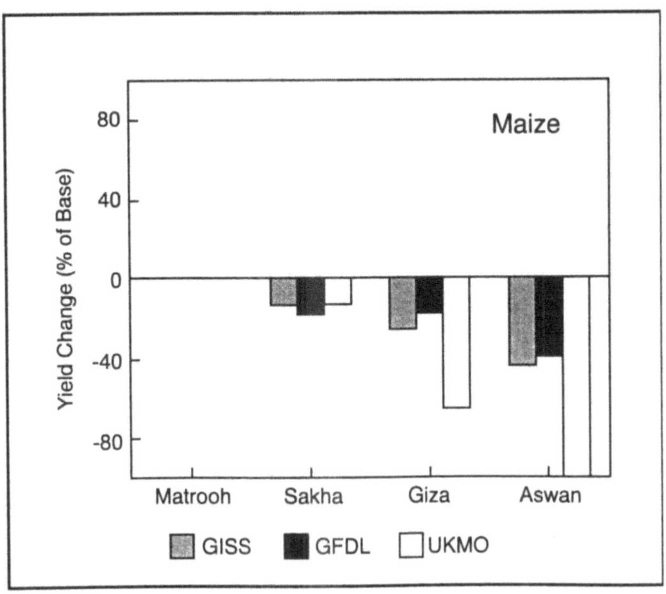

Figure 3. Simulated Changes in Crop Yield

climate at all sites. This increase was mitigated in part by the higher WUE of the crops in a CO_2-enriched atmosphere (Rosenzweig and Hillel 1993).

The simulation study considered changes in maize and wheat cultivars, as well as planting dates, as possible adaptive strategies to climate change (Figure 3). The cultivars tested were described by Eid (1994). Planting dates were adjusted to an optimum that would minimize yield losses in the warmer climate (one month in advance). Nevertheless, for the locations considered, these on-farm adjustments did not significantly reduce the negative effects of climate change on the crops (compare Figure 3 with Table 4). Furthermore, the agronomic WUE did not improve with the adaptive changes considered.

Discussion

Vulnerability of Egyptian Agricultural Regions and Systems

The following analysis of the vulnerability to climate change of Egyptian agriculture is based on the results of the crop simulation study, changes in resources that affect the agricultural system that were not included in the study (e.g., sea-level rise), and the problems and limitations of the current system (Rosenzweig and Hillel 1994).

On the basis of previous simulation studies, and extending and integrating site-based results to national yield and water consumption changes, Eid and collaborators (Eid and El-Sergany 1993; Eid et al. 1993, 1995) projected that climate change could decrease national production of all crops (ranging from 11% for barley to 28% for soybeans), while increasing crop demand for water (up to 16%). The results of this study are somewhat more pessimistic, with crop yield decreases of more than 20%.

This simulation study assumed that the crops received sufficient water from irrigation at all times. Under the tradition of infrequent irrigation in Egypt, however, sensitive crops are more likely to suffer from increased moisture stress and additional salt stress due to lack of water than the simulations indicate; therefore, yields may suffer further. Heat-sensitive crops already near the limit of their tolerance are especially vulnerable.

Higher evaporation rates may also worsen the tendency toward soil salinization by speeding the transport of damaging salts to the soil surface. Changes in the agronomic WUE may also be more negative than the simulations suggest because of excessive irrigation, poor drainage, salinization, and nonoptional management (e.g., insufficient or inappropriate fertilization, poor pest control, or poor choice of crop or poor germination). Thus, the agronomic WUE may continue to be much below optimum.

Especially vulnerable to the progressive degradation of land and water resources are the ill-drained areas of the lower Nile Delta that are already subject to land subsidence, water-table rise, and saltwater intrusion. Combating these processes will require large investments in expensive drainage, as well as greater government intervention and regulation. If investments, interventions, and regulations are lacking or haphazardly implemented, these lands will become unusable for agriculture. The strain on the coastal and delta system may also lead to clashes among competing interests (e.g., agricultural, urban, industrial, and tourist sectors).

Most vulnerable to sea-level rise are the low-lying lands along the Northern Delta, where the surface elevation is less than 1 m above sea level. Owing to land subsidence (perhaps 0.1 m in 50 years), as well as potential sea-level rise (estimated to total 0.2-0.5 m in the same period), the coastal fringe of land may reach 20 km or more. Within this strip, it will become progressively more difficult to maintain agriculture, and eventually, much land will be retired from production. Urban and industrial development will also be problematic because waterlogging, and the ecology and economy of the lagoons, will be affected by saltwater intrusion. Drainage to control waterlogging will become increasingly expensive.

Adaptation: Improved Resource Management and Effective Adaptive Strategies

Much can be done to mitigate the potential dire consequences of climate change, and the earlier the task is recognized and undertaken, the more likely it is to succeed (Rosenzweig and Hillel 1994). A few essential changes in resource management that would lead not only to adaptation to climate change, but to the overall improvement of the Egyptian agricultural system, are discussed.

Water Management

The first imperative is to improve both the technical water application efficiency and the agronomic WUE. This measure involves revamping the entire system of water delivery and control. Ideally, water should be made available on demand (rather than on a fixed schedule) and delivered in fixed quantities in closed conduits subject to effective monitoring and regulation while avoiding losses. Although this policy will be difficult to achieve in the Old Lands, where traditional systems prevail, it can be implemented from the outset in the New Lands.

To encourage water conservation, authorities should provide farmers with explicit guidance on optimal crop selection, irrigation, and fertilization and institute strong incentives for avoiding excessive water use (including the often suggested, but seldom implemented, pricing of water in increasing proportion to the amount used). Modern methods of irrigation based on the high-frequency, low-volume application of water and fertilizers directly on the plants must be adapted to the scale of operation and local practicalities of Egyptian farming. Fortunately, such systems are flexible and lend themselves readily to downsizing so as to accommodate the small-scale Egyptian farming units. Moreover, such systems can be applied to sandy and even gravelly desert soils (potential New Lands) that are not considered irrigable by the traditional surface-irrigation methods.

Improvements in WUE are not easy to implement. A strong system of rewards and penalties is needed to create incentives for conserving water and installing modern irrigation technology. Water metering and pricing must be instituted; water must be made available on demand or at high frequency (rather on a fixed schedule at infrequent intervals); and credit, as well as training, should be offered to farmers willing to modernize their irrigation.

Land Management

Another adaptive measure involves managing low-lying lands on the northern fringe of the delta, where the consequences of sea-level rise (submergence and salinization) are certain to wreak their greatest damage. Some of these lands must be retired from agriculture, and the water made available should be used to irrigate the New Lands outside the Nile Valley and Delta.

Adapted Crops

Another set of measures involves the careful selection and/or breeding of heat-tolerant, salinity-tolerant, water-conserving crops, as well as developing controlled-environmental production methods that minimize water use while maximizing the production of high-value crops (e.g., all-season vegetables, fruits, spices, medicinals). In addition, efforts should be made to promote the adoption of high-return, specialized, water-conserving crops instead of the water-profligate crops now grown, such as rice and sugar cane.

As an example, the simulation study considered on-farm adaptive techniques, such as planting alternative varieties and optimizing the planting date. However, these on-farm techniques, which imply no additional cost to the agricultural system, did not compensate for the yield losses under a warmer climate or improve the crop WUE. Egypt must develop new cultivars that are more adapted to higher temperatures. The crop models can be used to identify appropriate crops, varieties, and management strategies to maximize benefits and minimize risks associated with future climate change. Therefore, further simulation studies would be valuable in assessing the risks associated with given production strategies.

Crop changes can be considered as adaptive measures. For example, cereals may be a better alternative than cotton in some areas. The recent crop liberalization policy allows farmers to adapt to more suitable crops in each area. As result of this policy, the cotton-growing area has sharply decreased, while the cereal-growing area has increased (CAPMAS 1994). Irrigation water could be saved by minimizing the cultivation of rice and sugar cane in areas with other crop alternatives.

El-Shaer (1995) listed and evaluated the potential impacts of climate change on different aspects of agriculture, animal breeding, fisheries, and epidemiology (from germplasm availability and genetic engineering to national policies); their possibility of adaptation to climate change; and a ranking of their adaptation potential (from very high to very low). This evaluation may serve as a basis for a more detailed adaptation study.

The overall effect of the measures discussed here, with or without climate change, will be to raise the potential and actual productivity of Egyptian agriculture. Thus, climate change may not thwart progress toward the goal of providing sufficiently for the Egyptian people. However, much depends on whether the rate of population growth in Egypt continues to decline fast enough to allow agricultural productivity to keep pace with the country's growing needs.

Conclusions

Egypt's vulnerability to climate change is acute. Rapid increases in population and urbanization will only exacerbate this problem. Because of the linkages among the Nile River, the delta, coastal resources, and surrounding deserts, potential impacts of climate change must be addressed in an integrated way by joining the disciplines of hydrology, agronomy, and coastal zone geography, as Strzepek et al. (1994) did. Such an approach will aid in understanding critical environmental processes and thus improve the future of the Egyptian people.

Opportunities exist for improving water-use efficiency and crop management. Egypt must conserve water and reduce drainage requirements, while raising crop yields in both the Old and New Lands. Although yields in the fertile lands of the Nile Valley and Delta are already high, water-use efficiency can be improved. The potential increase in productivity from improved irrigation in the Old Lands probably exceeds that for reclaiming New Lands in the desert outside the Nile Valley and Delta. The latter undertaking is not to be precluded, however, and will be enhanced by water conservation in the Old Lands.

References

Abdel Wahed, A., 1983, *Research Report on Soil Survey,* Soil Survey EMCIP Research Extension Centers, Publication No. 62, Consortium for International Development, ARC, Cairo, Egypt.

Acock, B., 1990, "Effects of Carbon Dioxide on Photosynthesis, Plant Growth and Other Processes," in B.A. Kimball, N. Rosenberg, and H. Allen (eds.), *Impact of Carbon Dioxide, Trace Gases, and Climate Change on Global Agriculture,* ASA Special Publication No. 53, American Society of Agronomy, Inc., Madison, Wisc., USA.

CAPMAS, 1994, *Statistical Yearbook: Arab Republic of Egypt 1952-1993,* Central Agency for the Public Mobilization and Statistics, Cairo, Egypt.

Carter, T.R., M.L. Parry, H. Harasawa, and S. Nishioka, 1994, *IPCC Technical Guidelines for Assessing Climate Change Impacts and Adaptations,* Center for Global Environmental Research, University College of London, United Kingdom.

Eid, H.M, 1994, "Impact of Climate Change on Simulated Wheat and Maize Yields in Egypt," in C. Rosenzweig and A. Iglesias (eds.), *Implications of Climate Change for International Agriculture: Crop Modeling Study*, U.S. Environmental Protection Agency, Washington, D.C., USA.

Eid, M.H., and D.Z. El-Sergany, 1993, "Impact of Climate Change on Simulated Soybean Yield and Water Needs," *Proceedings of the 1st Conference on the Environment of Egypt*, pp. 313-316.

Eid, H.M., M.I. Bashir, N.G. Ainer, and M.A. Rady, 1993, "Climate Change Crop Modelling Study on Sorghum," *Annals of Agricultural Science* 1:219-234.

Eid, H.M., N.M. El-Mowelhi, M.A. Metwally, N.G. Alner, F.A. Abbas, and M.A. Abd El-Ghaffar, 1995, "Climate Change and its Expected Impacts on Yield and Water Needs of Some Major Crops," *Proceedings of the 2nd ARC Field Irrigation and Agroclimatology Conference,* Paper 17.

El-Raey, M., 1995, personal communication, University of Alexandria, Egypt.

El-Shaer, M.H., 1994, *Some Features of Egyptian Agriculture*, University of Cairo, Cairo, Egypt.

El-Shaer, M.H., 1995, *Adaptation to Climate Change in Egyptian Agriculture*, Technical Report, University of Cairo, Egypt (in preparation).

Houghton, J.T., G.J. Jenkins, and J.J. Ephraums (eds.), 1990, *Climate Change: The IPCC Scientific Assessment*, Cambridge University Press, Cambridge, United Kingdom.

Houghton, J.T., B.A. Callander, and S.K. Varney (eds.), 1992, *Climate Change 1992, Supplementary Report to the IPCC Scientific Assessment*, Cambridge University Press, Cambridge, United Kingdom.

Nicholls, R.J., and S.P. Leatherman, 1995, "Sea-Level Rise," in K.M. Strzepek and J.B. Smith (eds.), *As Climate Changes: International Impacts and Implications,* Cambridge University Press, Cambridge, United Kingdom (in press).

Peart, R.M., J.W. Jones, R.B. Curry, K. Boote, and L.H. Allen, Jr., 1989, "Impact of Climate Change on Crop Yield in the Southeastern U.S.A," in J.B. Smith and D. Tirpak (eds.), *The Potential Effects of Global Climate Change on the United States*, EPA-230-05-89-050, Appendix C-1, pp. 2.1-2.54, U.S. Environmental Protection Agency, Washington D.C., USA.

Rosenzweig, C., 1995, in K.M. Strzepek and J.B. Smith (eds.), *As Climate Changes: International Impacts and Implications,* Cambridge University Press, Cambridge, United Kingdom (in press).

Rosenzweig, C., and D. Hillel, 1993, "Agriculture in a Greenhouse World: Potential Consequences of Climate Change," *National Geographic Research and Exploration* 9(2):208-221.

Rosenzweig, C., and D. Hillel, 1994, *Egyptian Agriculture in the 21st Century,* Collaborative Paper CP-9412, International Institute for Applied Systems Analysis, Laxenburg, Austria.

Rosenzweig, C., and A. Iglesias (eds.), 1994, *Implications of Climate Change for International Agriculture: Crop Modeling Study,* U.S. Environmental Protection Agency, Washington D.C., USA.

Rosenzweig, C., and M.L. Parry, 1994, "Potential Impact of Climate Change on World Food Supply," *Nature* 367:133-138.

Sestini, G., 1992, "Implications of Climatic Changes for the Nile Delta," in L. Jeftic, J.D. Milliam, and G. Sestini (eds.), *Climatic Change and the Mediterranean*, Edward Arnold, London, United Kingdom, pp. 535-601.

Strzepek, K.M., D.N. Yates, S.C. Onyeji, and M. Saleh, 1995, "A Socio-Economic Analysis of Integrated Climate Change Impact on Egypt," in K.M. Strzepek and J.B. Smith (eds.), *As Climate Changes: International Impacts and Implications*, Cambridge University Press, Cambridge, United Kingdom (in press).

Tegart, W.J. McG., G.W. Sheldon, and D.C. Grifflths (eds.), 1990, *Climate Change: The IPCC Impacts Assessment,* Intergovernmental Panel on Climate Change, World Meteorological Organization and United Nations Environmental Programme, Bracknell, United Kingdom.

Tsuji, G.Y., J.W. Jones, G. Uehara, and S. Balas (eds.), 1995, *Decision Support System for Agrotechnology Transfer, V3.0*, Vols. 1-3, IBSNAT, University of Hawaii, Honolulu, Hawaii, USA.

Adaptive Measures for Zimbabwe's Agricultural Sector

C.H. Matarira and F.C. Mwamuka
Scientific and Industrial Research and Development Centre
Environment and Remote Sensing Institute
P.O. Box 6640
Harare, Zimbabwe

J.M. Makadho
Agritex
P.O. Box 8117, Causeway
Harare, Zimbabwe

Abstract

Zimbabwe is aware that climate change and variability, due either to natural cycles or shifts, will substantially affect the national economy and ecology. Recurrent droughts have had adverse effects on all sectors of the economy during the past 15 years. The vulnerability of the agricultural sector to the impacts of climate change is evaluated. Measures that can be taken, at both the farm and national levels, to adapt Zimbabwe's agricultural sector to such impacts are examined. Sustained scientific research is recommended to enable more reliable predictions of the impacts of climate change, especially at the national and local levels.

Introduction

Agroclimatic Regions

Zimbabwe is a land-locked country between latitudes 16.5 and 22.5°S and longitudes 25°5′ and 33°E. Zimbabwe's tropical climate is modified by a large range in altitude. The country is divided into five Natural Regions (Figure 1) based on the linkages among altitude, rainfall, and temperature patterns. Table 1 outlines the attributes of these regions.

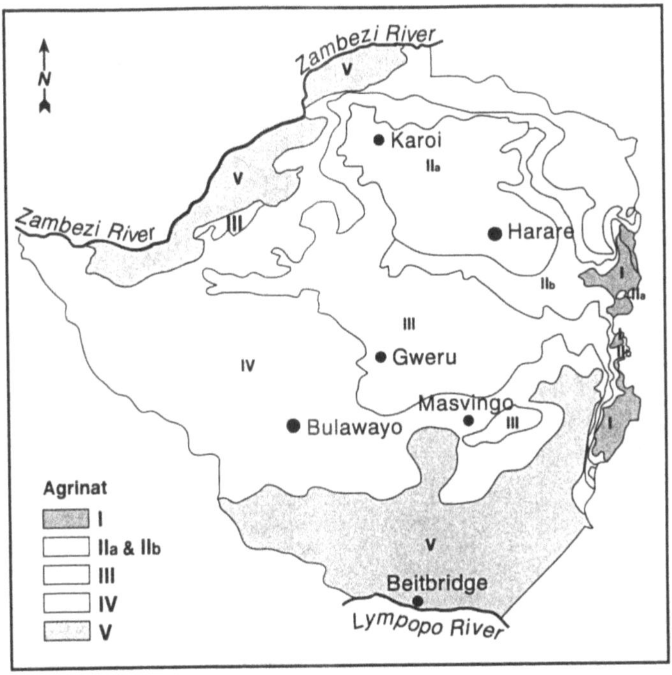

Figure 1. Zimbabwe's Five Natural Regions

The country is generally dry. Nearly 50% receives an annual rainfall of less than 600 mm; a few places in the southern region receive less than 400 mm. Only 37% of the country receives more than 700 mm of rain per year. The Central Plateau and the northern parts of the Zambezi Valley receive an annual rainfall of 600-1,000 mm, whereas the Eastern Highlands annually receive 1,000-1,400+ mm.

The mean annual temperatures vary between 14 and 28°C, and daily temperatures range from 5 to 33 °C. The lowest mean annual temperatures occur at the highest altitudes of the Eastern Highlands (14°C), and the highest temperatures occur in the lowlands of the extreme southeast, northeast, and northwest (28°C).

Agricultural Production

Although Zimbabwe has one of the most highly developed industrial sectors in Africa, agriculture remains a very important economic activity, contributing more than 40% of the country's annual exports and 14% of the gross domestic product (Government of Zimbabwe [GOZ] 1991). Frequent droughts over the

Table 1. Major Characteristics of Zimbabwe's Natural Regions

Natural Region	Altitude and Topography	Climate	Details on the Agricultural Season	Recommended Farming System	Need for Irrigation
I	Mostly highveld 2,000 m above sea level	• Rainfall 900-1,000 + mm/yr • Precipitation every month of the year. • Low temperature	• Rainfall is highly effective • Frost-free valleys	• Specialized and diversified farming • High-value crops: coffee, tea, deciduous fruit, forest products for paper, and timber	• Low need: irrigation • Supplements, minor variations in rainfall
II(a)	Most highveld and plateau up to 1,800 m above sea level	• Rainfall 750-1,000 mm/yr • 18 rainy pentads[a] per season. Very reliable rainfall	• Rainfall confined to summer • Rarely experiences dry spells in summer	• Intensive farming system • Major crops: maize, soya-beans, tobacco, ground-nuts, wheat, and cotton	• Low need: irrigation reduces effects of the midseason dry spells • Irrigation also lengthens growing season for many crops • Wheat is fully irrigated
II(b)		• 16-18 rain pentads per season[a]	• Susceptible to severe dry spells during the rainy season		
III	Midlevel, 1,000-1,600 m above sea level, open and undulating terrain	• Moderate rainfall 650-800 mm/yr in form of infrequent heavy falls • 14-16 rainy pentads • Unreliable start of rainy season	• Subject to severe midseason dry spells and periodic droughts • Marginal for major crops (i.e., maize, soya, and tobacco)	• Semi-intensive farming under good management • Maize, cotton, groundnuts, but under irrigation • Suited to livestock production	• Irrigation sustains crop production
IV	Low-lying plains 600-1,000 m above sea level	• Fairly low rainfall 450-600 mm/yr • High temperatures • High evapotranspiration	• Subject to periodic seasonal droughts and dry spells during the rainy season • Uncertain rainfall for cash crops	• Semi-extensive farming • Suited to livestock produc-tion and drought-resistant crops	• Great need: irrigation greatly enhances reliability of food crop production
V	Low-lying valleys below 600 m above sea level	• Low and erratic rainfall • Very high temperatures and evapotran-spirative demand	• Rainfall too low and erratic for reliable production of even drought-resistant fodder and grain	• Extensive farming system • Suitable for utilization of the veld alone (i.e., cattle and/or game ranching)	• Need for irrigation is greatest • No production of crops is possible without irrigation.

[a] Rainfall pentad is the center of five-day periods (pentad) that together receive more than 40 mm of rainfall and two days of which receive at least 8 mm.

past 15 years (in particular, the drought of 1991-1992) have contributed to the slow growth rate (2.8% per year) of the agricultural sector. This sector is significant in the development of the overall economy of the country for the following reasons:

- It is a source of livelihood for about 80% of the population (GOZ 1991).
- It provides direct and indirect employment to a large proportion of the country's labor force.
- It has made Zimbabwe self-sufficient in food requirements.
- It supplies the manufacturing sector with agricultural raw materials.
- It is a ready market for agricultural inputs manufactured by the industrial sector, that is, agricultural equipment and agrochemicals.

Zimbabwe has about 33 million ha of land under agricultural production, 18.9% of which lies in Natural Regions IV and V. Agricultural production is mainly located in regions that receive an annual rainfall of 800 mm and above. Intensive agricultural production occurs in Regions I and II, while the moderate rainfall of Region IV (~525 mm) makes it more suitable for extensive livestock and game ranching. Zimbabwe produces a wide range of crops and livestock. Most crops, except for winter crops, such as wheat and barley, and perennial crops, such as sugarcane, coffee, and tea, are produced under conditions of natural rainfall. Irrigation is used only to supplement rainfall. Currently, 190,000 ha (42.2%) of potentially irrigable land (450,000 ha) is under irrigation: (1) large-scale commercial farms (80%); (2) state farms (12%); and (3) small-scale farms, communal areas, and resettlement areas (8%) (GOZ 1991).

A total of 68.2% of Zimbabwe's cattle herd is held in communal areas. However, beef off-take from communal areas to the Cold Storage Company (formerly the Cold Storage Commission, a parastatal that had a monopoly on the processing of livestock products) remains at 1.5%, while commercial farms contribute 17% (GOZ 1991). The commercial herd supplies beef for the local and export markets. Low off-takes from communal areas occur mainly because the communal farmers rely on their livestock for draft power, for milk and manure, and as a form of social security.

The development of the agricultural sector has been promoted through three kinds of support services: (1) research and extension, (2) marketing, and (3) financial services. The Department of Veterinary Services provides farmers with an environment in which livestock can thrive and multiply free of major animal diseases. The Department of Agricultural Technical Extension Services (Agritex) provides professional and technical advice, as well as extension services, to the farming community. Agritex advises farmers in crop and animal production, land-use planning, agricultural management, soil and water conservation, agricultural engineering, and irrigation. Financial support for the agricultural sector comes from nongovernmental organizations, the Agricultural Finance Corporation, and commercial banks.

Assessment of Vulnerability to Climate Change

Agricultural production is very dependent on climate conditions, which leaves the sector extremely vulnerable. However, it is important that agricultural production processes are based on three broad types of resources that include, but go beyond, climate conditions (Riebsame 1989):

- Natural resources (e.g., climate, soil, topography, genetic endowments);
- Capital resources (e.g., fertilizers, animals, and machinery); and
- Human resources (e.g., labor inputs, management practices, and market conditions).

Climate can also affect human and capital resources. This factor increases the vulnerability of agricultural activities to climate changes. Matarira et al. (1995) used the CERES-Maize model (IBSNAT 1989) to simulate crop responses to changes in climate and management variables at four sites in Natural Regions II (Karoi), III (Gweru), IV (Masvingo), and V (Beitbridge).

Methodology

Baseline Climate Data

Daily observed climate data (precipitation, solar radiation, and maximum and minimum air temperatures) at each of the stations, collated by the Department of Meteorology for 1951-1991, were used in the simulation.

Climate Change Scenarios

By using General Circulation Models, the observed climate data were modified to create climate change scenarios for each site. The Canadian Climate Center Model (CCCM) (Boer et al. 1992) and the Geophysical Fluid Dynamics Laboratory (GFDL R-30) model (Mitchell et al. 1990) established climate change scenarios for the vulnerability assessment. Because the water supply in Zimbabwe depends entirely on climate conditions, any decrease in precipitation would be most significant with regard to water availability for crop irrigation.

Crop Model Inputs and Simulations

Potential changes in maize physiological responses (yields, season length, evapotranspiration, irrigation demand) to the major factors of climate (daily solar radiation, maximum and minimum temperature, precipitation), soils, and management (cultivar, planting date, plant population, row spacing, sowing depth) were estimated by means of the CERES-Maize model under different climate scenarios. The following assumptions were made in applying the crop model:

- Nutrients are nonlimiting.
- Pests are controlled.

- Soil conditions are not problematic.
- Catastrophic weather events do not occur.
- Technology and the climate tolerance of cultivars do not change under conditions of climate change.

Cultivar and Management Variables

A short-season maize variety, R201, commonly grown under dryland conditions, was used for the four sites. R201 would perform in both high- and low-rainfall areas.

The potential changes were simulated under dryland and irrigated conditions to provide a range of possible scenarios and to analyze the production changes. Because it is not possible to determine the amounts of irrigation water required for each region, irrigation was simulated under the automatic option to provide the crop with a hypothetical nonlimiting situation. For the irrigation simulation, the water demand was calculated assuming the following:

- The automatic irrigation system operates at 100% efficiency.
- The irrigation management depth is 300 mm.
- Irrigation is automatic when the available soil moisture is 50% or less of capacity.
- Soil moisture for each layer is reinitialized to 100% capacity at the start of each growing season.
- The plant population is kept the same in both dryland and irrigated conditions at 4.4 plant m^{-2}.

Soils

Soils in Zimbabwe are predominantly derived from granite and are often sandy and light-textured, with low agricultural potential due to low nutrient content, particularly nitrogen and phosphorus. Nevertheless, significant amounts of soils in all regions have a heavier clay content more suitable for crop growth. The representative soils in Karoi and Gweru are medium sandy loams; Masvingo and Beitbridge have sandy clay loams (Nyamapfene 1991).

Effects of Carbon Dioxide on Plant Physiology

The CERES-Maize model includes an option that simulates the physiological effects of carbon dioxide (CO_2) on photosynthesis and water-use efficiency, which produces higher crop yields (Acock and Allen 1985). For all climate scenarios included in this study, maize was simulated under normal climate conditions and then under conditions of climate change. These simulations also included the simulated physiological effects of CO_2 on crop growth and yield.

Validation of the Crop Model

The CERES-Maize model was validated by means of local experimental crop data. The experimental data included aspects such as cultivar, planting date, growth analysis, fertilizer application, harvesting date, and final yield components. Experimental crop and climate data were used for the 1988-1989 season at Harare Research Station and for the 1986-1987 season at Gweru. At Harare Research Station, the observed yield was 9.5% lower than the simulated yield, and the observed season length was 2.3% shorter than the simulated season length. In Gweru, the mean observed yield was 3% lower than the simulated yield, and the observed season length was 1.6% longer than the simulated season length. From these results, the CERES-Maize model was presumed to be an adequate tool to simulate maize growth, particularly to evaluate relative changes in crop yield in relation to planting date.

Results

The simulation results reveal significant variations in maize yields at the different sites (Tables 2, 3, and 4). However, some of the assumptions made in applying the crop model tend to overestimate the simulated yields. In particular, the amount of irrigation water used is overestimated, and consequently, so are the yields obtained. Nevertheless, this approach allows comparison between relative changes at each site. If arbitrary amounts of irrigation water were applied, the results would be more uncertain, and in-built errors would occur when comparing results from different sites.

Maize production at all stations is more consistent under a normal climate than it is under climate change conditions. Climate change introduces greater variability in maize yields, making maize production more risky. At all sites, maize planted late will not give the necessary yields that make maize production a viable activity under climate change conditions (Table 2). Although irrigation will increase maize production in all areas, the yields will be lower under climate changes than yields under normal climate.

The simulated changes in crop yields are driven by two factors: CO_2 enrichment and changes in climate. In Natural Region IV, for example, it is probable that climate change will turn the region into a non-maize-producing area, as exemplified by reduced maize production at Masvingo. If climate change becomes a reality, all of Natural Region IV, which represents 42% of communal areas, will not have adequate supplies of staple food crop.

Decreases in yield are caused primarily by the increase in temperature, which shortens the duration of the crop growth stages, particularly the anthesis and grain-fill periods. The length of the growing season is, thus, greatly reduced as a result of climate change, irrespective of planting dates (Table 5). A change in the length of the growing season limits maize production to short-season varieties. Decreases in yield due to the shortened growing season are somewhat compensated for by the physiological effects of CO_2, as simulated in this study; however, these effects are not sufficient to offset yield decreases induced by climate change, particularly in an irrigation environment. Except for the GFDL

Table 2. Effect of Climate Change on Dryland Maize Planted on Different Dates at Different Locations

Planting Date/ Climate Scenario	Average Maize Yield over 40 Seasons (kg ha^{-1})			
	Beitbridge	Masvingo	Gweru	Karoi
October 15				
Normal	738	3,006	3,006	3,727
CCCM($2XCO_2$)[a]	514	3,493	5,011	2,634
GFDL ($2XCO_2$)	1,640	3,097	5,446	2,940
November 1				
Normal	1,136	2,779	2,567	3,654
CCCM($2XCO_2$)	838	2,725	4,260	4,641
GFDL ($2XCO_2$)	1,740	2,402	3,697	4,630
November 15				
Normal	514	2,592	2,507	3,531
CCCM($2XCO_2$)	1,092	58	3,444	3,512
GFDL ($2XCO_2$)	1,422	47	2,815	3,507
December 1				
Normal	1,203	2,417	2,047	3,225
CCCM($2XCO_2$)	1,304	47	3,063	2,956
GFDL ($2XCO_2$)	1,453	45	2,640	2,940
December 15				
Normal	1,213	2,339	1,121	3,143
CCCM($2XCO_2$)	713	43	770	41
GFDL ($2XCO_2$)	725	40	735	41

[a] $2XCO_2$ = doubling of carbon dioxide.

Table 3. Effects of Irrigation on Yields for Maize Grown under Dryland Conditions at Different Locations

	Percentage Increase in Yield Due to Irrigation			
Climate Scenario	Beitbridge	Masvingo	Gweru	Karoi
Normal climate	1,655	353	420	255
CCCM ($2XCO_2$)	626	16,928	296	207
GFDL ($2XCO_2$)	462	21,375	298	215

Table 4. Effects of Climate Change on the Yields of Maize Grown under Irrigation at Different Locations

	Decrease in Yield from Base (%)			
Climate Scenario	Beitbridge	Masvingo	Gweru	Karoi
CCCM ($2XCO_2$)	12	17	17	14
GFDL ($2XCO_2$)	11	14	14	12

Table 5. Effects of Climate Change on Average Growing Season Length under Dryland Conditions at Different Locations

Scenario/ Planting Date	Average Season Length (days)			
	Beitbridge	Masvingo	Gweru	Karoi
Normal				
October 15	83	121	121	127
November 1	85	121	122	129
November 15	88	121	121	103
December 1	85	124	121	132
December 15	84	127	113	135
CCCM ($2XCO_2$)				
October 15	77	101	110	105
November 1	79	101	107	108
November 15	78	101	103	108
December 1	78	102	104	109
December 15	83	102	112	109
GFDL ($2XCO_2$)				
October 15	87	104	111	107
November 1	84	103	106	110
November 15	81	103	103	111
December 1	82	104	105	111
December 15	84	105	113	112

R-30 model, which gives an average increase of 8% in precipitation at Beitbridge for all planting dates, the models indicated a reduction in the amount of available precipitation with climate change (Table 6).

Zimbabwe has experienced reduced precipitation and recurrent drought conditions for the past 15 years. A reduction in mean seasonal precipitation under climate changes also implies reduced capacity for irrigation. Although these effects are indicated for maize, it is likely that climate change will affect other crops in more or less the same way. Agricultural production that depends on irrigation, such as sugarcane, winter crops, and the horticultural subsector,

Table 6. Effects of Climate Change on Available Precipitation at Different Locations

Planting Date/ Climate Scenario	Base precipitation (mm) (average available precipitation over 40 seasons under normal climate)			
	Beitbridge	Masvingo	Gweru	Karoi
October 15	138	451	506	501
November 1	164	501	536	571
November 15	175	498	531	584
December 1	170	480	510	577
December 15	166	433	445	531
	Change in Available Precipitation (Percent of Base)			
CCCM ($2XCO_2$)	Beitbridge	Masvingo	Gweru	Karoi
October 15	-36	-33	-29	-36
November 1	-31	-28	-22	-29
November 15	-29	-19	-23	-28
December 1	-18	-23	-20	-27
December 15	-9	-16	-12	-27
GFDL ($2XCO_2$)	Beitbridge	Masvingo	Gweru	Karoi
October 15	+15	-16	-15	-29
November 1	+7	-13	-12	-21
November 15	+2	-9	-14	-16
December 1	+11	-8	-11	-12
December 15	+7	-6	-3	-9

would be seriously affected by reduced irrigation (an example is the effect of the 1991-1992 drought on sugarcane production in the lowveld). Thus, climate changes that may occur as a result of increased atmospheric CO_2 will, thus,require different levels of resource surveillance and management than are applied today and/or the development of different management strategies for the agricultural sector that will minimize the negative socioeconomic impacts of climate change.

The simulation results obtained in this study are supported by the findings of Downing (1992), who noted the following:

- With a temperature increase of 2°C, the wet zones of Zimbabwe (areas with a water surplus) decrease by one-third, from 9% to about 2.5%.
- The drier zones double in area.
- An additional increase in temperature of +4°C reduces the summer water-surplus zones to less than 2% of the country's area, approximately corresponding to the 1991-1992 drought.
- In addition to a decrease in the agricultural area, crop yields in marginal zones become more variable.

Simulations conducted by Muchena (1991) also indicate that if temperatures increase 2°C, yields currently expected 70% of the time would decrease and be expected only 40% of the time. Such studies indicate that small-scale or subsistence farmers in the marginal semiarid regions of Zimbabwe are the most vulnerable to climate change.

Adaptations to Climate Change

For the agricultural sector, adaptations to climate change can occur at two levels: the farm level and the national level, as reflected in government policy.

Although agriculture is very sensitive to climate, it may be flexible to climate change. As a unit exposed to impact, agriculture is a moving target, continually adjusting to both perceived climatic and nonclimatic conditions (Parry and Duinker 1990). Adaptive measures to climate change can be either reactive or anticipatory. Reactive adaptive measures are taken after, or in response to, climate change. Adaptive responses would seem more appropriate, given the uncertainties in our understanding of climate change and its effects. However, reactive approaches may not be satisfactory and may prove to be too costly.

Nevertheless, it is also necessary to examine anticipatory approaches to adaptation. The goal of anticipatory measures is to minimize the impact of climate change by reducing vulnerability to its effects or enabling more efficient reactive adaptation, faster and at lower cost (Smith and Mueller-Vollmer 1993).

Reactive Adaptive Measures

The potential for agricultural adaptation at the farm level is very promising. Farm-level adaptations tend to be reactive and arise from the farmer's perception of changed or changing conditions. Farmers are already operating in an environment in which climate conditions vary from place to place and from season to season. Thus, a vast wealth of knowledge exists on how to adjust to, and cope with, these variations within farming communities. Table 7, for example, shows the relative frequencies of some adaptive measures and coping mechanisms commonly observed in Zimbabwe during climate stress. Zimbabwe's recurrent droughts of varying severity have alerted the government and farming communities of the need to re-examine land-use and management practices, as well as on-farm infrastructure.

Land-Use Changes

Parry and Duinker (1990) identified the Southern African region as one of the areas that appears most vulnerable to climate change. In Zimbabwe, climate change is likely to increase the constraints on agricultural production, with marginally productive areas shifting to nonagricultural use. For areas where crop farming is or becomes nonviable, livestock and dairy production can be developed as the major agricultural activities. Some farmers have already switched to game ranching because of recurrent droughts. This offers great opportunities to promote ecotourism. Farmers may also switch to different crops or change to more drought- and disease-tolerant crops.

In areas where high temperatures and rates of evapotranspiration lead to reduced levels of available moisture, introducing irrigation systems will help sustain agricultural production. Switching from monocultures (which are more vulnerable to climate change, pests, and disease) to more diversified agricultural production systems will also help farmers cope with climate changes. Using supplementary feeds and livestock breeds adaptable to drought gives farmers greater flexibility in adapting to climate change.

Management Changes

Changes in management practices can offset many potentially negative impacts of climate change (Smith and Mueller-Vollmer 1993). The timing of farming operations (e.g., planting dates; application of fertilizers, insecticides, and herbicides) will become more critical if farmers must reduce their vulnerability to the impacts of climate change. Changing planting densities and application rates of agrochemicals and fertilizers will also help farmers to cope with the impacts of climate change. Practices, such as conservation tillage, intercropping, and crop rotation, will enhance the long-term sustainability of soils and improve the resilience of crops to climate changes (U.S. Environmental Protection Agency

1992). Farmers should also seriously consider using greenhouses to produce a wider range of crops than they currently do.

Climate change could lead to more efficient irrigation systems because of the need for stricter water management practices to counter increased demand. For

Table 7. Relative Frequencies of Adaptive Measures and Coping Mechanisms Commonly Observed in Zimbabwe during Climate Stress

Adaptive Measures / Coping Mechanisms	Relative Frequency		
	High	Moderate	Low
Accept self-insured loss			
Work for food	*	*	
Work for wages to buy food		*	
Sell cattle to buy food			*
Use savings to buy food	*	*	
Store more than one season's food when crop			*
is good	*	*	
Work for crop seed			
Distribute and share loss			
Ask friends and relatives to help		*	*
Ask the government to help			*
Eliminate moisture waste			
Weed plots	*	*	
Stop planting when rains are insufficient		*	
Change moisture requirements			
Plant drought-resistant crops	*	*	
Affect source			
Hold rainmaking ceremonies	*	*	
Pray for rains	*	*	
Change location			
Plant in wet places		*	*
Improve moisture storage and distribution			
Implement ridging	*	*	
Install irrigation		*	
Schedule for optimal moisture			
Plant without rain		*	*
Plant only when enough rain comes	*	*	
Stagger planting	*	*	

orchards and vines, drip-irrigation systems can be used to conserve water. Water lost through seepage and evaporation in canal- and flood-irrigation systems can be minimized by lining the canals with cement or changing to pipe-irrigation systems. The significantly higher costs of production related to irrigation systems will probably result in shifts to less water demanding uses in areas with higher rates of moisture loss.

Livestock and dairy farmers can use supplementary feeds and fodder trees as low-cost grazing systems become less sustainable in marginal areas. Farmers can also explore improving pastures by using municipal wastewater.

Infrastructural Changes

Changes in the types of agricultural production and irrigation systems require significant changes in farm layout and the types of capital equipment used. Areas requiring irrigation systems may need additional water reservoirs or boreholes. In reaction to recent droughts, the government has embarked on the construction of a number of medium- to large-sized dams throughout the country. Parry and Duinker (1990) noted that because of the large costs involved in infrastructural changes (at the farm level), only small incremental adjustments may occur without changes in government policy.

Anticipatory Adaptive Measures

At the national level, adaptive measures are more anticipatory. The measures taken at this level are longer term and tend to affect the community at large. Agriculture is affected in many ways by a wide range of government policies that influence input costs, product pricing, and marketing arrangements. Government policies pertaining to land and water resources (the basic foundation for agricultural production) significantly impact agricultural production. The government has little leeway to act directly in promoting adaptation to anticipated climate change because of the uncertainties about the magnitude and rate of change (especially at finer scales). Thus, it is imperative that any anticipatory measures allow for great flexibility to permit these measures to be revised as new information about the magnitude and direction of climate change becomes available.

Through its policies on infrastructural development, research and development, education, water resources management, and product pricing, the government can put both reactive and anticipatory adaptive measures into place. Ideally, a policy-relevant research program could help to identify appropriate actions as the current state of knowledge evolves (Office of Technology Assessment 1993).

Infrastructural Development

The government has an ongoing program of constructing medium- to large-sized dams throughout the country. Marginal increases in the capacities and numbers of these dams will enhance the availability of water resources in the future. Their construction costs may also be significantly less now than in future years. Construction of these dams also allows policymakers to establish irrigation plans. Such plans are already operational in some areas. Rukuni (1994) notes increasing evidence of high rates of return from investments in small-scale/subsistence irrigation plans.

It is necessary for the government to undertake a major review of land-use planning and to consider an integrated resources management approach. The scope is great for expanding the Communal Areas Management Programme for Indigenous Resources into areas that could become marginal for agricultural production. If the anticipated impacts of climate change are considered, the ongoing resettlement programs, primarily targeted at relieving population pressure from the marginal communal areas, can become more efficient and enhance the sustainability of agricultural production in these areas. If more areas become marginal, more intensive agricultural production will shift to the more favorable locations. Hence, if such locations can be identified, the supporting infrastructure (e.g., transportation and communication networks, markets) can be improved. Although the setup of such infrastructure may not be critical at this stage, it can still be fully utilized and significantly improve the efficiency of agricultural production. However, its most critical significance will become more apparent as adaptive measures become more efficiently implemented and the impacts of climate change are minimized.

Research and Development

Government policy and support for research and development significantly affect the agricultural production sector. The availability of facilities, the level of funding, and the outlook on private-sector initiatives greatly influence the rate at which crop varieties, livestock breeds, agricultural technologies, and management systems adaptable to climate change can be implemented. An intense research program is needed to study crops and livestock that are more tolerant to disease and drought. Continued research on short-season, high-yield crop varieties and livestock breeds is of paramount importance to adaptation. The government should also support research efforts aimed to do the following:

- Ascertain ways to increase and sustain agricultural production in the country's marginal regions without causing detrimental effects on the environment;
- Increase support to develop agrochemicals to counter diseases and pests that would probably increase with climate change;
- Develop more appropriate fertilizers for the new crop varieties;

- Improve pastures and tree fodder crops to enhance the sustainability of livestock and dairy production if climate changes occur;
- Implement research on irrigation systems and low-input permaculture and agroforestry to enhance water-use efficiency and conservation systems; and
- Invest in research on diseases and pests, both of which currently afflict the agricultural sector, and those diseases and pests that may occur as a result of climate change.

Research is also needed to provide effective storage systems for agricultural products. The government recognizes the utility of improvements to storage, processing, and preservation techniques in overcoming production shortages (GOZ 1991). However, it has not made a firm commitment to undertaking such improvements.

In addition, the government should seriously consider supporting research to find a more decentralized method for storing the country's strategic food reserves. Increased local participation should be encouraged. Recent events prove that the rural majority is hardest hit during persistent droughts. An enabling environment and government support would encourage the private sector to invest more resources into these areas of research. Private-sector participation would result in more rapid application of research output within the agricultural sector. The government should also establish and maintain seed banks for crop varieties adaptable to changeable climates. A regional approach could be used here, as was done with the Southern Africa Development Community Tree Seed Centre Network.

Through regional cooperation, the government can promote the exchange of advisory services and experiences, as well as the transfer of environmental management and rehabilitation technologies that affect the agricultural sector. It is also important that the government fully use research and development information when it formulates and/or reformulates policies that affect the agricultural sector. The government should carefully examine the inadvertent damage to the capacities of research and development institutions as a result of budgetary and staff cuts under the Economic and Structural Adjustment Programme. Finally, the government needs to improve incentives to attract and retain outstanding scientists in these research and development institutions.

Education

In Zimbabwe, a fairly significant amount of agricultural produce comes from small-scale and subsistence farmers. The greatest challenge to the government lies in sensitizing subsistence farmers to the impacts of climate change. These farmers already operate in the most marginal areas and are the most vulnerable group. A more intense approach should be adopted to make these farmers aware of the need for crop diversification, crop switching, conservation tillage, and water conservation. In particular, the government needs to promote the use of traditional small grains that seem to adapt more easily to harsh environments.

In most areas, livestock owned by subsistence farmers far exceeds the capacity of the marginal land they occupy. Intensive government awareness programs are needed to enhance the sustainability of these marginal areas. The government can actively encourage crop switching and diversification, as well as the use of appropriate fertilizers when they assist small-scale and subsistence farmers through the drought recovery program. The government should also consider the possibility of setting up a high-level interagency task force to develop a coherent national drought policy. Such a task force would examine the viability of establishing a national drought insurance scheme and spearhead an awareness campaign on matters related to drought and other impacts of climate change.

An "enabling environment" is needed for small-scale and subsistence farmers to organize into unions, commodity groups, and cooperatives, which would provide the government with a balanced view of the rural majority. Hence, with an enhanced awareness of their rights and an enabling environment, consistent with human rights and democratic governance, it would be legally and institutionally easier for farmers to form such groupings. Such groupings would enhance the flow of information to farmers, making it easier to communicate research results with respect to adaptations to climate change.

Water Resources Management

The government should review its current policy on water rights. A new policy should reflect the need to conserve and use water resources more efficiently. It should consider the increasing need to share equitably a diminishing resource. The government's recent move (*The Herald* 1995) to reserve a guaranteed amount of water in all government and Regional Water Authority dams to be used for irrigation by communal, small-scale, and resettlement farmers is long overdue. Until this move, commercial farmers have been the primary beneficiaries of water for irrigation. Therefore, it is essential that the government assist small-scale farmers in developing the basic infrastructure because most of these farmers may not have easy access to the necessary capital resources. The government can also support and offer incentives to farmers who help to develop infrastructure that will lead to the conservation or increased availability of water resources.

The government should explore the possibilities of interbasin water transfers to enhance the sustainability of areas that become intensively used for agricultural production or marginal as a result of climate change. The formulation and implementation of catchment management policies would enhance mitigation of erosion and, thus, siltation of river systems and dams.

Input Costs and Product Pricing

Until recently, most of Zimbabwe's agricultural produce was marketed through five main marketing boards: the Grain Marketing Board (GMB), the Cotton Marketing Board, the Dairy Marketing Board, the Tobacco Marketing Board, and the Cold Storage Commission. However, the Economic Structural Adjustment Programme has resulted in major parastatal reform and commercialization, particularly with respect to market and trade liberalization. The marketing boards no longer have a monopoly on buying and supplying agricultural produce, although the GMB, for example, remains a major buyer of grain products and maintains its traditional role of setting price ceilings for grain products.

This system puts farmers at a disadvantage because they cannot negotiate a better price for their products, since manufacturers will opt to obtain supplies from the GMB. Consequently, farmers may switch to more financially rewarding activities, which would threaten food security. Recurrent droughts and poor prices for grain products in recent years have led to new ventures for many commercial farmers: horticulture, game ranching, and ostrich farming. Market liberalization should not result in the marginalization of small-scale and subsistence farmers. Therefore, the government should establish and strengthen existing institutions geared toward extending credit to small-scale and subsistence farmers, and facilitate cost-effective ways to market their produce. National public agricultural research shows that many small-scale farmers can seize market opportunities in a favorable macroeconomic environment (Rukuni 1994).

Input costs and product pricing can function as incentives or disincentives in determining which agricultural crops to produce. Thus, pricing policy can steer the agricultural sector in a direction more adaptable to climate change. Through pricing policy, the government can actively influence crop changes, water conservation measures, and a host of other management activities, thus making the agricultural sector adaptable to climate change.

Conclusions

As a country highly dependent on the agricultural production sector, Zimbabwe could see a rapid deterioration in the livelihood of its citizens as a result of climate change. Without the appropriate policies or adaptive strategies in place, small-scale farmers will find it extremely difficult to operate sustainable agricultural production systems in an environment experiencing climate changes. The potential solutions to problems resulting from climate change in the agricultural sector require increased financial resources and greater commitment to research and development efforts. However, the scientific community should not concentrate only on developing new technologies and management systems, but should also draw and improve on existing traditional systems.

For Zimbabwe to meet the growing demands for food (locally and regionally), national development programs should give high priority to sustainable growth of the agricultural production sector. Expanding the diversity of crops and available farm technologies will improve the chances for successful adaptation to a future in which existing farming systems are threatened by climate change. Thus, anticipatory measures will enhance the ability of farmers to adapt to climate change. Such measures will speed up the rate at which farming systems can be adapted to climate change and will significantly lower the potentially high costs associated with adjustment. Even without climate change, the adaptive measures suggested are considered to be beneficial, given the prevailing climate conditions, which are characterized by recurrent droughts and decreased precipitation.

Because of the considerable uncertainties about the magnitude and extent of the possible effects of climate change, it is relatively difficult to plan appropriate responses (policies and strategies). These uncertainties dictate that any anticipatory measures should be of maximum flexibility for them to be beneficial to the agricultural sector even without climate changes, as well as allow adjustments as more information about climate change becomes available.

The direction, magnitude, timing, and path of the impacts of climate change are neither fully understood nor accurately predictable. Thus, sustained scientific research is needed to enable more confident prediction of the impacts of climate change, especially at the national and local levels.

References

Acock, B., and L.H. Allen, 1985, "Crop Responses to Elevated Carbon Dioxide Concentrations," in B.R. Strain and J.D. Cure (eds.), *Direct Effects of Increasing Carbon Dioxide on Vegetation*, DOE/ER-0238, pp. 33-97, U.S. Department of Energy, Washington, D.C., USA.

Boer, G.J., N.A. McFarlane, and M. Lazare, 1992, "Greenhouse Gas-Induced Climate Change Simulated with the CCC Second-Generation General Circulation Model," *Bulletin of the American Meteorological Society* 5:1045-1077.

Downing, T.E., 1992, *Climate Change and Vulnerable Places: Global Food Security and Country Studies in Zimbabwe, Kenya, Senegal and Chile*, Research Report No. 1, Environmental Change Unit, University of Oxford, Oxford, United Kingdom.

Government of Zimbabwe (GOZ), 1991, "Sectoral Development — Agriculture," in *Second Five-Year National Development Plan 1991-1995*, pp. 23-33, Government Printers, Harare, Zimbabwe.

The Herald, 1995, "Water Is Reserved for Small Farmers," Harare, Zimbabwe.

IBSNAT, 1989, *Decision Support System for Agrotechnology Transfer Version 2.1 (DSSAT V2.1)*, Department of Agronomy and Soil Science, College of Tropical Agriculture and Human Resources, University of Hawaii, Honolulu, Hawaii, USA.

Matarira, C.H., J.M. Makadho, and F.C. Mwamuka, 1995, "Zimbabwe: Climate Change Impacts on Maize Production and Adaptive Measures for the Agricultural Sector," *Interim Report on Climate Change Country Studies*, in C. Ramos-Mañé and

R. Benioff (eds.), DOE/PO-0032, pp. 105-114, U.S. Country Studies Program, Washington, D.C., USA.

Mitchell, J.F.B., S. Manabe, T. Tokioka, and V. Maleshko, 1990, "Equilibrium Change," in *Climate Change: The IPCC Scientific Assessment*, J.T. Houghton, G.J. Jenkins, and J.J. Ephraums (eds.), Cambridge University Press, New York, N.Y., USA.

Muchena, P., 1991, *Implications of Climate Change for Maize Yields in Zimbabwe*, Plant pp. 1-9, Protection Research Institute, Department of Research and Specialist Services, Harare, Zimbabwe.

Nyamapfene, K., 1991, *Soils of Zimbabwe*, Nehanda Publishers, Harare, Zimbabwe.

Office of Technology Assessment (OTA), 1993, *Preparing for an Uncertain Climate: Summary*, OTA-O-563, United States Congress, Washington, D.C., USA.

Parry, M.L., and P.N. Duinker, 1990, "Agriculture and Forestry," in W.J. McG. Tegart, G.W. Sheldon, and D.C. Griffiths, *Climate Change — The IPPC Impacts Assessment*, World Meteorological Organization/U.N. Environmental Programme, Intergovernmental Panel on Climate Change, Australian Government Publishing Service, Canberra, Australia.

Riebsame, W.E., 1989, *Assessing the Social Implications of Climate Fluctuations. A Guide to Climate Impact Studies*, World Climate Impacts Program, U.N. Environment Programme, Nairobi, Kenya.

Rukuni, M., 1994, "Getting Agriculture Moving in East and Southern Africa and Framework for Action," *East and Southern Africa Conference of Agricultural Ministers*, April 12-14, 1994, Harare, Zimbabwe, World Bank, Washington, D.C., USA.

Smith, J.B., and J. Mueller-Vollmer, 1993, *Setting Priorities for Adapting to Climate Change*, RCG/Haigler, Bailly, Inc., Boulder, Colo., USA.

U.S. Environmental Protection Agency (EPA), 1992, *Agriculture. Climate Change Discussion Series*, Office of Policy, Planning, and Evaluation, Washington, D.C., USA.

Model-Based Climate Change Vulnerability and Adaptation Assessment for Wheat Yields in Kazakhstan

S.V. Mizina, I.B. Eserkepova, O.V. Pilifosova, and S.A. Dolgih
Kazakh Hydrometeorological Science Research Institute, Seifullin pr., 597
480032 Almaty, Kazakhstan

E.F. Gossen
Kazakh Agricultural Academy, Ablai Han pr., 79
480091 Almaty, Kazakhstan

Abstract

The CERES-Wheat crop growth model and General Circulation Model–based climate change scenarios assessed possible impacts of climate change on wheat production in Kazakhstan. Strategies to reduce the negative effects of regional climate change are suggested. These strategies include partially switching the crop from spring wheat to winter wheat; planting drought-resistant wheat varieties; changing planting dates; changing tillage practices; implementing snow reserving; planting forests; planting perennial vegetation in vacant areas; and reducing cereal production.

Introduction

Kazakhstan covers an area of approximately 270 million ha. Its climate is arid, with extreme temperatures in summer and winter. Diverse soil types, climates, and vegetation cover greatly influence agricultural production. Kazakhstan has six natural and economic regions based on soil type, climate, and economic conditions (Abugaliev 1985). Two of these regions are important in wheat production: the semi-arid and arid steppe zones. The semi-arid steppe zone has typical and southern chernozem soils and complexes of these two soils. This zone includes North Kazakhstan, most of the Kokshetau oblast, a large part of the Kostanai oblast, and the northern parts of the Aqmola and Pavlodar oblasts. The arid-steppe zone region is dominated by dark-chestnut soils, and chernozem and chestnut soils are common in the north and south. This zone includes the northern areas of West Kazakhstan, the Aktobe and Torgai oblasts, a large part of the Kostanai oblast, and most of the Aqmola, Pavlodar, and Karagandy oblasts.

More than 90% of the total crop area is planted in hard varieties of wheat. After soil protection measures were implemented in North Kazakhstan, the average wheat yield at scientific institutes amounted to 1.6-1.7 Mg ha^{-1} compared with 0.5-0.6 Mg ha^{-1} when the soil was first developed. Recently, however, the average wheat yield has been 0.92 Mg ha^{-1} because of economic conditions.

The possibility of global warming necessitates developing adaptive measures for different economic sectors. Spring and winter wheat production is particularly vulnerable to potential global warming, and adaptation strategies are crucial for sustainable development of agriculture. The main goals of this study are to evaluate the vulnerability of wheat production and analyze adaptation strategies for agriculture in Kazakhstan.

Methods

The framework for developing an adaptation strategy is composed of seven steps (Carter et al. 1994). The first step is to define the objective. Our main objective is to develop recommended adaptive measures to preserve spring and winter wheat production in Kazakhstan under global warming.

The second step in developing an adaptation strategy is to specify the climate impacts. This step includes a detailed description of the magnitude and extent of future changes (climate scenarios) that may affect the exposed unit (wheat production). Vulnerability assessment is an essential part of this step. Vulnerability is defined as the degree to which an exposed unit is affected by climate change. The change in wheat yield expected under a doubling of the atmospheric concentration ($2XCO_2$) was determined.

Crop Model

The CERES-Wheat model (Ritchie and Otter 1985), based on the Decision Support System for Agrotechnology Transfer (DSSAT) (Tsuji et al. 1994) developed by International Benchmark Sites Agrotechnology Transfer, assessed the vulnerability and adaptation of spring and winter wheat yields in Kazakhstan. The DSSAT models describe the development, growth, and yield of crops on homogenous areas of soil exposed to certain climate and weather conditions. This system allows one to run and validate models, conduct sensitivity analyses, and evaluate the variability and risks of different management strategies for a range of locations specified by soil and weather data. The Russian version of the SOIL-module of DSSAT 2.1 was used to create a soil database.

Climate Change Scenarios

General Circulation Models (GCMs) take into account a wide range of physical processes on land, in the atmosphere, and in the upper mixed ocean layer. Such models estimate climate change due to increased concentrations of greenhouse gases in a physically consistent manner. GCM-based scenarios are preferable to incremental scenarios, which are only hypothetical.

Long-term climate change scenarios for Kazakhstan ($2XCO_2$ by 2050-2075) were prepared. Three GCMs were used: the Canadian Climate Centre (CCC) model (Boer et al. 1991), Geophysical Fluid Dynamics Laboratory (GFDL) model (Manabe and Wetherald 1987), and transitional version of the GFDL model (GFDL-T). Climate change scenarios were obtained from model results based on the GRADS data bank provided by the National Center for Atmospheric Research.

A baseline climate scenario was also used to evaluate future climate changes. This scenario represents the current climate for the base period 1951-1980 without a warming trend. No other environmental and socioeconomic scenarios were considered.

All three models predict temperature increases under $2XCO_2$ in Kazakhstan, but the magnitude of this warming varies by season and region. The CCC model predicts the greatest warming. On average, the temperature change under the CCC scenario is 4-6°C higher than that predicted with the other two models. Minimum changes occur in summer, maximum in winter. The GFDL-T model gives the smallest change, and the GFDL R-0 results are inbetween. The GFDL R-30 model also gives a better estimate of current (observed) conditions. Changes in precipitation patterns are consistent among the three models; their relative changes are within the range of natural variability or slightly higher.

Input Information

Climate Data

Observed daily maximum and minimum temperatures and precipitation rates were obtained from the Kazakh Hydrometeorological Science Research Institute data bank (Antonov and Smirnova 1995). Data on daily solar radiation were generated by using long-term monthly norms (obtained from the same source) and the solar radiation generator in DSSAT 3.

Climate Change Data

The study area covers multiple GCM grid cells. Estimates of meteorological variables were obtained and averaged within the Kazakhstan latitude zones.

Tables 1 and 2 show estimated temperature and precipitation changes averaged over point values between 45° and 51°N latitude (zone 1) and 51° and 57°N (zone 2) in Kazakhstan.

Table 2 shows that the precipitation changes are small and close to norms. Moreover, under current climate conditions, summer precipitation in zone 1 is about 130 mm, and the potential evaporation is about 400 mm. In zone 2, these values are 76 and 500-700 mm, respectively. The seasonal distribution of precipitation change is more important than the total annual precipitation change for the region.

Soil Data and Initial Soil Conditions

Data on typical chernozem, southern chernozem, and dark-chestnut soil were obtained from *Agrohydrological Properties...* (1980) and *Soils of Kazakh...* (1966). Table 3 lists sites according to soil type.

Initial conditions for water and nitrogen in the soil profiles were obtained from the results of field expeditions by the agrometeorology department of the Kazakh Hydrometeorological Center. Such information was not available for every site. Standard values according to geographic proximity and environmental and climatic conditions were used in those cases.

Wheat Varieties

The Bezostaya-1 and Mironovskaya-808 winter wheat varieties used in this study are the most common varieties in Kazakhstan (more than 60% of the

Table 1. Latitude-Averaged Monthly and Annual Temperature Change under GCM-Based Climate Change Scenarios in Kazakhstan

	Temperature Change (°C)												
Model	Jan	Feb	Mar	Apr	May	Jun	Jul	Aug	Sep	Oct	Nov	Dec	Annual
Zone 1													
GFDL R-30	3.5	3.8	7.7	8.2	5.4	5.4	6.4	6.6	5.5	5.7	6.5	4.6	5.8
CCC	7.8	7.7	6.8	8.4	10.9	5.1	3.7	3.8	3.3	3.1	3.1	8.8	6.0
GFDL-T	5.6	5.1	6.0	4.9	3.7	3.9	4.2	3.9	5.3	6.0	4.8	6.2	4.4
Zone 2													
GFDL R-30	3.1	3.5	6.4	6.0	4.9	5.1	5.8	5.8	5.6	5.1	4.3	2.7	4.6
CCC	8.5	9.8	9.3	11.2	9.3	6.4	6.3	5.9	4.8	3.4	3.9	7.9	7.1
GFDL-T	4.8	4.6	4.9	4.4	4.4	4.3	3.9	4.8	4.9	6.6	4.5	5.3	4.8

Table 2. Latitude-Averaged Monthly and Annual Precipitation Change under GCM-Based Climate Change Scenarios in Kazakhstan

Model	Precipitation Change (%)												
	Jan	Feb	Mar	Apr	May	Jun	Jul	Aug	Sep	Oct	Nov	Dec	Annual
Zone 1													
GFDL R-30	118	121	135	93	89	108	112	99	117	148	125	124	116
CCC	117	125	128	151	134	104	108	101	103	108	100	120	120
GFDL-T	142	150	135	91	101	139	94	110	132	135	100	121	121
Zone 2													
GFDL R-30	110	114	94	77	89	144	169	210	140	149	103	94	123
CCC	111	120	105	121	94	64	65	71	93	106	94	105	96
GFDL-T	115	134	115	107	124	103	79	95	119	111	93	121	120

Table 3. Soil Types Used in the Simulation and Vulnerability and Adaptation Assessment

Site	Soil Type
Kotyrkol	Typical chernozem
Schuchinsk	Southern chernozem
Ruzaevka	Typical chernozem
Kostanai	Typical chernozem, southern chernozem
Pavlodar	Dark-chestnut
Petropavlovsk	Typical chernozem
Karagandy	Dark-chestnut

winter wheat crop area). The Saratovskaya-29 and Omskaya-19 varieties were used for spring wheat. All these varieties produce good-quality wheat.

Calibration

The CERES-Wheat model was calibrated and validated for several wheat varieties (Ritchie and Otter 1985). This study used specific varieties planted in Kazakhstan. The winter varieties Bezostaya-1 and Mironovskaya-808 were characterized by a specific genetic coefficient set of the CERES-Wheat model (Tsuji et al. 1994). Earlier investigations calculated genetic coefficients for some varieties of spring wheat grown in Kazakhstan, but this information was not available (Menzhulin et al. 1994). Therefore, genetic coefficients for spring varieties were determined on the basis of experimental data provided by the Kazakh Hydrometeorological Institute and the Kazakh Academy of Agricultural Sciences. The model was validated by comparing observed experimental data on spring wheat (Saratovskaya-29) yield over 17 years and simulated values.

Table 4 presents the observed data (*Recommendations*... 1982), and Table 5 presents the predicted values. The model was validated by using agrometeorological forecasts, a method used by Kazhydromet (*Guidelines*... 1983). The relative error P_i for each simulation result was calculated as follows:

$$P_i = |(Y_{ob} - Y_{sim}) / Y_{ob}| \times 1{*}00\%,$$

where

P_i = relative error,
Y_{ob} = observed wheat yield, and
Y_{sim} = simulated wheat yield.

Table 4. Observed Spring Wheat Yields (Saratovskaya-29 and Schuchinsk) by Planting Date

	Observed Wheat Yield by Planting Date (t/ha)						
Year	May 5	May 10	May 15	May 20	May 25	May 30	June 5
1961	1.03	1.23	1.38	1.88	1.80	1.25	1.02
1962	1.77	1.54	1.54	1.68	1.49	1.55	1.31
1963	0.55	0.57	0.80	0.94	1.11	1.32	1.38
1964	-	-	3.06	2.99	2.31	2.07	0.91
1965	0.55	0.70	0.72	0.73	0.81	0.83	0.95
1966	-	-	-	1.48	1.73	2.18	1.90
1967	0.65	0.70	0.74	0.72	0.92	1.00	0.98
1968	1.24	1.61	1.76	1.79	1.54	-	1.59
1969	0.89	1.07	1.28	1.55	1.67	1.75	1.72
1970	2.06	2.25	2.27	2.27	1.95	2.26	1.96
1971	1.59	-	1.77	1.83	2.16	2.09	1.93
1972	1.62	1.63	1.67	2.03	1.33	1.24	0.66
1973	1.55	1.48	1.57	2.0	1.54	1.91	1.30
1974	1.10	1.16	1.34	1.44	1.55	1.52	1.44
1975	0.77	0.88	0.88	0.81	1.04	1.22	1.18
1976	1.80	1.74	1.88	1.70	1.59	1.87	1.87
1977	0.84	0.94	1.20	1.18	1.26	1.55	1.73

Table 5. Simulated Spring Wheat Yields (Saratovskaya-29 and Schuchinsk) by Planting Date

	Simulated Spring Wheat Yields by Planting Date (Mg ha^{-1})						
Year	May 5	May 10	May 15	May 20	May 25	May 30	June 5
1961	1.19	1.51	1.12	1.15	1.40	1.00	1.35
1962	1.95	1.93	2.33	1.39	2.21	2.31	1.66
1963	0.72	0.68	1.12	0.88	1.73	1.10	0.97
1964	-	-	0.73	1.64	4.00	1.90	0.51
1965	0.55	0.54	1.16	0.93	1.13	1.04	0.59
1966	-	-	-	1.61	2.42	1.92	2.39
1967	0.74	0.83	0.42	0.80	0.61	0.83	0.87
1968	1.53	2.06	1.16	1.74	0.99	-	1.49
1969	0.77	0.89	0.78	1.78	1.12	2.19	1.60
1970	1.77	1.96	1.73	2.72	1.68	2.10	2.23
1971	1.40	-	1.89	1.65	1.92	1.71	2.41
1972	1.88	1.74	1.69	1.91	1.20	1.14	0.76
1973	1.47	1.63	1.73	1.94	1.72	2.08	1.37
1974	1.18	1.28	1.34	1.34	1.36	1.72	1.35
1975	0.59	0.95	0.79	0.73	1.20	1.18	1.14
1976	1.67	1.48	1.94	1.65	1.78	1.76	1.89
1977	0.66	0.82	1.14	1.06	1.39	1.47	1.80

Table 6 presents the relative error. The total error P was calculated as follows:

$$P = (P_1 + P_2 + \dots + P_N) / N,$$

where

P = total error,
$P_1, P_2, \dots, P_N$ = relative errors in 1, 2, ..., N simulations, and
N = number of simulations.

The reliability of simulation R was calculated as follows: $R = 100\% - D$. The accuracy of the agrometeorological forecasts was evaluated as follows:

- If $R \geq 91\%$, the mark is 5;
- If $81\% \leq R \leq 90\%$, the mark is 4;
- If $70\% \leq R \leq 80\%$, the mark is 3; and
- If $R < 70\%$, the mark is 0.

In this case, $N = 112$, $P = 18.8\%$, $R = 81.2\%$, and the mark of the quality of the simulations is 4.

Table 6. Relative Errors in Simulations of Spring Wheat Yields (Saratovskaya-29 and Schuchinsk) by Planting Date

	Relative Errors in Spring Wheat Yields by Planting Date (%)						
Year	May 5	May 10	May 15	May 20	May 25	May 30	June 5
1961	16	23	19	39	22	20	32
1962	10	25	51	17	48	49	27
1963	30	19	53	6	56	17	30
1964	-	-	76	45	73	8	44
1965	0	23	61	28	64	25	38
1966	-	-	-	9	40	12	26
1967	14	18	43	11	34	17	11
1968	23	28	34	3	36	-	6
1969	13	17	39	15	27	25	7
1970	14	13	24	20	14	7	14
1971	12	-	7	10	11	18	25
1972	16	7	1	6	10	8	15
1973	5	10	10	3	12	9	5
1974	7	10	-	7	12	13	6
1975	23	8	10	10	15	3	3
1976	7	15	3	3	12	6	1
1977	21	13	5	10	10	5	4

Results

Changes in the Wheat Crop Yields

Six sites located in zone 1 (Kotyrkol, Schuchinsk, Ruzaevka, Kostanai, Pavlodar, and Petropavlovsk) and one site in zone 2 (Karagandy) were chosen (Figure 1). The magnitude of changes in wheat yield under different climate scenarios strongly depends on the following:

- Temperature change and seasonality of this change,
- Precipitation change and how this change is distributed through the year, and
- Beneficial direct effects of CO_2 on crop growth.

Solar radiation will change minimally (±1-4%) and will hardly affect wheat yield change.

Table 7 shows possible changes in wheat yield calculated by the CERES-Wheat model. The GFDL-T climate change scenario is the most favorable; the other scenarios are quite unfavorable.

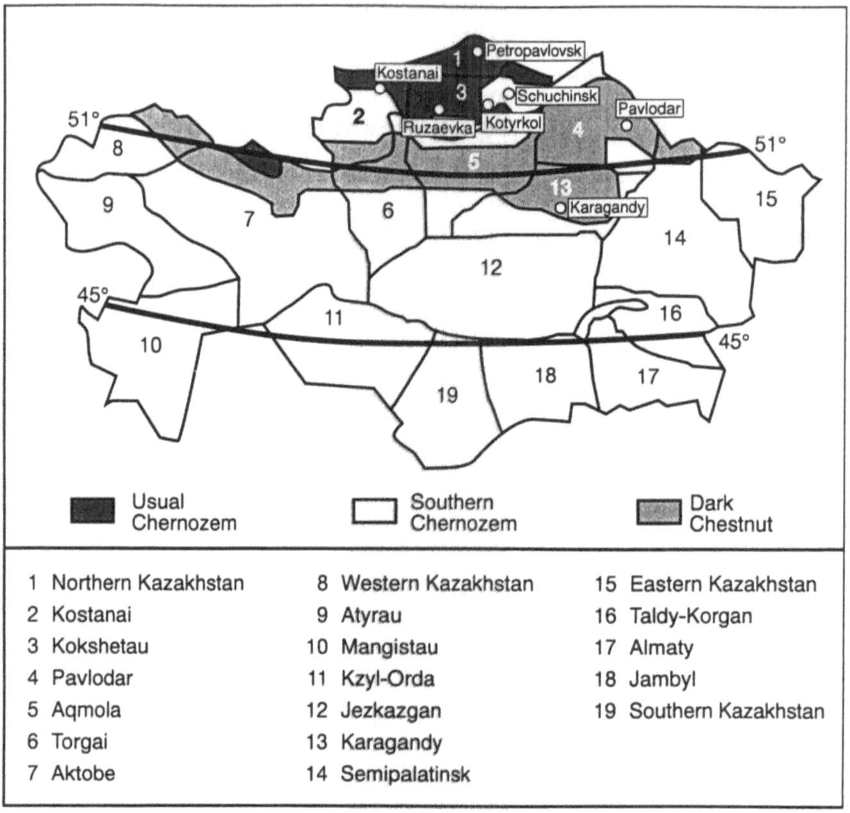

Figure 1. Latitude Zones, Soils, and Sites of Kazakhstan

Table 7. Simulated Change in Wheat Yield under GFDL R-30, CCC, and GFDL-T Climate Change Scenarios

	GFDL R-30		CCC		GFDL-T	
Site	Spring	Winter	Spring	Winter	Spring	Winter
Kotyrkol	51	121	44	125	71	132
Schuchinsk	47	120	40	120	68	127
Ruzaevka	50	116	38	121	68	125
Kostanai	48	113	42	124	73	120
Pavlodar	43	107	37	110	63	115
Petropavlovsk	57	125	61	101	93	109
Zone 1 average	49	117	44	117	88	121
Karagandy	114	159	32	104	71	109

Spring wheat yield will decrease at all sites under all scenarios. The CCC and GFDL R-30 scenarios show the greatest change (44 and 49% of current yield, respectively); the GFDL-T scenario shows less change (88% of current yield). Although the spring wheat yield will decrease, the winter wheat yield will increase under all climate scenarios. The GFDL-T scenario is more favorable for winter wheat yield (121% of current yield). The impact under the GFDL R-30 and CCC scenarios will be the same (117% of current yield).

Experiments have simulated changes in wheat yield under $2XCO_2$ with temperature and precipitation at current levels. Spring and winter wheat yield would increase by 45-74% and 57-79%, respectively.

Table 7 shows combined effects of climatic change and direct effects of CO_2 on wheat production. The lowest spring wheat yield was obtained under the CCC scenario. During the vegetative period (approximately mid-May to mid-August) (*Mean Long-Term...* 1982), warming is predicted to be 10.9-3.7°C. Precipitation will increase 34% in May and 1-8% in the summer months in zone 1 (Tables 1 and 2). Therefore, spring wheat will undergo warming and water stress in the most important vegetative phases (heading to wax maturity phases) (Abugaliev 1985). As a result, the spring wheat yield will be reduced to 56% of current yield.

Additional precipitation (51% in April and 34% in May) will increase the winter wheat yield, which means that the winter wheat yield would be 17% higher than the current yield. Winter wheat undergoes the vegetative phases earlier than spring wheat does.

The temperature change in May under the GFDL R-30 scenario (5.4°C) is smaller than that under the CCC scenario (Table 7). In summer, the temperature change will be 5.4-6°C. The precipitation will be 89-112% of the norm in zone 1. Therefore, the spring wheat yield will be reduced by 51% on average. The heading phase of spring wheat must coincide with the maximum precipitation in July. Precipitation will increase 12%, which corresponds to 3 mm of precipitation for zone 1; this amount is not sufficient for the spring wheat yield. However, winter wheat yield will increase by 17% on average, because it can better use soil moisture stored during fall and winter (Abugaliev 1985) when precipitation is higher than normal, according to this scenario.

The GFDL-T scenario results in the smallest temperature change (3.7-4.2°C) from May to August. Precipitation will be 94% of the norm in July. Therefore, spring wheat yield will be reduced by 12%. However, the winter wheat yield will be 21% higher than its current value because of a 39% precipitation increase in June, when heading takes place.

In zone 2, spring wheat yield increases only under the GFDL R-30 scenario because of increased precipitation in June, July, and August (44, 69, and 110%, respectively). For the other scenarios (Table 7), the changes in wheat yield in zone 2 are essentially the same as those in zone 1. But results for only one site are not sufficient for a thorough evaluation.

Adaptation

As the foregoing vulnerability assessment shows, climate change will have a significant detrimental effect on spring wheat yield in Kazakhstan. According to the Kazakh Agricultural Academy research results (*Conceptual Program* 1994), areas that now give low yields will no longer support crops. The current average wheat yield is about 0.92 Mg ha^{-1}. In some oblasts (Semipalatinsk, Pavlodar, Karagandy, Jezkazgan, Torgai, and Aktobe), however, wheat is grown not only in the steppe and arid-steppe zones, but even in the desert-steppe zone on light-chestnut and gray-brown desert soils. The soil and climate of that area allow a wheat yield of no more than 0.5-0.6 Mg ha^{-1}.

The cost of wheat production in Kazakhstan is currently $130/ha. Under measures (fertilizers, pesticides, weed control, etc.) to increase wheat yield to 1.49 t/ha, the cost will increase to $160/ha (*Conceptual Program* 1994). This figure was calculated on the basis of the cost of fertilizers, pesticides, workers' salaries, etc. Critical wheat yields that ensure a given profitable level were defined for different areas. The critical wheat yields are 0.8 t/ha with current farming technology and 1.0 Mg ha^{-1} under intensive technology (assuming the price of wheat is about $160 mg). Under these considerations, adaptive measures will work only in areas with rich soils (typical chernozems, southern chernozems, and dark-chestnut soils).

Some adaptive measures should be taken before a climate change occurs, and others can be taken later. Adaptive measures can be applied at two levels: the farm level and the national level. The costs and benefits of these measures must be considered. This study analyzes anticipatory farm-level adaptive measures and gives preliminary results of a cost-benefit assessment. A complete economic analysis will be conducted later.

To reduce the effect of climate change on wheat yield in Kazakhstan, several approaches to adaptive measures can be offered. This study obtained estimates of adaptive responses to climate change mainly at local sites or individual farms. The results can be generalized for application to the national level.

National-level anticipatory measures require government support. The Kazakhstan Country Study Program is only the first step in this direction. Other recommendations are as follows: (1) as soon as possible, *educate the population* of Kazakhstan regarding the possible climate change and the vulnerability and adaptation of agriculture (particularly the wheat crop); (2) support further *study of agricultural vulnerability and adaptation* (particularly for the wheat yield); (3) improve the study of higher-yield, drought-resistant, earlier-maturing, disease- and pest-tolerant wheat varieties; (4) *plant perennial vegetation* in areas available after fields are reduced (In 1991, Kazakhstan planted 24 million ha of cereals. In 1995, 18-19 million ha will be planted, and the agricultural area will be reduced to 16-18 million ha. The available areas should be planted in perennial grassy and bushy vegetation, which will absorb CO_2. This action will reduce aridity caused by climate change and protect soils.); and (5) *plant forests*.

This last recommendation may be important for advance warning of soil and water erosion, coping with droughts, and maintaining the hydrological regime of rivers. Forests can also reduce aridity caused by climate change. Forest covers 3.7% of Kazakhstan now and could be increased to 5.1%. The Kazakh State Project Institute of Forest developed the $3.5 million Forests of Kazakhstan program in 1994. Forested area will be increased to 4.6% by 2010 according to proposals from the Ministry of Agriculture and the Kazakh Scientific Research of Agriculture and Forestry Melioration Institute. Protecting forests in agricultural areas requires about $1.2 million

The following measures should be applied as a response to climate change.

Use more irrigation. Irrigation will significantly increase wheat yield. But in North and Central Kazakhstan, which have undeveloped potential, irrigation may be too expensive. Under the GFDL-T scenario, water resources will remain unchanged. Under the GFDL R-30 and CCC scenarios, water resources will be reduced by 20-30%. Irrigation will not be economical under a potential climate change.

Change planting dates. In the moderate-dry climate of North Kazakhstan, the July maximum precipitation is important to the spring wheat yield. Planting dates should be chosen so that the heading phase will coincide with the maximum precipitation. Under a $2XCO_2$ climate change, the planting dates could be shifted to reduce the loss due to warming. Planting dates would be 15-20 days earlier than those under current climate conditions (Table 8).

Switch areas from spring wheat to winter wheat. As shown in Table 7, winter wheat yield increases as a whole, and spring wheat yield decreases; therefore, some land should be switched from spring to winter wheat. To maintain spring and winter wheat production in Kazakhstan at the current level, changes

Table 8. Spring Wheat Yield Response to Changes in Planting Dates under $2XCO_2$ Climate Change Scenarios

Site	Mean Planting Date	Optimal Planting Date/Percentage Change in Yield		
		GFDL R-30	CCC	GFDL-T
Kotyrkol	17.05	28.04/+34	26.04/+31	23.04/+7
Schuchinsk	17.05	28.04/+32	26.04/+31	24.04/+10
Ruzaevka	10.05	25.04/+33	23.04/+30	23.04/+9
Kostanai	10.05	25.04/+30	23.04/+28	23.04/+7
Pavlodar	11.05	26.04/+27	27.04/+26	24.04/+5
Petropavlovsk	16.05	28.04/+34	25.04/+30	24.04+7
Karagandy	04.05	30.04/+15	20.04/+5	20.04/+3

in spring and winter wheat yield averaged over natural and economic regions can be used to calculate the redistribution of the crop areas. In zone 1, arithmetic averages of the site values were used, but this is a rough approach because there are few sites. Spring wheat occupies most of the agricultural area in zone 1 (99.83% in 1994). According to a simple calculation, 70% of the area should be switched to winter wheat under the CCC scenario, 68% under GFDL R-30, and 35% under GFDL-T. However, it is difficult to make this redistribution. At most, 35% of the area may be redistributed. Winter wheat has not been grown in the Karagandy oblast recently, but about 5% of zone 2 is planted in winter wheat. However, it is impossible to make recommendations on the basis of one calculation.

Increase the percentage of fallow land. Kazakhstan was assumed to have 4 million ha of fallow land, but in reality, almost no fallow land is available because of the economic situation. Three areas in which the percentage of fallow land could be increased were chosen according to the annual distribution of precipitation. Zone 1, which has typical chernozems, needs 20% fallow land when annual precipitation is about 450-500 mm. Zone 2, which has southern chernozems, needs no less than 25% fallow land when annual precipitation is about 300-350 mm. Areas with dark-chestnut soils need more than 33% fallow land when annual precipitation is less than 300 mm. More fallow land increases wheat yield by 0.5-0.6 t/ha (Abugaliev 1985).

Table 9. Characteristics of Selected National Anticipatory Adaptive Measures

Adaptive Measure	Cost (in millions of U.S. $)	Possibility
Conduct a public education program.	-	Possible through Kazakhstan efforts
Support further study of agricultural vulnerability and adaptation.	-	Possible through Kazakhstan efforts
Improve studies of new wheat varieties.	-	Possible through Kazakhstan efforts
Plant forests.[a]	3.5	Impossible without international subsidies
Plant perennial vegetation on areas available after reduction of fields.[a]	1.2	If funding is not available, areas will be overgrown in several years anyway.

[a] May also be done on the farm level.

Table 10. Characteristics of Selected Farm-Level Intensive Adaptive Measures in Response to Climate Change

Adaptive Measure	Cost ($\$ Mg^{-1}$)	Yield Increase[a] ($Mg\ ha^{-1}$)	Beneficial Effect ($\$ Mg^{-1}$)	Possibility
Irrigation	-	0.43	-	Impossible
Change sowing dates	-	0.1 (CCC, GFDL R-30) 0.021 (GFDL-T)		Possible through Kazakhstan efforts
Switch crops from spring wheat to winter wheat	-	0 (for all scenarios)	-	Possible through Kazakhstan efforts
Increase fallow percentage	15.62	0.5-0.6	-15.2	Possible through Kazakhstan efforts
Practice snow reserving	-	0.3	-	Possible under general economic improvement
Introduce new wheat varieties	1.19	0.115	+16.63	Possible under general economic improvement
Apply fertilizers, pesticides, and weed control	25	0.21-0.25	(+8)-(+13)	Possible under general economic improvement
Use zonal growing technology	70.1	0.098	+8.18	Possible under general economic improvement

[a] Average for the entire area.

Practice snow reserving. Snow reserving increases wheat yield by about 0.3 t/ha (Abugaliev 1985). Snow reserving is not a new tillage practice in Kazakhstan, but it has not been used since 1993 because of economic considerations.

Change varieties of wheat. Use higher-yielding, drought-resistant, earlier-maturing, disease- and pest-tolerant.

Apply fertilizers, pesticides, and weed control. This measure could increase wheat yield if combined with other farming operations and new crop cycles, but might not have a considerable effect by itself.

Tables 9 and 10 summarize these measures. Table 9 shows national anticipatory adaptive measures, and Table 10 shows farm-level measures. These measures can be used not only as a response to climate change, but as current practice.

Kazakhstan needs 16 million mg annually of wheat for its own consumption, 3 million tons of which is used for meal and cereals, 3 million tons for sowing, and the rest for other requirements: natural payment, insurance funds, cattle forage, etc. (5-7 million tons is needed as a reserve). Considering that winter wheat costs $80-100 mg and spring wheat costs $100-130 mg, adaptive measures are preferable to purchasing wheat abroad because purchasing can be costly (more than $1,600 million every year), and a significant reduction of cropland would result in unemployment and social instability.

Conclusions

The main conclusions of this study are as follows:

- In spite of uncertainties, especially on a regional level, GCM outputs can be used to create climate change scenarios for vulnerability and adaptation assessments of Kazakhstan.
- The DSSAT gives affirmative results, but its functions and opportunities have not all been applied in this study. Further studies will be done, and the use of DSSAT will be expanded.
- Climate change will have a significant detrimental effect on wheat yield in Kazakhstan. The GFDL-T climate change scenario is more favorable, but the CCC and GFDL R-30 scenarios are rather unfavorable.
- Low-yield areas will be taken out of production and therefore will not apply to vulnerability and adaptation analysis according to the Conceptual Program Agriculture and Industry Complex Development of the Republic of Kazakhstan.
- Any adaptive measures for Kazakhstan will be preferable to purchase of wheat abroad.

Thus, one can draw inferences about the possibilities of adaptation to climate change, but these estimates are preliminary, and further study is needed.

References

Abugaliev, I.A. (ed.), 1985, *Reference-book of Agronomist*, Kainar, Alma-Ata, Kazakhstan (in Russian).

Agrohydrological Properties of Soils in Kazakhstan, 1980, Alma-Ata, Kazakhstan (in Russian).

Antonov, A.G., and E.Yu. Smirnova, "On Main Principles of Development of 'Climate' Data Bank," *Hydrometeorologia and Ecologia,* No. 2 (in Russian).

Boer, G.J., N. MacFarlane, and M. Lazare, 1992, "Greenhouse Induced Climate Change Simulated with the Canadian Climate Center Second Generation General Circulation Model," *Bulletin of the American Meteorological Society* 5: 1045-1077.

Carter, T.R., M.L. Parry, H. Harasawa, and S. Nishioka, 1994, *IPCC Technical Guidelines for Assessing Climate Change Impacts and Adaptations*, Working Group II of the Intergovernmental Panel on Climate Change. University College of London, United Kingdom. Center for Global Environmental Research.

Conceptual Program Agriculture and Industry Complex Development of Republic of Kazakhstan on the 1993-1995 Period and till 2000, 1994, Bastau, Almaty, Kazakhstan (in Russian).

Guidelines on Reliability of Agrometeorological Forecasts, 1983, Hydrometeoizdat, Moscow, USSR (former Soviet Union) (in Russian).

Manabe, S., and R.T. Wetherald, 1987, "Large Scale Changes in Soil Wetness Induced by an Increase in Carbon Dioxide," *Atmospheric Science* 44:1211-1235.

Mean Long-Term and Probability Characteristics of Productive Water Stores under Winter and Early Spring Cereals, 1981, Vol. 4, Hydrometeoizdat, Leningrad, USSR (former Soviet Union) (in Russian).

Menzhulin, G.V., L.A. Koval, and A.L. Badenko, 1994, "Potential Effects of Global Warming and Carbon Dioxide on Wheat Production in the Former Soviet Union," in *Implication of Climate Change for International Agriculture: Crop Modeling Study*, Policy Planning and Evaluation (2122), EPA 230-B-94-003, FSU 1-36, U.S. Environmental Protection Agency, Washington, D.C., USA (in Russian).

Recommendations on the Agricultural Management System. Tselinograd Oblast, 1982, Kainar, Alma-Ata, Kazakhstan (in Russian).

Ritchie, J.T., and S. Otter, 1985, "Description and Performance of CERES-Wheat: A User-Oriented Wheat Yield Model," in W.O. Wills (ed.), *ARS Wheat Project*, U.S. Department of Agriculture, Agricultural Research Service, Washington, D.C., USA.

Soils of Kazakh SSR, 1966, Kazakhstan Academy of Sciences, Alma-Ata, Kazakhstan (in Russian).

Tsuji, G.Y., J.W. Jones, G. Uehara, and S. Balas (eds.), 1995, *Decision Support System for Agrotechnology Transfer, V3.0*, Vols. 1-3, IBSNAT, University of Hawaii, Honolulu, Hawaii, USA.

The Impact of Climate Change on Spring Wheat Yield in Mongolia and Its Adaptability

Sh. Bayasgalan, B. Bolortsetseg, D. Dagvadorj, and L. Natsagdorj
Hydrometeorological Research Institute
Khuldaldsany-5
Ministry of Nature and Environment, Ulaanbaatar, 11 Mongolia

Abstract

The characteristics of the current tendency toward climate change in Mongolia and the doubling of carbon dioxide (CO_2) levels (referred to as $2XCO_2$) scenario from the General Circulation Model (GCM) are briefly described. The effect of climate change on the production of spring wheat in Mongolia, given the nation's geographic and climatic conditions, was assessed. In general, the yield decreased 19-67% under the GCM-based $2XCO_2$ scenario. This scenario assumed that growing season temperatures and precipitation would increase, but that potential evapotranspiration would be higher. However, wheat yields for the actual climate change trend (i.e., reduced temperature during the growing season and increased precipitation) increased 10.4-70.2% at six of seven locations. Simulated adaptive measures, such as changing planting dates, using different varieties of spring wheat, and applying the ideal amount of nitrogen fertilizer at the optimum time, are potential responses that could modify the effects of climate change on wheat production.

Introduction

Mongolia is bordered on the north by the Siberian Taiga and permafrost zone and on the south by global semiarid and desert areas. Mongolia is also part of the Asian watershed. These facts suggest that climate change in Central Asia and Mongolia could have serious social, economic, and ecological effects, at both the national and regional level. In addition, environment and climate play a key role in the sustainable development of the country because of its unique geographic location and economic structure.

Agriculture plays an important role in Mongolia's economy. Each year, the country's main crop, spring wheat, is planted on 500,000-800,000 ha of land.

Climate change may affect Mongolia's spring wheat production. Scenarios from General Circulation Model (GCM)-based doubling of carbon dioxide (referred to as $2XCO_2$) and a scenario of historical climate change trends in Mongolia are examined. Several adaptation options are discussed as potential responses that could modify the effects of climate change on wheat production.

Current Climate Change Trends in Mongolia

Studies by Dagvadorj et al. (1994) show that global climate change has been observed in Mongolia. Since 1940, the annual mean air temperature of the territory has generally increased by 0.7°C. The intensity of the warming tendency differs in the various natural and climatic areas. For example, the temperature increased 1.8°C in the Great Lakes Basin, which is the coldest location in Mongolia; 1.0°C in the basin of the Selenge and Orhon Rivers, which is a major agricultural area; and 0.3°C in the Gobi and steppe areas. These findings suggest that parts of the country are experiencing warming that exceeds the global average. The intensity of temperature increases also varies seasonally. Temperature increases during the winter and decreases during the summer (Figure 1).

Changes in the amount of precipitation also vary in different areas of the country (Figure 2). The Gobi and desert areas have recorded decreases in precipitation, while other areas have shown small increases. This pattern seems to follow the conclusions of some scientists regarding precipitation decreases in the mid-latitudes of the Northern Hemisphere (Parry 1990; Parry and Zhang 1991).

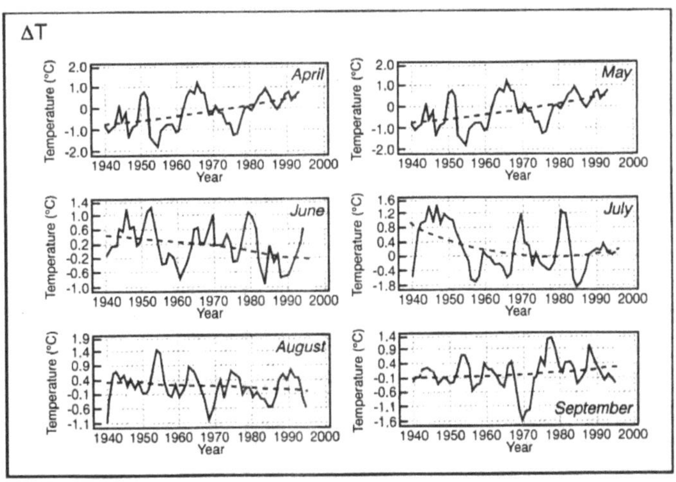

Figure 1. Climate Change Trends in Temperature for the Growing Season in Mongolia

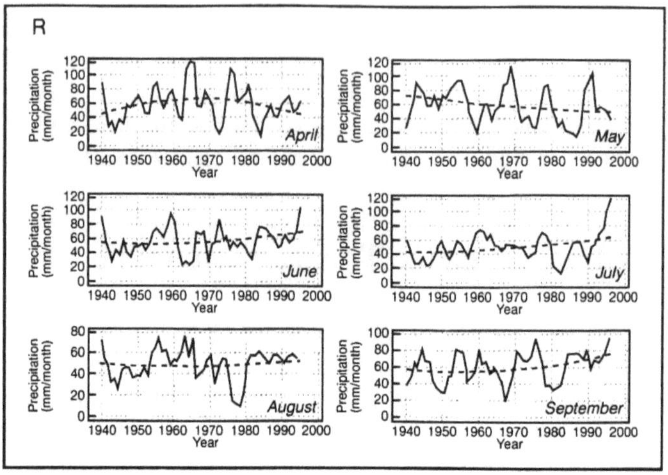

Figure 2. Climate Change Trends in Precipitation for the Growing Season in Mongolia

Methodology

The Decision Support System for Agrotechnology Transfer version 3 (DSSAT3) model was used to study the effects of climate change on spring wheat in Mongolia (U.S. Country Studies Program 1994). Nonlinear models and methods for time series analysis estimated climate change trends.

Impact Assessment

Scenarios

Three climate scenarios were examined: two Goddard Institute for Space Studies (GISS) and Geophysical Fluid Dynamics Laboratory (GFDL) GCM-based $2XCO_2$ scenarios, and one scenario of historical climate change trends in Mongolia. The climatological baseline was 1940-1990. Future changes in other environmental and socioeconomic factors were not considered.

The output from GCMs often fails to reproduce the seasonal pattern of present-day climate observed at a regional level. This fact casts doubt on the ability of GCMs to accurately estimate future regional climate, and, consequently, on the reliability of assessing the impact of climate change on crop production. Temperature increases obtained from GISS model simulations generally range from 2.9-13.4°C for all months. But GFDL model simulations predict temperature decreases of -3.0 to -5.9°C during the winter and temperature increases for the other months. Thus, the output from GCMs should be treated, at best, as broad-scale sets of future climatic conditions and its effects, and should not be regarded as predictions. The GISS and GFDL models give fairly accurate reproductions of the actual climate in a region. Therefore, the scenarios and current climate from these GCMs were used in this study.

Locations

To run the simulation models, researchers selected seven locations in the main grain crop area. The necessary climate and soil data for running the models for these locations were then collected. Table 1 gives the latitude and longitude, as well as the sea-level elevation for these locations.

Results

2XCO_2 Scenario

Figure 3 shows that spring wheat yields for 2XCO_2 scenarios at five of the seven locations decreased 19-67%; two locations experienced increases in yield. For example, in Baruunharaa, spring wheat yields increased 70 kg/ha under the GISS weather scenarios and 29 kg ha^{-1} under the GFDL weather scenarios. Therefore, in general, spring wheat yields could be reduced substantially by the end of the next century as a result of negative impacts of climate change in the region.

Clearly, increased atmospheric CO_2 concentrations can directly affect the photosynthesis and growth rate of crop plants (Parry 1990). In addition, as a result of ambient temperature increases, physiologic temperatures needed by

Table 1. Geographic Characteristics of the Locations

Location	Latitude (deg)	Longitude (deg)	Elevation (m)
Baruunharaa	48.91	106.06	807
Darhan	49.46	105.98	709
Erdenesant	47.14	104.14	1,363
Hutag	49.38	102.69	933
Orhon	49.14	105.39	748
Ugtaal	48.26	105.41	1,151
Zuunmod	47.71	106.94	1,530

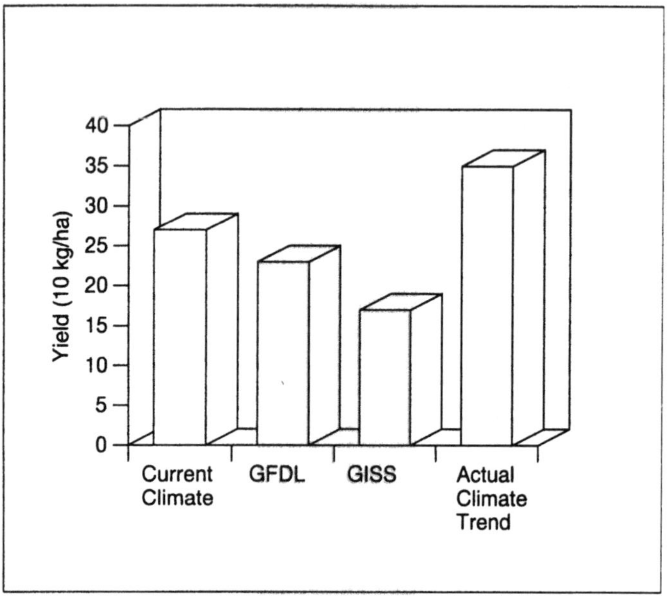

Figure 3. Spring Wheat Yields under Different Climate Change Scenarios

vegetation for ripening would occur more frequently; consequently, the time needed for crops to ripen would decrease. However, maximum temperature increases have negative effects on the normal rate of phenological phases of wheat maturity and on the accumulation process of organic matter in grain.

Under this scenario, the number of days with precipitation and cloud cover would increase because precipitation would increase. Also, the number and area of leaves in one spike of wheat, the number of grains per unit area (m^2), and the weight of grain could be reduced because of the decrease in photosynthesis in the crop. To determine how the photosynthesis rate of crops changes requires more detailed information and an advanced methodology.

Increases in temperature and precipitation affect the concentration of mineral substances in the soil. Table 2 gives the nitrogen requirements of spring wheat during the growing season under various climatic conditions. The requirement for nitrogen would be reduced by 12.1-15.6 kg ha^{-1} under the GISS and GFDL scenarios.

Historical Climate Change Trend Scenario

The results of this study show many differences between the historical climate change trend in Mongolia and the GCM scenarios. Therefore, the next step was to study how the actual climate change in the region affected grain crop production.

Table 2. Nitrogen Requirements of Spring Wheat (kg ha^{-1})

Location	Current Climate	Calculated by GCMs	
		GISS	GFDL
Baruunharaa	68.2	70.3	74.7
Darhan	78.2	78.0	60.3
Erdenesant	80.7	70.0	60.0
Ugtaal	94.7	59.6	50.0
Zuunmod	71.7	60.2	70.5
Mean	**78.7**	**67.6**	**63.1**

Scientists used quadratic polynomial equations for 1940-1994 to estimate the monthly climate change tendency and calculated its trend values for each month. The main peculiarity of these trends is a temperature decrease of –0.28 to –0.67°C for June, July, and August, and precipitation increases of 28% and 46% for June and July, respectively. Simulation results from DSSAT3 with this climate scenario show that the spring wheat yield could increase if the climate change trend continued. However, the data necessary to make a detailed investigation on the Mongolian climate change in the near future and to examine its effect on the environment are not available.

Adaptation Options

The analysis with GCM-based 2XCO_2 scenarios predicts reductions in the spring wheat yield as a result of the negative impact of possible climate changes in the region. According to these findings, Mongolia will face serious difficulties in providing much of its grain needs. As a result, the country needs to respond to modify these negative effects. Two broad responses have been identified: mitigation and adaptation.

Mitigation (or limitation) attempts to deal with the causes of climate change. It involves taking actions that prevent or retard the increase of atmospheric greenhouse gas (GHG) concentrations by limiting current and future emissions from sources of GHGs and enhancing potential sinks for GHGs. However, it is likely that realistic policies of mitigation can fully prevent climate changes and that alternative adaptive measures are needed.

Adaptation options are potential responses that could modify the effects of climate change on wheat production. Each option, planting, nitrogen, and cultivar, is discussed in the following subsections.

Planting

The main goal of considering this option is to estimate the optimum planting date for spring wheat under potential changes in climate conditions. Scientists analyzed wheat yield values using different planting dates at selected locations. Planting dates were moved ahead of the average planting date (May 11) because climate scenarios suggest that future climate would be warmer than current climate. Table 3 gives wheat yields under climate scenarios from the GISS and GFDL models at planting dates of April 25 and May 1.

The optimum planting date for spring wheat in Mongolia is April 25, which means that the grain crop should be planted 15 days earlier than the current planting date. Planting the wheat crop on April 25 could increase wheat yields 9.2-79.7% under the GISS scenario and 9.0-55.6% under the GFDL scenario.

Nitrogen

Changes in climate could greatly affect soils. Policymakers may need to review land-use policies in the light of this change, with regard to the possibility of changing soils for crops.

Table 3. Spring Wheat Yields by Different Planting Dates (10 kg ha^{-1}) as Percentage of Current Yields (100% = no change)

	GCMs and Planting Dates					
			GISS		GFCL	
Location	GISS	GFDL	May 1	April 25	May 1	April 25
Baruunharaa	28.7	24.6	ND [a]	ND	ND	ND
Darhan	24.1	12.5	ND	ND	14.5	14.7
Erdenesant	24.2	12.7	26.5	28.2	14.3	16.8
Hutag	18.4	15.6	20.6	20.1	19.4	19.5
Orhon	7.9	16.6	11.4	14.2	18.8	18.1
Ugtaal	20.2	8.1	23.0	24.3	10.4	12.6
Zuunmod	23.9	26.9	24.1	27.0	28.9	29.2

[a] ND = no data available.

In this option, scientists examined the optimum value of nitrogen fertilizer on 1 m^2 of land and the optimum soil depth for inserting nitrogen. The results of their estimate show that the optimum value of nitrogen fertilizer is 60 kg ha^{-1} to obtain maximum yields. It is important to choose the date to apply nitrogen in the soil. (DSSAT3 was used to examine different dates.) Data from Table 4 show that maximum yields could be obtained by applying nitrogen on the 70th day following the planting date. If nitrogen were applied when the grain is planted, the wheat yield would increase, but the increase would not be as high as it would be if nitrogen were applied on the later date. Also, the study shows that the most appropriate depth for inserting nitrogen is 7 cm. If Mongolia follows these nitrogen options, spring wheat yields could increase as much as 115 kg ha^{-1} under the GCM-based $2XCO_2$ scenarios.

Cultivar

The current variety of spring wheat, "Orhon," yields less under the $2XCO_2$ scenarios. Therefore, on the basis of experience, scientists estimated genetic coefficients for a "new" variety of spring wheat that might be more adaptable to changed climate conditions in the future. Table 5 gives the optimum values for genetic coefficients of a variety of spring wheat, as estimated by DSSAT3 simulation. This variety of spring wheat yields 200-250 kg ha^{-1}. The grain starch in this variety of wheat is 1.3-1.9 times higher than that in the current variety of wheat.

Table 4. Spring Wheat Yields (10 kg ha^{-1}) According to Dates of Nitrogen Insertion in Soil

				Insert Together When Grain Is Planted		At the 70th Day from the Planting Date	
Location	Current Climate	GFDL	GISS	GFDL	GISS	GFDL	GISS
Baruunharaa	21.7	24.6	28.7	ND[a]	ND	ND	ND
Darhan	23.3	12.5	24.1	14.2	ND	15.9	ND
Erdenesant	29.7	12.7	24.2	13.9	25.2	16.2	27.7
Hutag	25.9	15.6	18.4	16.9	20.4	19.1	21.9
Orhon	23.8	16.6	7.9	18.6	10.7	20.1	11.4
Ugtaal	31.5	8.1	20.1	10.6	21.1	11.6	23.4
Zuunmod	38.0	23.9	26.9	25.9	28.4	27.4	29.9

[a] ND = no data available.

Table 5 . Genetic Coefficients of a "New" Variety of Spring Wheat (see text for explanation)

P1D	P5	G1	G2	G3
3.4	4.0	4.6	3.0	4.2

In Table 5, P1V indicates the relative amount that development is slowed for each day of unfulfilled vernalization, assuming that 50 days of vernalization is sufficient for all cultivars. P1D is the relative amount that development is slowed when plants are grown in a photoperiod 1 hour shorter than the optimum, which is 20 hours. P5 is the relative grain filling duration based on thermal time (degree-days above a base temperature of 1°C), where each unit increase above zero adds 40 degree-days to an initial value of 300 degree-days. The value G1 denotes the kernel number per unit weight of stem (fewer leaf blades and sheaths) plus spike at anthesis (1/g); G2, the kernel filling rate under optimum conditions (mg day^{-1}); and G3, the nonstressed dry weight of a single stem (excluding leaf blades and sheaths) and spike when elongation ceases (g).

Discussion and Conclusions

The options discussed are some examples of many possible adaptive measures (Parry and Zhang 1991). It is reasonable to expect that numerous changes in land use (e.g., changes in farmed area, crop type, and crop location) and changes in management (e.g., changes in the use of irrigation and fertilizers, in the form of crop) would be adopted over time as the effects of climate change are observed.

It appears that these adaptive measures would be most favorable for Mongolia if climate shifted toward warmer and wetter conditions in the future. It is clear that any adaptive measure at the regional level is not one of climate change abatement, but rather one of optional adjustment to climate change (Carter et al. 1994).

Results obtained so far for crop yields have been encouraging because the scientific community is beginning to determine the importance of changes in climate variability in this area. Future research should help to refine knowledge of the effects of variability in climate. It would be useful if technological advances in irrigation efficiency and crop drought resistance continue, as well as improvements in a number of crop-specific characteristics, including harvest index, photosynthetic efficiency, and pest management.

References

Carter, T.R., M.L. Parry, S. Nishioka, and H. Harasawa, 1994, *IPCC Technical Guidelines for Assessing Climate Change Impacts and Adaptations*, Working Group II of the Intergovernmental Panel on Climate Change, University College, London, United Kingdom, and Center for Global Environmental Research, Tsukuba, Japan.

Dagvadorj, D., 1993, "Numerical Assessment of Climate Change Impacts on Grain Yield in Mongolia," presented at the *1st PRC-Mongolia Workshop on Climate Change in Arid and Semiarid Region over the Central Asia*, May 8-11, 1993, Beijing, People's Republic of China, pp. 104-108.

Dagvadorj, D., R. Mijiddorj, and L. Natsagdorj, 1994, "Climate Change and Variability Studies in Mongolia," *Annual Scientific Journal of Hydrometeorological Research*, 17: 3-10, Ulaanbaatar, Mongolia.

Parry, M., 1990, *Climate Change and World Agriculture*, Earthscan Publications Ltd., London, United Kingdom.

Parry, M., and J.-C. Zhang, 1991, "The Potential Effect of Climate Changes on Agriculture," in *Proceedings of the 2nd World Climate Conference: Climate Change: Science, Impacts and Policy*, Cambridge University Press, Cambridge, United Kingdom.

U.S. Country Studies Program, 1994, *Guidance for Vulnerability and Adaptation Assessments*, Version 1.0 (U.S. Country Studies Program, Washington, D.C. , USA).

Some Adaptations of the Tea Plant to Dry Environments

M.A. Wijeratne
Tea Research Institute of Sri Lanka
St. Jochim Estate
Ratnapura, Sri Lanka

Abstract

Agriculture plays a major role in the economy of most developing countries. Sri Lanka's main agricultural exports are tea, rubber, and coconut. The productivity of such crops largely depends on climate. Therefore, variations in the weather pattern or changes in climate affect agricultural programs, leading to severe economic losses. Moreover, debilitation or death of tree crops changes the local environment because of soil erosion, floods, or dried-up streams. Hence, the use of specific cultivars or species of crops that can adapt to many climate conditions has become a necessity. Prolonged dry spells significantly affect tea production, which means that identifying and assessing inherent adaptations of the tea plant to dry environments are of paramount importance for minimizing the impact of climate change on the economy. Experimental results have shown that under dry conditions, drought-tolerant tea clones maintain a favorable water status because they use soil moisture efficiently through stomatal control and water potential adjustments. The use of drought-tolerant tea clones is the most cost-effective and environmentally friendly measure in the tea industry for adapting to global climate change.

Introduction

Economic and social development in many developing countries depends largely on agriculture. In Sri Lanka, agricultural exports bring in about 25-35% of the country's foreign exchange. Variations in weather or changes in climate, such as recurrent and prolonged droughts or a global increase in temperature, affect the production and planned programs in the agricultural sector, leading to irreparable losses to the economy. In addition, debilitation or death of plantation crops (e.g., tea) that cover extensive lands, especially hilly regions, leads to sudden changes in the local environment because vegetation cover is destroyed, which results in soil erosion, floods, or dried-up streams. Moreover, such natural hazards directly and immediately affect the economy and the life of individuals employed in the agricultural sector. The use of improved cultivars or a species of agricultural crops that can adapt to a wide range of climate changes has become a necessity.

In Sri Lanka, tea ranks first in earning net foreign exchange. More inhabitants are employed in plantation agriculture than in any other sector of the economy. Tea plantations cover about 11% of agricultural lands (i.e., equivalent to about 3.4% of the total land area), mostly in the central hilly part of the country. About 15% of the country's workers are employed as manual laborers on plantations. Tea is a rain-fed plantation crop, and its yield is greatly determined by the environment. The productivity of many plantations in the mid-elevations (600-1,200 m) has become marginal, but could be become viable if crops were diversified. Bush debilitation and erosion of topsoil on slopes during the past few decades are responsible for increased casualties or death of bushes. Improper crop and soil management and adverse weather are the principal causes of such damage.

Generally, tea grows well within a temperature range of about 18-25°C and a minimum annual rainfall of about 1,200-1,250 m. However, an even distribution of rain is more important than the amount of rain (Carr 1972; Watson 1986). Differences among tea plant clones are evident by their performances under different environmental conditions. In low-country tea-growing regions (>600 m above mean sea level), daytime temperatures generally exceed optimal limits. Moreover, the annual dry period during the first quarter of the year is usually due to uneven distribution of rain. Observations and assessments on drought damage to tea have alerted scientists to grant high priority to study ways to minimize environmental hazards on tea plantations (Nawaratne 1992; Yatawatte 1992). As a result of a drought in early 1992, national tea production declined by 19%; simultaneously, the cost of production increased by 19% over the previous year, thus depriving the economy of a significant amount of foreign exchange (Central Bank 1992).

In the past decade, the government has spent more funds on disaster relief for farmers who have suffered damage from drought than it has for relief to other sectors of the economy. Significantly more families have been affected by drought than by other natural hazards. Climate change scenarios have predicted that Sri Lanka may experience more frequent severe droughts and warm seasons in the future as a result of global climate change. The agricultural sector is more vulnerable to climate change than are other sectors (Asian Development Bank 1992). Therefore, adverse weather would cause more frequent and severe crop losses in the coming years, resulting in irreversible harm to the economy. Research programs should aim to find acceptable socioeconomic measures to adapt to adverse climate changes. Such measures would secure the role of the tea industry in the economic and social development of Sri Lanka.

Potential Adaptive Measures

Initially, seedling tea was planted on all estates. However, high-yield clonal tea is now commonly used. Most high-yield clones are vulnerable to adverse weather conditions because they require large amounts of nutrients and water. Studies and observations have shown that seedling tea and drought-tolerant tea clones adapt more easily to adverse climate conditions and suffer less damage during dry spells. Hence, planting drought-tolerant clones is a possible adaptive measure in response to climate change. Preliminary studies that identify and assess inherent drought-tolerant properties and the degree of drought tolerance in tea are very important if this type of adaptive strategy is used.

Methodology

To test the feasibility of introducing drought-tolerant clones as an adaptive measure for dry environments, a greenhouse experiment was conducted at the Tea Research Institute of Sri Lanka low-country station at Ratnapura (6°40'N, 80°25'E; 60 m above mean sea level). The purpose of the experiment was to identify inherent drought-tolerant characteristics and study the capability of hardy tea clones that could withstand dry environments.

A group of about seven-month-old vegetatively propagated plants — TRI 2025 (drought-tolerant clone) and TRI 2023 (drought-susceptible clone) — was selected for the experiment. These plants were raised in 41 plastic pots and placed in a greenhouse.

All potted plants were thoroughly watered and allowed to drain for 24 hours before treatments. Two clones (TRI 2025 and TRI 2023) and two moisture regimes (well watered and moisture stressed) were randomized factorially in three replicates, each having 80 plants in four rows (plots). Plants given well-watered treatment were watered daily in the morning; the other plants were allowed to dry out until they were permanently wilted. To estimate the bare soil evaporation, the main stems of six potted plants were cut immediately below the soil surface and placed in the three replicates. One-half of these pots were maintained at each moisture regime.

Assessments

The ability of plants to withstand dry conditions was tested by comparing the diurnal variations of plant-water relations under dry and wet soil conditions inside the greenhouse. Four areas were studied: soil moisture, transpiration and diffusive resistance, relative water content (*RWC*), and leaf-water potential (*LWP*). These areas are discussed in the following sections.

Measurement of Soil Moisture

One plant/treatment/replicate was tagged and weighed six times daily at 2-hour intervals between 0700 and 1700.

Measurement of Transpiration and Diffusive Resistance

Transpiration *TR* and diffusive resistance *DR* of one mature leaf of each plant tagged for weighing were measured, along with photosynthetically active radiation (PAR). Measurements were made by using a steady-state porometer (LI-1600, Li Cor, Inc. Ltd., United States) at time intervals similar to those given in the above section.

Relative Water Content

Three mature leaves of similar age were excised from three plants/treatment/replicate and measured. The fresh weight *FW*, turgid weight *TW* after floating them on distilled water for three hours at room temperature under a light intensity of 90 μmol $m^2 s^{-1}$, and dry weight *DW* after oven drying at 90°C to constant weight were also recorded (Sandanam et al. 1981). The relative water content *RWC* was estimated as:

$$RWC = (FW - DW)/(TW - DW) .$$

Measurements were taken for the above assessments at 2-hour intervals during the first two days (from 0700 to1700) and three times daily (0700, 1300, and 1700) during the subsequent period.

Leaf-Water Potential

Leaf-water potential was measured on a similar set of leaves at comparable time intervals as for relative water content (Scholander et al. 1965). A plant water status console (Model 3005, Soil Moisture Equipment, United States) was used in the measurements.

The water potential of the drying soil was estimated from a calibrated curve for water potential versus moisture content (Reeve and Carter 1991). Root-water potential *RWP* was also estimated (Jones 1990):

$$RWP\ (D) = LWP\ (D) - [LWP\ C\ g_S\ (D)]\ /\ g_S C\ ,$$

where *LWP*, *RWP*, and g_s are the leaf-water potential, root-water potential, and stomatal conductance (*1/DR*), respectively, of plants exposed to drought *D* and well-watered *C* plants.

In addition to the above assessments, the moisture content of soil from permanently wilted plants was determined by oven drying a soil sample of about 200 g from the root zone at 105°C to constant weight. At the end of the study, the fresh and dry weight of plant and soil, and the leaf area of each plant used for weighing, was recorded. Moreover, air temperature and saturation vapor pressure deficit (SVPD) inside the greenhouse were also monitored at hourly intervals.

Results

Diurnal Variations of Water Relations

The diurnal variations of PAR, air temperature, and SVPD are shown in Figure 1. The highest temperature and SVPD, together with the highest PAR, were experienced around midday.

Variations in the diurnal and drying cycle in soil and plant water potentials, namely, for leaves and roots, are illustrated in Figure 2. These variations are similar to those described by Slatyer (1967) and Jackson (1989). The soil-water potential in the root zone was inferred from the predawn *RWP* when all three water potentials were usually very close. Water potentials consistently showed

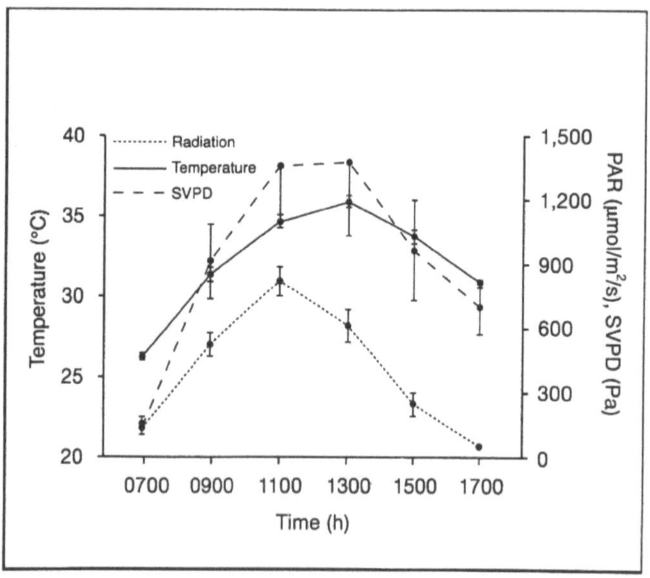

Figure 1. Diurnal Variation in Temperature, SVPD, and Solar Radiation in the Greenhouse (means are given for the experimental period)

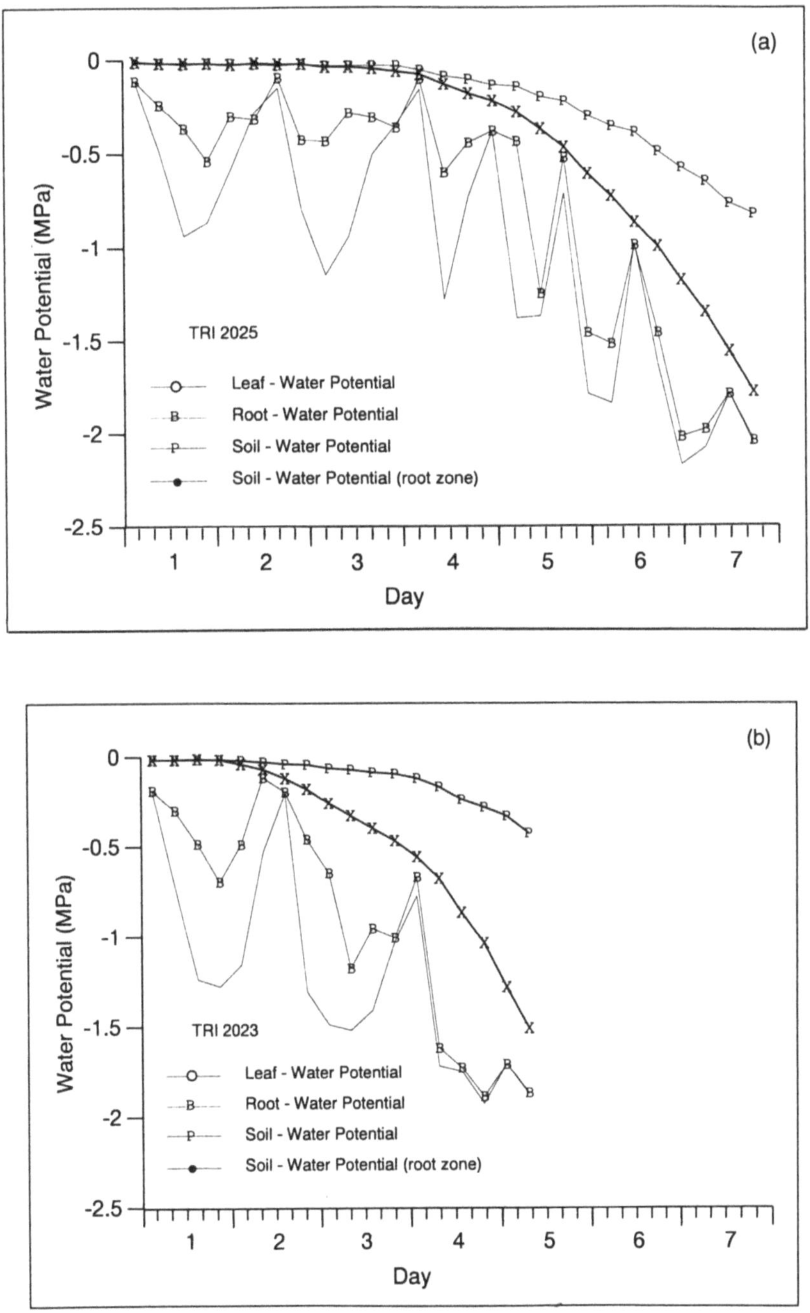

Figure 2. Soil Moisture Depletion over the Experimental Period: (a) TRI 2025 and (b) TRI 2023 (Data points for each day 0700, 0900, 1100, 1300, 1500, and 1700 hours.)

an opposite trend to that of environmental factors and reduced as soil dried. Higher water potentials were observed in the morning and evening, while lower potentials occurred midday. The period needed to complete a drying cycle (i.e., from field capacity to permanent wilting point [PWP]) was shorter for TRI 2023 plants (four days) than for TRI 2025 plants (seven days).

The soil-water potential in the root zone at PWP, estimated by the soil samples collected from plants reaching PWP, was -1.47 (±0.19) MPa for TRI 2023 plants and -2.78(±0.16) MPa for TRI 2025 plants. Corresponding gravimetric soil moisture contents at these water potentials were 5.9 (±0.2)% and 5.1 (±0.1)%, respectively. However, the soil-water potentials estimated by pot weights were higher than the above values because the moisture content was overestimated for the entire pot, whereas root growth was restricted to the top two-thirds of the soil.

When compared with the changes in environmental factors, it is evident that the diurnal variations of transpiration and diffusive resistance (Figure 3) are largely influenced by SVPD, temperature, and solar radiation (Squire 1979; Callander and Woodhead 1981; Sandanam et al. 1981). Diurnal fluctuations of transpiration and diffusive resistance were in the opposite order. In both TRI 2025 and TRI 2023, the higher values of diffusive resistance, combined with the lower values of transpiration, were measured in the morning and evening, while higher transpiration and lower diffusive resistance values were

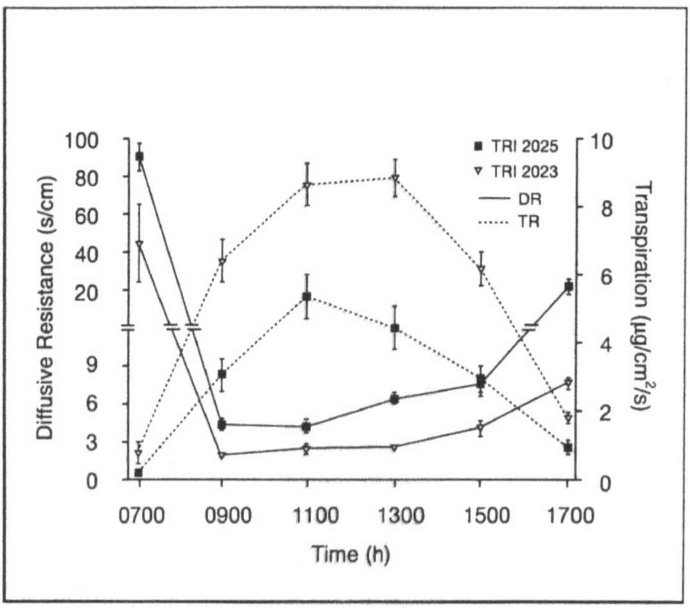

Figure 3. Diurnal Variation in Diffusive Resistance *DR* and Transpiration *TR* of Well-Watered Plants

observed around midday. Moreover, TRI 2025 leaves had lower transpiration than those of TRI 2023. The opposite was observed for diffusive resistance. The difference in transpiration and diffusive resistance could be attributed to the lower stomatal density in TRI 2025 compared with that of TRI 2023 (Wadasinghe and Wijeratne 1989). Transpiration in TRI 2025 appeared to decrease earlier in the day than in the other clone, that is, after about 1100 for TRI 2025 and after about 1300 for TRI 2023. These characteristics of the drought-tolerant clone would have contributed to its ability to conserve moisture, thus partly making it drought tolerant, as characterized by many other plant species adapted to dry environments (Chaney 1981).

The diurnal fluctuation in *RWC* and *LWP* of well-watered plants of both clones (Figure 4) follows a similar pattern. That is, high *RWC* is accompanied by high *LWP* and vice versa. The *RWC* and *LWP* of TRI 2025 leaves were significantly higher than those of TRI 2023. TRI 2025 recovered from moisture stress earlier than did TRI 2023 leaves, which supports the changes in transpiration discussed earlier. Other researchers have also shown that drought-tolerant plants have low transpiration and high diffusive resistance, *RWC*, and *LWP* (Hurd 1976; Kaufmann 1981; Sandanam et al. 1981; Handique and Manivel 1986; Planchon 1987).

When the relationship between transpiration and SVPD was considered, the rate of increase in transpiration in response to increasing SVPD was higher for

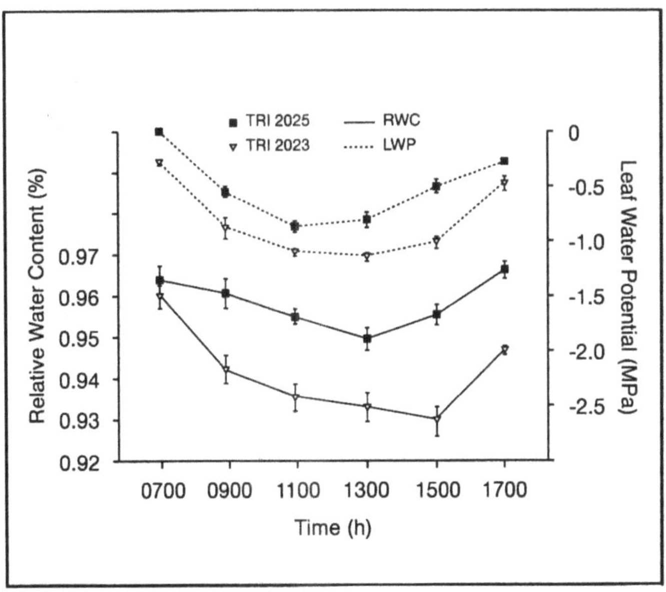

Figure 4. Diurnal Variation in Relative Water Content *RWC* and Leaf-Water Potential *LWP* of Well-Watered Plants (mean values are given for days one and two)

TRI 2023 than for TRI 2025 leaves. As a result of increased water loss in TRI 2023, the rate of reduction in *RWC* and *LWP* of TRI 2023 was higher than that of TRI 2025. These characteristics in TRI 2025 plants may help to minimize the adverse effects of dry environments on physiological functions, such as photosynthesis. Such mechanisms are related to drought tolerance (Hurd 1976; Planchon 1987).

The relationships among SVPD, transpiration (*TR*), *RWC*, and *LWP* for both clones are presented below:

$TR = 0.38\ (\pm 0.35) + 1.76\ (\pm 0.29)$ SVPD — $R^2 = 59\%$ --- TRI 2025 $p < 0.001$

$TR = 0.56\ (\pm 0.60) + 3.60\ (\pm 0.49)$ SVPD — $R^2 = 70\%$ --- TRI 2023 $p < 0.001$

$RWC = 0.968\ (\pm 0.002) - 0.009\ (\pm 0.002)$ SVPD — $R^2 = 48\%$ -- TRI 2025 $p < 0.001$

$RWC = 0.967\ (\pm 0.004) - 0.019\ (\pm 0.004)$ SVPD — $R^2 = 54\%$ -- TRI 2023 $p < 0.001$

$LWP = -0.135\ (\pm 0.049) - 0.256\ (\pm 0.040)$ SVPD — $R^2 = 63\%$ - TRI 2025 $p < 0.001$

$LWP = -0.202\ (\pm 0.071) - 0.422\ (\pm 0.059)$ SVPD — $R^2 = 67\%$ - TRI 2023 $p < 0.001$

The relationship between *LWP* and *RWC* for well-watered conditions is shown in Figure 5. It reveals that at low *LWP*, TRI 2025 plants can maintain a higher *RWC* than TRI 2023. Furthermore, TRI 2023 leaves transpire more water; thus, their *LWP* and *RWC* are smaller than those of the other clone. During the drying cycle, the two clones experience changes in leaf-water relations (Figure 6). Accordingly, TRI 2023 plants lost more water than TRI 2025 plants. The effects of soil-moisture stress on *RWC* and *LWP* of TRI 2025 started about three days after withholding water. However, corresponding data for TRI 2023 showed a significant effect by the second day. As the leaf area of plants was not significantly different between the clones ($50 \pm 3.9\ cm^2$ for TRI 2023 and $52 \pm 2.7\ cm^2$ for TRI 2025), the difference in water loss would have been caused by their inherent differences in transpiration. This finding implies that the drought-tolerant clone requires less water than is needed by the drought-susceptible clone.

To study the characteristics of water relations between the two clones in detail, researchers used pressure volume curves to estimate *RWC* and *LWP* at the turgor loss point (Figures 7 and 8). Results were based on the daily averages of *RWC* and *LWP* recorded on both well-watered plants and plants submitted to drought. The osmotic potential at full turgor was not significantly different between the two clones (i.e., -1.06 and -0.99 MPa for TRI 2025 and TRI 2023, respectively). The values were comparable with those reported for tea by Othieno (1978). However, the osmotic potential of the TRI 2025 at turgor loss

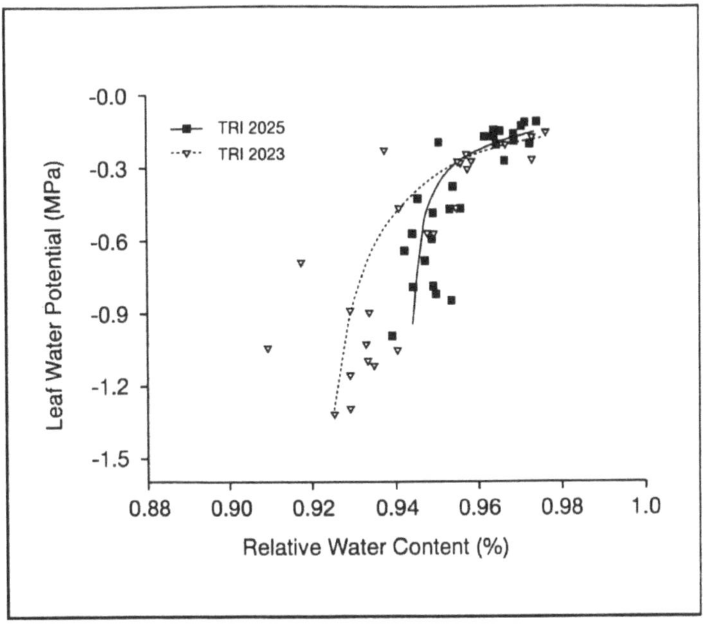

Figure 5. Relationship between Relative Water Content *RWC* and Leaf-Water Potential *LWP* of Well-Watered Plants

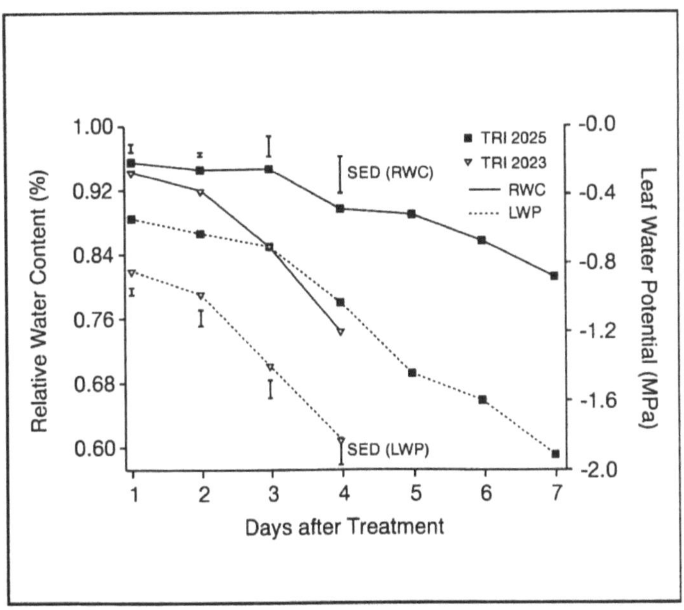

Figure 6. Daily Averages of Relative Water Content *RWC* and Leaf-Water Potential *LWP* of Plants Suffering from Drought

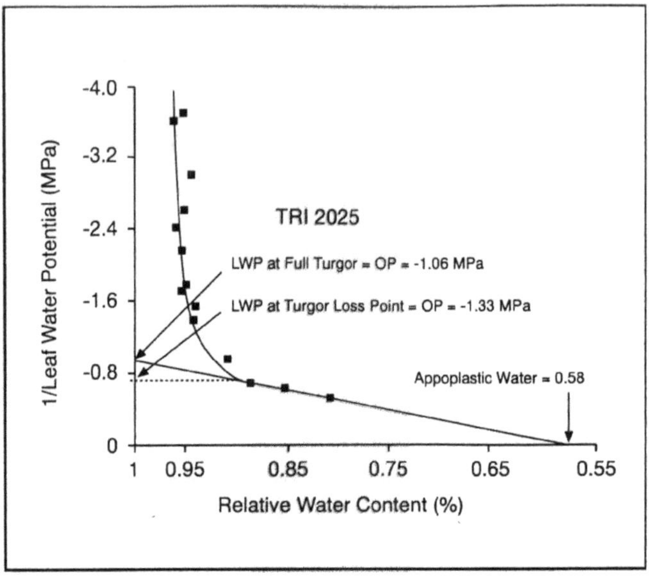

Figure 7. Estimation of Osmotic Potential (OP) at Full Turgor and Turgor Loss Point for TRI 2025 (*LWP* = leaf-water potential)

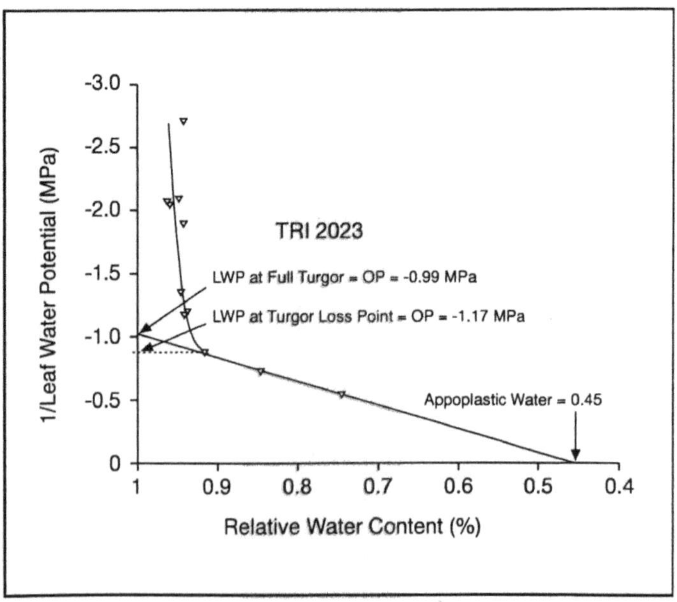

Figure 8. Estimation of Osmotic Potential (OP) at Full Turgor and Turgor Loss Point for TRI 2023 (*LWP* = leaf-water potential)

point was lower than that of TRI 2023 (i.e., -1.33 MPa and -1.17 MPa, respectively). The proportion of appoplastic water content of the leaves of TRI 2025 (58%) was lower than that of TRI 2023 (45%) and indicated drought tolerance (Ryadnova and Lebedeva 1971).

These results indicate that the drought-tolerant clone had a higher *LWP* under well-watered conditions and withstood soil drying by lowering its plant water potentials. Using such clones could be a feasible adaptive measure for envisaged climate changes. Drought-tolerant tea clones proved to be more adaptable to dry environments because of the following reasons:

- Low and earlier reduction of transpiration because of high stomatal resistance,
- Maintenance of high *RWC* and *LWP*,
- Low rate of increase in transpiration in response to increasing dryness of the air (SVPD), and
- Maintenance of high *LWP* under well-watered conditions and reduction in *LWP* to smaller values in response to soil drying.

These findings show that using drought-tolerant tea clones would reduce the risk of environmental hazards on tea plantations. Moreover, this study provides scientific evidence for past observations on drought casualties in tea fields (i.e., that TRI 2025 and moderately yielding tea clones are less affected by dry periods) (Nawaratne 1992). Therefore, drought-tolerant tea clones could be an adaptive measure against possible climate changes in Sri Lanka. However, some tea planters are reluctant to use hardy tea clones because of their low yield.

Hence, it is necessary to educate and encourage tea planters to select tea clones to suit the environment for their ongoing planting programs. In addition, availability of those planting materials should be increased. At present, tea planters are unaware of potential global climate change and its impact on productivity. However, damages incurred recently have led planters to realize the importance of using hardy tea clones. The small tea planter sector in Sri Lanka is expanding, and intercropping systems, such as planting tea with rubber to reduce the risk of monoculture, are being introduced. Therefore, it is more likely that the extent of tea lands with hardy clones will increase as an adaptive measure for climate change.

An efficient advisory and extension service is in place for tea cultivation. Therefore, it should be easy to educate tea growers about the effects of global climate changes on the productivity of tea plantations. Moreover, once the clones suitable for use in adverse environmental conditions are screened, they could be distributed to estates and commercial nurseries for propagation. If so, adequate quantities would be available for growers without a delay.

Conclusions

The results of this study show that both drought-tolerant and drought-susceptible cultivars suffer, in varying degrees, from adverse weather and changes in climate. However, the adverse effects of dry environments on drought-tolerant cultivars would be significantly less than those on drought-susceptible ones. The delay in developing plant water deficit in drought-tolerant cultivars is attributable to the moisture conservation through efficient stomatal control and uptake of soil moisture at lower water potentials by adjusting plant water potentials. Hence, drought-tolerant cultivars can be selected through planned breeding and selection programs and should be introduced to the growers for infilling, replanting, and new planting. Replacing tea clones vulnerable to dry environments with those more adaptable to varied climates, especially droughts, is the most cost-effective and environmentally friendly adaptive measure for the tea industry to minimize the adverse effects of climate change.

Acknowledgments

I wish to thank R. Fordham (Wye College, University of London, London, United Kingdom), A. Anandacoomaraswamy (Tea Research Institute, Sri Lanka), and J. Ratnasiri (Coordinator, U.S. Country Studies Program, Sri Lanka) for useful discussions and critical comments.

References

Asian Development Bank, 1994, "Climate Change in Asia," *Regional Study on Global Environmental Issues*, Manila, Philippines.

Callander, B.A., and T. Woodhead, 1981, "Canopy Conductance of Estate Tea in Kenya," *Agricultural Meteorology* 23:151-167.

Carr, M.K.V., 1972, "The Climatic Requirements of the Tea Plant: A Review," *Experimental Agriculture* 8:1-14.

Central Bank, 1992, *Annual Report*, Central Bank of Sri Lanka.

Chaney, W.R., 1981, "Woody Plant Communities," in T.T. Kozlowski (ed.), *Water Deficit and Plant Growth*, Vol. VI, pp. 1-47, Academic Press, Inc., London, United Kingdom.

Handique, A.C., and L. Manivel, 1986, "Shoot Water Potential in Tea, II, Screening Toklai Cultivars for Drought Tolerance," *Two and a Bud* 33:39-42.

Hurd, E.A., 1976, "Plant Breeding for Drought Resistance," in T.T. Kozlowski (ed.), *Water Deficit and Plant Growth*, Vol. IV, pp. 317-353, Academic Press, Inc., London, United Kingdom.

Jackson, I.J., 1989, *Climate, Water and Agriculture in the Tropics*, 2nd ed., Longman, London, United Kingdom.

Jones, H.G., 1990, "Physiological Aspects of the Control of Water Status in Horticultural Crops," *Hortscience* 25:19-26.

Kaufman, M.R., 1981, "Water Relations during Drought," in C.G. Paleg and D. Aspinall (eds.), *The Physiology and Biochemistry of Drought Resistance in Plants*, pp. 55-70, Academic Press, Inc., Sydney, Australia.

Nawaratne, D.K., 1992, "A Survey of Drought Damage in the Ratnapura Region in 1992," *Tea Bulletin* 12:34-42.

Othieno, C.O., 1978, "Supplementary Irrigation of Young Clonal Tea in Kenya. II. Internal Water Status," *Experimental Agriculture* 14:309-316.

Planchon, C., 1987, "Drought Avoidance and Drought Resistance in Crop Plants: Inter and Intra Specific Variability," in L. Monti and E. Proceddu (eds.), *Drought Resistance in Plants: Physiological and Genetic Aspects*, pp. 79-94, Commission of the European Communities, Luxembourg.

Reeve, M.J., and A.D. Carter, 1991, "Water Release Characteristics," in K.A. Smith and C.E. Mullins (eds.), *Soil Analysis: Physical Methods*, pp. 111-116, Marcel Dekker, Inc., New York, N.Y., USA.

Ryadnova, I.M., and T.A. Lebedeva, 1971, "The Indices of Drought Resistance in Some Peach Cultivars," *Trudy Krymskaya opytno Seleltsionnayam Stantsiya VNII Rastenievodstva* 6:141-150.

Sandanam, S., G.W. Gee, and R.B. Mapa, 1981, "Leaf-Water Diffusion Resistance in Clonal Tea: Effect of Water Stress, Leaf Age and Clones," *Annals of Botany* 47:339-349.

Scholander, P.F., H.T. Hammel, E.D. Bradstreet, and E.A. Hemmingsen, 1965, "Sap Pressure in Vascular Plants," *Science* 148:339-346.

Slatyer, R.O., 1967, *Plant Water Relationships*, Academic Press, Inc., New York, N.Y., USA.

Squire, G.R., 1979, "Weather Physiology and Seasonality of Tea (*Camellia sinensis*) Yields in Malawi," *Experimental Agriculture* 15:321-330.

Wadasinghe, G., and M.A. Wijeratne, 1989, "Effects of Potassium on Water Use Efficiency of Young Tea (*Camellia sinensis L.*)," *Journal of Soil Science Society of Sri Lanka* 6:46-55.

Watson, M., 1986, "Soil and Climatic Requirements," in P. Sivapalan, S. Kulasegaram, and A. Kathiravetpiliai (eds.), *Handbook on Tea*, pp. 3-5, Tea Research Institute of Sri Lanka, Talawakelle, Sri Lanka.

Yatawatte, S.T., 1992, "A Survey of Drought Damage in the Mid Country in 1992," *Tea Bulletin* 2:8-27.

Climate Change Impacts on Agriculture and Global Food Production: Options for Adaptive Strategies

M.I. Budyko and G.V. Menzhulin
State Hydrological Institute,
23 Second Line
St. Petersburg, 199053, Russian Federation

Abstract

The possible effects of global warming on agriculture may have economic, ecological, and social repercussions. Major factors in analyzing these effects include expected population growth, possible increases in agricultural production, improved agrotechnology, and changes in crop productivity. Favorable climate changes, especially the effect of increased carbon dioxide (CO_2)on crop productivity, can help to solve the food supply problem expected in the 21st Century. Techniques related to agrotechnology, climate, and crop productivity contribute to a balanced strategy for adapting to future global changes.

Introduction

The effects of global warming on agriculture will determine both the ecological and economic effects of climate change and the way the world will adapt. Agriculture will probably suffer the greatest consequences of global climate change, both favorable and disastrous. Increased agricultural production is of primary importance in solving global food problems. Estimates of global climate effects on agriculture deserve particular attention (Smith and Tirpak 1989; Tegart et al. 1990; Reilly 1994; Rosenzweig and Parry 1994).

According to estimates for the United States (Thompson 1976), almost the half the total losses in all economic sectors are caused by unfavorable weather conditions; therefore, reliable estimates of the agricultural consequences of climatic change are essential. However, many estimates of the sensitivity of agriculture to climatic factors reflect only the average effect of meteorologically anomalous years. Global warming, which would change not only the frequency of anomalous phenomena but also the climatic norms, could have considerably greater consequences. However, the relative sensitivity of various economic sectors to climate change would remain the same.

This study of the effect of expected climate changes on future global food production analyzes the following factors:

- Potential population growth rates,
- Development of global agriculture by cultivating all available arable land,
- Increased crop yield by using modern intense agrotechnology,
- Effects of increased carbon dioxide (CO_2) concentration on crop productivity, and
- Changes in crop productivity due to global warming.

World Population Growth

On the basis of an analysis of current and future demographic situations, the global population will be 8.3-10 billion in 2040 (Zachariah and Vu 1988; *World Population Prospects 1988*). Estimates for 2040 are used in this study because the atmospheric CO_2 concentration is expected to double by then.

Prospects of Expanding Agricultural Lands

Is it possible to increase crop area by cultivating all the land that meets current climatic requirements? About 10 years ago, some specialists suggested this possibility for countries in Africa and South America. However, it proved unrealistic (*Agriculture for Global Climate Change* 1992). According to statistical data from the Food and Agriculture Organization ([FAO] 1961-1993) (Figure 1), the total area used for growing 13 major food crops increased at the end of the 1960s, but it stabilized at about 800 million ha and decreased slightly during the last 25 years. The per capita amount of land used for these crops decreased from 0.204 ha in 1960 to 0.142 ha in 1993. Therefore, the overall increase in global crop production over the last 30 years resulted from an increase in yield due to improved agrotechnology. The mean value of this increase has been about 45% (75% for wheat, 60% for corn, and 55% for soybeans and rice). Because of population growth, however, per capita production of these crops increased by only 8% (Figure 2).

One reason the amount of arable land has not increased is the poor technological base of agriculture in some countries, which makes it unprofitable to cultivate crops on large, low-productivity lands. It is more profitable to increase food production through intensive agrotechnology, particularly when crop yields are low (i.e., when productivity is more responsive to the use of fertilizers, mechanization, new regional crop species, and other techniques).

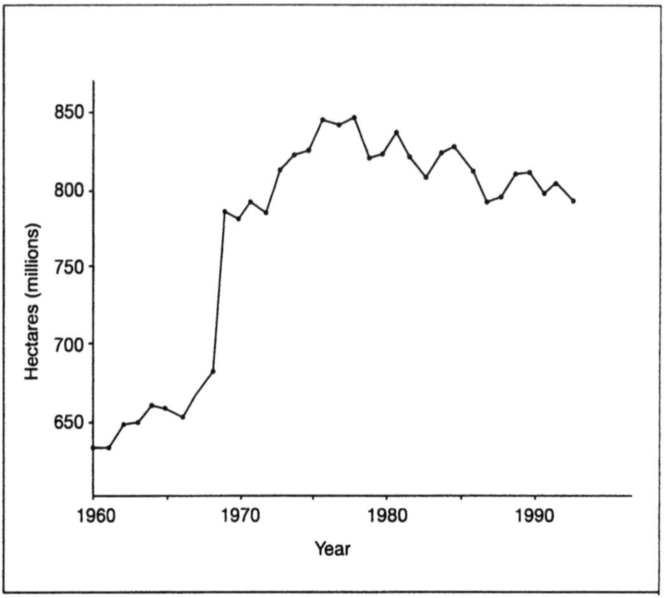

Figure 1. World Land Use for the Production of 13 Food Crops

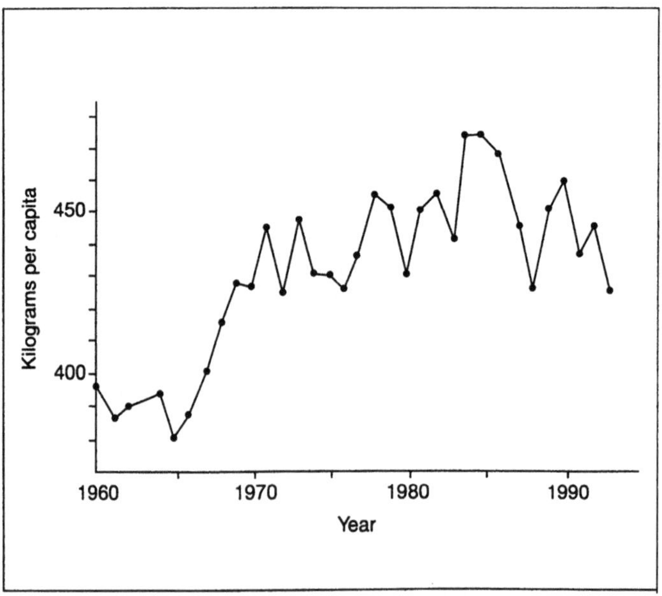

Figure 2. World Per Capita Production of 13 Principal Food Crops

What are the prospects for using lands whose temperature and precipitation regime could change favorably because of future climatic changes? Global warming would significantly affect on high-latitude temperature. A longer warm season would make it possible to expand agriculture into northern regions. It may be possible to use intensive agricultural techniques in such areas; however, the problem of preserving boreal forests and the forestry industry could limit the northward advancement of agriculture.

Agrotechnological Factors

Is it possible to increase crop productivity solely through advances in agrotechnology? The answer depends on how sensitive crop productivity is to technological factors, such as fertilizers, new highly productive species, and irrigation. The answer can be obtained by using a simple model based on two assumptions (Menzhulin 1995):

- Growth rates of agricultural expenditures will remain the same on average.
- The current highest yield for each crop is its physiological limit.

In determining the necessary level of crop productivity for 2040, it is assumed that the current mean value of per capita food production (about 440 kg for the principal food crops [Figure 2]) should be maintained. The mean yield per hectare should increase from the current 3.10 mg to about 5 t for a population of 8.3 billion and about 6 t for a population of 10 billion. This model is quite reliable, but it is possible that the world community will spend ever-increasing resources on agriculture, or that current yields are not the physiological productivity limits for the main crops.

The model calculations show that mean productivity for some crops, primarily low productivity crops, is expected to increase considerably (even double) in the next five decades. Among these crops are oats, sorghum, millet, rye, sunflowers, soybeans, and groundnuts. Five crops (wheat, rice, corn, potatoes, and barley) provide more than 80% of total food production. Estimated productivity increases for these crops are much lower, about 50% on average (43% for rice, 53% for barley).

The effects of agrotechnological factors on crop yields can be determined from an empirical analysis of historical data on mineral fertilizer consumption for different crops in different countries (Figure 3). The effect of fertilizer is limited by crop biological productivity, as are other agrotechnological factors (Menzhulin and Nikolaev 1987).

Estimates show that it will not be possible to produce enough food solely on the basis of improved agrotechnology, if the forecasted population growth for 2040 is accurate. A comparison of these estimates with data on the current crop productivity in countries with highly developed agricultural sectors supports this

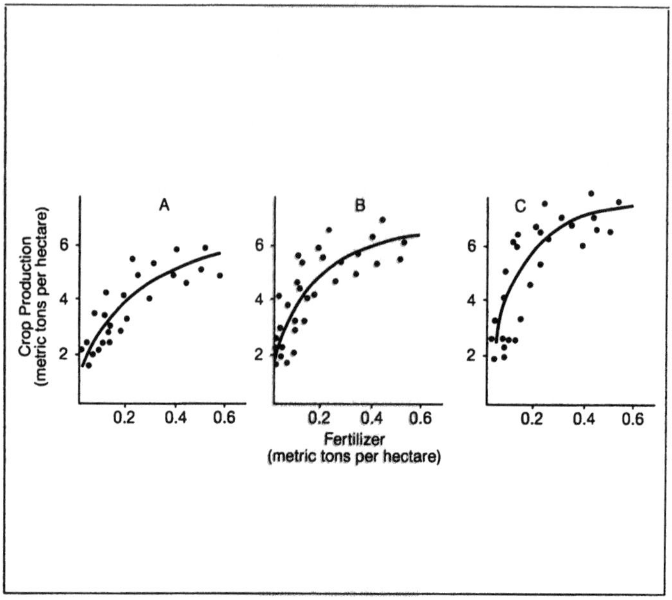

Figure 3. Effect of Mineral Fertilizers on (a) Wheat, (b) Rice, and (c) Maize Yields

conclusion (Table 1). Only five countries have achieved corn yields of 6 Mg ha^{-1}; only the Republic of Korea, Japan, and Spain have achieved equivalent yields of rice; and only a few Western European countries have achieved this yield for soft, highly productive forage wheat species. The productivity of other crops is much lower in many countries.

A proportional increase in the main food crops (wheat, rice, and corn) through agrotechnology would not be easy to obtain. All countries would have to achieve crop yields close to the maximums achieved by countries with favorable climates and highly developed agricultural sectors. It would be easier to increase the yields of currently less productive crops.

Direct Effect of Increased CO_2 on Crop Productivity

What will solve the problem of future food supply? The anthropogenic increase in atmospheric CO_2 concentration and related global climate change may help. Experimental studies show increased productivity for crops grown under elevated CO_2 levels (Rose 1989; Kimball 1993; Rogers and Dahlman 1993). According to estimates based on experiments with a doubled CO_2 concentration, the productivity of C_3 plants (wheat, rice, barley, and oats) was 33% higher (Kimball 1983). The leaf surface area and thickness increased in experimental

Table 1. Mean Wheat, Rice, and Maize Yields (Mg ha^{-1}, averaged for 1979-1993)

Wheat		Rice		Maize	
Algeria	0.64	Campuchea		Mozambique	0.52
Tunisia	0.90	Tanzania	1.45	Morocco	0.73
Iran	1.08	Brazil (9)[a]	5.59	Zaire	0.79
Brazil	1.09	Laos	1.68	Philippines	1.00
Ethiopia	1.10	Madagascar	1.82	India	1.15
Portugal	1.15	Nepal	1.83	Nigeria	1.15
South Africa	1.22	Thailand (5)	1.96	Nicaragua	1.16
Uruguay	1.31	Bangladesh (4)	2.04	Pakistan	1.27
Australia (8)	1.35	India (2)	2.07	Colombia	1.38
Pakistan (9)	1.55	Philippines (10)	2.31	Guatemala	1.41
Former USSR (1)	1.56	Vietnam (7)	2.38	Kenya	1.49
India (4)	1.68	Pakistan	2.49	Portugal	1.52
Kenya	1.73	Burma (8)	2.62	Indonesia	1.52
Argentina	1.76	Sri Lanka	2.64	Mexico (5)	1.53
Canada (6)	1.86	Romania	2.79	Venezuela	1.64
Turkey (7)	1.87	Malaysia	2.83	Brazil (3)	1.69
Spain	1.99	Guyana	2.93	South Africa (9)	1.71
China (2)	2.22	Venezuela	2.95	Zimbabwe	1.87
United States (3)	2.3[b]	Iraq (2)	2.96	Peru	1.88
Greece	2.54	Ecuador	2.99	Thailand	2.26
Romania	2.73	Mexico	3.19	Turkey	2.72
Italy	2.74	Argentina	3.59[b]	Australia	2.86
Japan	3.15	Iran	3.70	Argentina (8)	3.18
Poland	3.19	Indonesia (3)	3.72	China (2)	3.33[b]
Egypt	3.52	Dominican	3.75	Egypt	4.23
Bulgaria	3.79	Cuba	4.17	Yugoslavia (7)	4.25
Mexico	3.88	Portugal	4.31	Bulgaria	4.34
New Zealand	4.01	Peru	4.48	Romania (4)	4.35
Austria	4.21	China (1)	4.50	Chile	4.56
Hungary	4.34	Colombia	4.50	Czech	4.83
Czech Republic	4.37	Italy	5.40	Spain	5.26
Switzerland	4.80	United States	5.46	Hungary (10)	5.57
Sweden	4.87	Greece	5.52	France (6)	5.70
France (5)	5.12	Egypt	5.53	Canada	5.80
Be-Ne-Lux	5.14	Japan (6)	5.96	Germany	6.05
Germany	5.38	Australia	6.09	United States (1)	6.34
Denmark	5.75	South Korea	6.10	Italy	6.85
United Kingdom (10)	5.78	Spain	6.14	Austria	7.25
Ireland	5.84[c]		6.60[c]	Greece	7.39[c]

[a] Number in parentheses indicates position among the major producers of this crop.

[b] Current mean world yield of this crop.

[c] Future mean world crop yield needed by 2040 for a population of 10 billion.

plants (Chaudhuri et al. 1987; Cure et al. 1987). The photosynthetic productivity of C_4 plants (maize and sorghum) also increased with doubled CO_2, although transpiration was also greatly affected in some plants (Morison and Gifford 1983, 1984; Rogers et al. 1983; King and Greer 1986).

Other experiments studied the feedback that inhibits the positive effects of increased CO_2 on plant productivity (DeLucia et al. 1985; Sasek et al. 1985; Idso et al. 1987a). With increased CO_2, maximum productivity was seen in plants that transported photosynthesis products from the chloroplasts where they were formed to organs where they accumulated, and in plants genetically predisposed to the formation of new sink organs (Rutty and Huber 1983; Stitt et al. 1984a,b).

An increased CO_2 concentration can also change other physiological functions of plants, including transpiration. Experiments with C_3 and C_4 plants show that increased atmospheric CO_2 increases the coefficient of transpiration effectiveness (Rogers et al. 1983; Idso et al. 1987b; Leuning et al. 1993). One practical effect of this phenomenon may be to increase the resistance to aridity in some crops (Jones 1985; Kimball et al. 1986; Schonfeld 1987).

Maximum estimated C_3 crop yields increased with proper mineral nutrition, agrotechnology, and environmental conditions. For some species, these yield increases could be close to 100% if the atmospheric CO_2 concentration doubled (Kimball et al. 1992).

Climate Change Scenarios

The results of climate change calculations obtained with General Circulation Models (GCMs) can be used for different purposes (Houghton et al. 1990, 1992). Regional temperature changes estimated on the basis of future global climate scenarios generated by different GCMs correlate well (Boer et al. 1992; Budyko et al. 1992). However, estimates of regional changes in precipitation obtained by different models continue to be a weak point. Because of the significant uncertainties in current climate change scenarios, other information must be used in impact and adaptation assessments. Modern paleoclimatic reconstructions can provide such information (Budyko and Izrael 1987; MacCracken et al. 1990; Budyko et al. 1992).

Crop Models

A new class of crop models can be used for assessing the effect of climate change on crop yields and substantiating adaptation strategies. The CERES-Wheat Model (Ritchie and Otter 1985; Tsuji and Balas 1993) was used for the results presented in Figures 4 and 5 and Table 2. The model was validated for

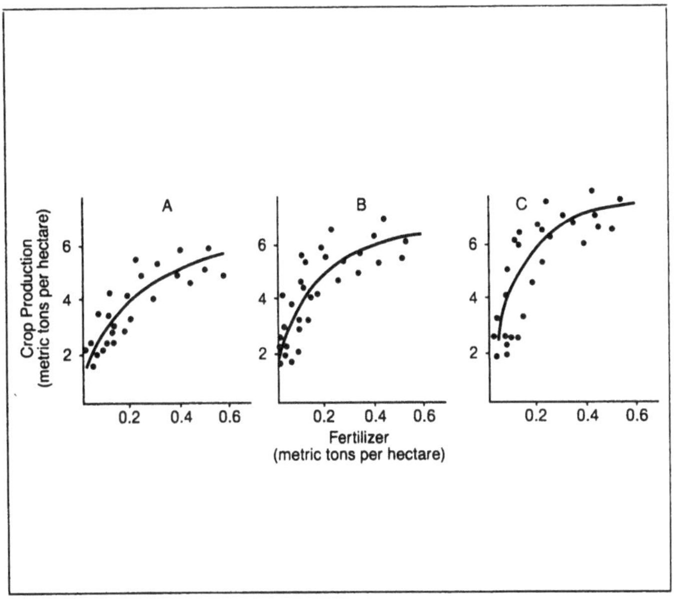

Figure 4. Changes in Spring and Winter Yields in the Former Soviet Union Predicted with GISS Climate Change Scenario for 2040 (percentage of current values)

Table 2. Aggregated Estimates of Changes in Former Soviet Union Wheat Production According to GCM-Generated and Paleoanalog Climate Change Scenarios (% of current production)

Scenario	Year	CO_2 Concentration (ppm)	Wheat Spring	Wheat Winter
GISS transient	2010	405	+ 8	+16
GISS transient	2030	460	+12	+24
GISS transient	2050	530	+18	+38
GISS equilibrium	2050	550	+21	+41
GFDL equilibrium	2050	550	- 4	+12
UKMO equilibrium	2050	550	-18	+ 9
Pliocene optimum	2040-2050	550	+34	+51

the environmental conditions of the former Soviet Union and the wheat cultivars grown in that region (Menzhulin et al. 1994).

These crop models require many input parameters that may be difficult to obtain. Therefore, it is often necessary to use artificial expedients (input data

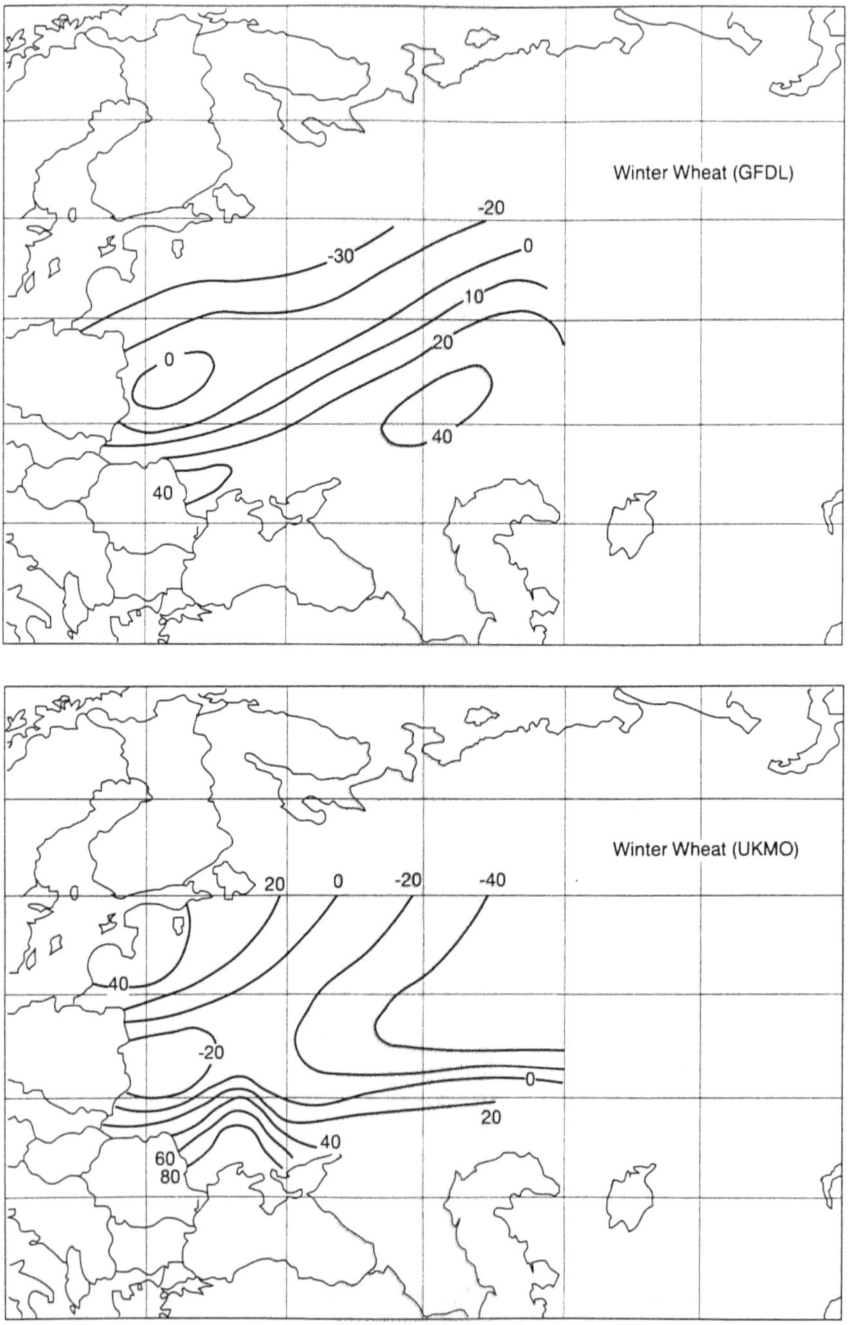

Figure 5. Changes in Winter Wheat Yields in the Former Soviet Union Predicted with GFDL and UKMO Climate Change Scenarios for 2040 (percentage of current values)

generation, extrapolation of soil parameters and genetic coefficients). For this reason, special parameterized quasi-static crop models are more convenient for large-scale calculations. In these models, agroclimatic indices highly correlated with crop yields are used as output predicted parameters. Five versions of these models were used by the Russian State Hydrological Institute (Menzhulin and Savvateyev 1980; Menzhulin et al. 1983, 1987a,b; Menzhulin 1984; Menzhulin and Nikolaev 1987; Budyko et al. 1994). In general, the models use the same input parameters as the IBSNAT crop models, but they are averaged over several growth seasons. The atmospheric parameters include temperature, precipitation, air humidity, solar radiation, and CO_2 concentration; the soil parameters include the initial water and nitrogen contents in the upper soil layer. Required physiological data include the light and CO_2 dependences of the photosynthesis and photorespiration rates and the temperature dependence of dark respiration. The parameterized model was able to estimate the effect of increased use of nitrogen fertilizers.

Table 3 presents the results of calculations of climate change impact on wheat, barley, oats, rice, and maize productivity indices in different agricultural regions of the world.

Results and Discussion

Climate change scenarios were generated by equilibrium GCMs developed at the Goddard Institute of Space Studies (GISS), Geophysical Fluid Dynamics Laboratory (GFDL), and United Kingdom Meteorological Office (UKMO) as well as the GISS transient scenario (Hansen at al. 1983, 1988; Manabe and Wetherald 1987; Wilson and Mitchell 1987). A Pliocene optimum paleoanalog scenario, corresponding to 3-4°C of global warming by 2040, was also used. Other analog scenarios were developed by paleoclimatologists from the Russian State Hydrological Institute (Borzenkova 1987, 1990; Budyko and Izrael 1987; MacCracken et al. 1990; Zubakov and Borzenkova 1990; Borzenkova et al. 1992; Budyko et al. 1992).

A noticeable discrepancy exists among the estimates obtained by the different GCM scenarios. For example, there are three patterns (Figures 4 and 5) for winter wheat in the former Soviet Union. Even some consistency is found among these patterns, it is difficult to determine the future productivity distribution of this area. Nevertheless, on the basis of the calculations, it can be concluded that the expected climate change will have a positive effect on winter wheat production in Russia and other states of the former Soviet Union. This conclusion is substantiated in Table 2. The results of the GISS equilibrium scenario show a difference of 20% between estimates for winter and spring wheat. Approximately the same difference is characteristic for other scenarios for 2040 (GISS transient, GFDL, and UKMO equilibrium) For the Pliocene optimum

Table 3. Increase in Productivity of Grain Crops by 2040 Due to Increased CO_2 Concentration (% of current values)

Region	Averaged over Region	Wheat	Barley	Oats	Rice	Maize
Europe	33	35.8	37.2	39.5	56.3	36.3
Former USSR	14	43.8	48.6	51.7	64.1	71.3
North America	6	33.3	36.0	39.2	35.0	27.3
Latin America	26	52.1	57.3	60.8	59.8	53.1
Africa	43	52.4	66.5	64.8	36.6	55.1
Far East	14	48.1	49.5	55.6	60.8	50.7
Near and Middle East (including India)	15	62.2	65.4	65.5	36.2	60.7
Australia	2	49.0	54.0	57.0	61.0	52.8

scenario, productivity will increase more for winter wheat than for spring wheat. One way to adapt grain production to possible global warming is to plant winter wheat in areas now used for spring wheat.

Table 2 shows that estimates of changes in wheat yields based on the Pliocene optimum scenario substantially exceed the model estimates, even in the case of the favorable GISS equilibrium model. According to the paleoanalog scenario, the forecasted changes in precipitation through the middle of the 21st Century are favorable for agriculture. Calculations were carried out for countries from eight large geographic regions of the world (Europe, former Soviet Union, North America, Latin America, Africa, Far East, Near and Middle East including India, and Australia) for five principal grain crops. The results are presented in Table 3.

The calculations show that under a CO_2 concentration increase to 550 ppm and temperature and precipitation changes corresponding to the Pliocene optimum scenario, the productivity of C_3 crops (wheat, barley, oats, and rice) can increase up to 30-65% of current values. The lowest increase in total wheat production occurs in North America, the highest in Africa. For maize, the highest increase is expected in the former Soviet Union because the current mean corn yields in Russia, Ukraine, and Moldova, used as the reference level, are low. The low increase in corn yields for North America is partly a result of the high current yield.

These calculations show that the increased CO_2 concentration and global warming expected by 2040 will improve moisture conditions in basic crop regions. According to the paleoanalog scenario, the mid-latitude regions of Eurasia and North America and the grain belts of Australia and Argentina will be the most productive regions.

Calculations using information on population growth show a disproportionate increase in per capita food production in developed countries in the temperate latitudes of the Northern Hemisphere as compared with developing countries. This factor will influence the global adaptive strategy for solving the international food problem. The world community should attempt to decrease this disproportion in food production by means of international trade and aid to developing counties.

Conclusion

An increased CO_2 concentration and global warming should be considered a positive influence on the food supply for a growing world population. World population has reached the point where current agrotechnology cannot ensure enough food for everyone. The favorable effects of an increased CO_2 concentration and global warming may enhance the food supply during the next five decades, after which the global population should stabilize. However, even though the agricultural effects will be favorable, adaptive strategies cannot be delayed. Without well-developed adaptive strategies tailored to the economical, ecological, and climatic conditions of each region, these auspicious changes in agroclimatic potential will not be realized.

References

Agriculture for Global Climate Change, 1992, Council for Agricultural Science and Technology, Ames, Iowa, USA.

Boer, G.J., K. Arpe, M. Blackburn, M. Deque, W.L. Gates, T.L. Hart, H. Le Treut, E. Roecker, D.A. Sheinin, I. Simmonds, R.N.B. Smith, T. Tokioka, R.T. Wetherald, and D. Williamson, 1992, "Some Results from an Intercomparison of the Climates Simulated by 14 Atmospheric General Circulation Models," *Journal of Geophysical Research* 97:12771-12786.

Borzenkova, I.I., 1987, "The Features of the Moisture Conditions on the Land of the Northern Hemisphere during the Different Geological Epochs," *Soviet Meteorology and Hydrology* (10):53-61.

Borzenkova, I.I., 1990, "Climate Changes in the Late Warm - Holocene (the Last 20 ka)," *Trudy GGI* (348):70-7 (in Russian).

Borzenkova, I.I., V.A. Zubakov, and A.G. Lapenis, 1992, "Global Climate Changes During the Warm Epochs of the Past," *Soviet Meteorology and Hydrology.*

Budyko, M.I., and Yu.A. Izrael (eds.), 1987, *Anthropogenic Climatic Changes*, Gidrometeoizdat, Leningrad (in Russian) (English translation: Arizona University Press, 1990, Tucson, Arizona, USA).

Budyko, M.I., I.I. Borzenkova, G.V. Menzhulin, and K.I. Selyakov, 1992, "The Forthcoming Regional Climate Changes," *Seriya Geograph.* (4):36-52.

Budyko, M.I., I.I. Borzenkova, G.V. Menzhulin, and I.A. Shiklomanov, 1994, *Cambios Antropogenicos del Clima en America del Sur*, Series No. 19, Academia Nacional de Agronomia y Veterinaria, Buenos Aires, Argentina.

Chaudhuri, U.N., R.B. Burnett, E.T. Kanemasu, and M.B. Kirkham, 1987, *Effect of Elevated Levels of CO_2 on Winter Wheat Under Two Moisture Regimes*, No. 040, Series: Response of Vegetation to Carbon Dioxide, U.S. Departments of Energy and Agriculture., Washington, D.C., USA.

Cure, J.D., T.W. Rutty, and D.W. Israel, 1987, "Assimilate Utilization in the Leaf Canopy and Whole Plant Growth of Soybeans during Acclimating to Elevated CO_2," *Bot. Gaz.* 148:67-72.

DeLucia, E.H., T.W. Sasek, and B.R. Strain, 1985, "Photosynthetic Inhibition after Long-Term Exposure to Elevated Levels of Atmospheric Carbon Dioxide," *Photosynt. Res.* 7:175-184.

FAO, 1961-1993, *Monthly Bulletin of Agricultural Economic and Statistics*, U.N. Food and Agricultural Organization.

Hansen, J., G. Russell, D. Ring, P. Stone, A. Lacis, S. Lebedeff, R. Ruedy, and L. Travis, 1983, "Efficient Three-Dimensional Global Model for Climate Studies: Models 1 and 2," *Monthly Weather Review* 3(4):609-662.

Hansen, J., I. Fung, A. Lacis, D. Ring, S. Lebedeff, R. Ruedy, and G. Russell, 1988, "Global Climate Changes as Forecasted by Goddard Institute for Space Studies, Three-Dimensional Model," *Journal of Geophysical Research* 92: 9341-9364.

Houghton, J.T., G.J. Jenkins, and J.J. Ephraums (eds.), 1990, *The IPCC Scientific Assessment*, Cambridge University Press, Cambridge, United Kingdom.

Houghton, J.T., B.A. Callander, and S.K. Varney, 1992, Climate Change 1992, *The Supplementary Report to the IPCC Scientific Assessment*, Cambridge University Press, Cambridge, United Kingdom.

Idso, S.B., B.A. Kimball, M.G. Anderson, and J.R. Mauney, 1987a, "Effects of Atmospheric CO_2 Enrichment on Plant Growth: The Interactive Role of Air Temperature," *Agric. Ecosystems and Envir.* 20:1-10.

Idso, S.B., B.A. Kimball, and J.R. Mauney, 1987b, "Atmospheric Carbon Dioxide Enrichment Effects on Cotton Midday Foliage Temperature: Implication for Plant Water Use and Crop Yield," *Agron. J.* 79: 667-672.

Jones, P.H., L.H. Allen, Jr., and J.W. Jones, 1985, "Responses of Soybean Canopy Photosynthesis and Transpiration to Whole-Day Temperature Changes in Different CO_2 Environments," *Agron. J.* 77:242-249.

Kimball, B.A., 1983, "Carbon Dioxide and Agricultural Yield: An Assemblage and Analysis of 430 Prior Observations," *Agron J.* 75: 779-788.

Kimball, B.A., 1993, "Ecology of Crops in Changing CO_2 Concentration," *Journal of Agricultural Meteorology* 48:559-567.

Kimball, B.A., J.R. Mayney, J.W. Radin, F.S. Nakayama, S.B. Idso, D.L. Hendrix, D.H. Akey, S.G. Allen, M.G. Anderson, and W. Hartung, 1986, "Effects of Increasing Atmospheric CO_2 on the Yield and Water Use of Crops," No. 039, Series: Response of Vegetation to Carbon Dioxide, U.S. Departments of Energy and Agriculture, Washington, D.C., USA.

Kimball, B.A., J.R. Mauney, and E.A. Lakatos, 1992, *Carbon Dioxide Enrichment: Data on the Response of Cotton to Varying CO_2, Irrigation, and Nitrogen*, CDIAC-44, Carbon Dioxide Information Analysis Center, Oak Ridge National Laboratory, Oak Ridge, Tenn., USA.

King, K.M., and D.H. Greer, 1986, "Effects of Carbon Dioxide Enrichment and Soil Water on Maize," *Agron. J.* 78:515-521.

Leuning, R., Y.P. Wang, D. de Pury, O.T. Denmead, F.X. Dunin, A.G. Gordon, S. Nonhebel, and J. Goudriaan, 1993, "Growth and Water Use of Wheat under Present and Future Levels of CO_2," *Journal of Agricultural Meteorology* 48:807-810.

MacCracken, M., A.D. Hecht, M.I. Budyko, and Yu. A. Izrael (eds.), 1990, *Prospects for Future Climate: A Special US/USSR Report on Climate and Climate Change*, Lewis Publishers.

Manabe, S., and R.T. Wetherald, 1987, "Large Scale Changes in Soil Wetness Induced by the Increase in Carbon Dioxide," *Journal of Atmospheric Sciences* 44:1211-1235.

Menzhulin, G.V., 1984, "The Influence of Contemporary Climate Changes and CO_2 Content Growth on the Crops Productivity," *Soviet Meteorology and Hydrology* (4):95-101.

Menzhulin, G.V., 1995, "Forecasting of Regional Climate Changes and Estimating of their Principal Consequences," in *Climate Change Assessments in the Commonwealth of Independent States*, U.N. Environmental Programme (in press).

Menzhulin, G.V., and Nikolaev, M.V., 1987, "Methods for Study Year-to-Year Variability of Cereal Crop Yields," *Trans. State Hydrol. Inst.* 327:113-131 (in Russian).

Menzhulin, G.V., and Savvateyev, S.P., 1980, "The Impact of Expected Climate Change on Crop Productivity," in *The Problems of Atmospheric Carbon Dioxide, Transactions of Soviet-American Symposium*, pp. 186-197, Gidrometeoizdat, Leningrad, USSR (former Soviet Union) (in Russian).

Menzhulin, G.V., L.A. Koval, and S.P. Savvateyev, 1983, "The Principles on Constructions of Crop Productivity Parameterized Models," *Trans. State Hydrol. Inst.* 280:119-129 (in Russian).

Menzhulin, G.V., S.P. Savvateyev, L.A. Koval, and M.V. Nikolaev, 1987a, "The Agroclimatic Consequences of Modern Global Climate Changes," in *The Problems of Agroclimatic Providing for the USSR Foodstuff Programme*, pp. 72-81, Gidrometeoizdat, Leningrad, USSR (former Soviet Union).

Menzhulin, G.V., S.P. Savvateyev, L.A. Koval, and M.V. Nikolaev, 1987b, "Estimating Climatic Changes Agricultural Consequences: Scenario for North America," *Trans. State Hydrol. Inst.* 327:132-146 (in Russian).

Menzhulin, G.V., L.A. Koval, and A.L. Badenko, 1994, "Potential Effects of Global Warming and Carbon Dioxide on Wheat Production in the Former Soviet Union," in *Implication of Climate Change for International Agriculture: Crop Modeling Study*, C. Rosenzweig and A. Iglesias (eds.), EPA 230-B-94-003, U.S. Environmental Protection Agency, Office of Policy, Planning, and Evaluation, Washington, D.C., USA.

Morison, J.I.L., and R.M. Gifford, 1983, "Stomatal Sensitivity to Carbon Dioxide and Humidity: A Comparison of Two C_3 and C_4 Grass Species," *Plant Physiol.* 71:789-796.

Morison, J.I.L., and R.M. Gifford, 1984, "Plant Growth and Water Use with Limited Water Supply in High CO_2 Concentration," *Austral. J. Plant Physiol.* 11:361-374.

Reilly, J., 1994, "Crops and Climate Change," 1994, *Monthly Nature* 2(1):31-32.

Ritchie, J.T., and S. Otter, 1985, "Description and Performance of CERES-Wheat: A User-Oriented Wheat Yield Model," in *ARS Wheat Yield Project*, W.O. Wilis (ed.), ARS-38, pp. 159-175, U.S. Department of Agriculture, Agricultural Research Service, Washington, D.C., USA.

Rogers, H.H., G.E. Bingham, J.D. Cure, J.M. Smith, and R.A. Surano, 1983, "Responses of Selected Plant Species to Elevated Carbon Dioxide in the Field," *Journal of Environmental Quality*. 12:569-574.

Rogers, H.H., and R. Dahlman, 1993, "Crop Responses to CO_2 Enrichment," *Vegetatio* 104:117-131.

Rose, E., 1989, "Direct (Physiological) Effects of Increasing CO_2 on Crop Plants and Their Interactions with Indirect (Climatic) Effects," in J.B. Smith and D.A. Tirpak (eds.), *The Potential Effects of Global Climate Change on the United States, Appendix C: Agriculture*, Vol. 2, U.S. Environmental Protection Agency, Office of Policy, Planning, and Evaluation, Washington, D.C., USA.

Rosenzweig, C., and M. Parry, 1994, "Potential Impact of Climate Change on World Food Supply," *Monthly Nature* 2(1):65-70.

Rutty, T.W., and S.C. Huber, 1983, "Changes in Starch Formation and Activities of Sucrose-Phosphate Syntheses and Cytoplasmic Fructose-1,6-bisphosphatase in Response to Source-Sink Alteration," *Plant Physiology*. 72:474-480.

Sasek, T.W., E.H. DeLucia, and B.R. Strain, 1985, "Reversibility of Photosynthetic Inhibition in Cotton after Long-term Exposure to Elevated CO_2 Concentration," *Plant Physiol.* 78:619-622.

Schonfeld, M., 1987, *Drought Sensitivity in Winter Wheat Population under Field Condition and Elevated Atmospheric CO_2*, Doctoral Thesis, Oklahoma State University, Tulsa, Okla., USA.

Smith, J.B., and D.A. Tirpak (eds.), 1989, *The Potential Effects of Global Climate Change on the United States, Appendix C: Agriculture*, Vol. 2, U.S. Environmental Protection Agency, Office of Policy, Planning, and Evaluation Washington, D.C., USA.

Stitt, M., B. Herzog, and H.W. Heldt, 1984a, "Control of Photosynthetic Sucrose Synthesis by Fructose 2.6-bisphosphate. I. Coordination of CO_2 Fixation and Sucrose Synthesis," *Plant Physiology*. 75:548-553.

Stitt, M., B. Kurzel, and H.W. Heldt, 1984b, "Control of Photosynthetic Sucrose Synthesis by Fructose 2,6-bisphosphate. II. Partitioning between Sucrose and Starch," *Plant Physiol.* 75:554-560.

Tegart, W.J. McG., G.M. Shelton, and D.C. Griffiths (eds.), 1990, *The IPCC Impact Assessment*, Australian Government Publishing Service, Canberra, Australia.

Thompson, I., 1976, *Living with Climate Change: Phase II*, Symposium Report, Mitre Corporation.

Tsuji, G.Y., and S. Balas (eds.), 1993, *The IBSNAT Decade: Ten Years of Endeavor at the Frontier of Science and Technology*, International Benchmark Sites Network for Agrotechnology Transfer, Department of Agronomy and Soil Science, College of Tropical Agriculture and Human Resources, University of Hawaii, Honolulu, Hawaii, USA.

Wilson, C.A., and J.F.B. Mitchell, 1987, "A Doubled-CO_2 Climate Sensitivity Experiment with a Global Model Including a Simple Ocean," *Journal of Geophysical Research* 92:13315-13343.

World Population Prospects 1988, 1989, Population Studies No. 106, Department of International Economic and Social Affairs, United Nations, New York, N.Y., USA.

Zachariah, K.C., and M.T. Vu, 1988, *World Bank World Population Projection, 1987-1988: Short- and Long-Term Estimates*, Johns Hopkins University Press, Baltimore, Maryland, USA.

Zubakov, V.A., and I.I. Borzenkova, 1990, *Global Climate throughout the Cenozoic*, Elsevier Publishers, Amsterdam, the Netherlands.

3

WATER RESOURCES

Rapporteur's Statement

Ron Benioff
U.S. Country Studies Program
1000 Independence Ave., PO-63
Washington, DC, 20585, USA

Adaptation of water resource management is one of the major issues in dealing with climate change. Important issues have been identified and are discussed in this chapter. The following paragraphs highlight the issues discussed.

Suitability of Adaptation Assessment Methods for Water Resources. Several important considerations must be included in selecting and designing methods to evaluate climate change adaptation options for the water resources sector:

- In evaluating adaptative measures, countries should decide whether to (1) delay action until more information is available about impacts, (2) apply a "no regrets" approach, or (3) apply optimality rules to select measures that will help to cope with climate change impacts under different scenarios, while satisfying other water resource objectives.
- The evaluation process should foster communication and active participation of all persons and organizations affected.
- Assessing adaptive measures should be integrated with analyzing measures for interrelated systems (e.g., agriculture, coastal resources, forests, health).
- Evaluation of adaptive measures should consider all environmental, economic, and social costs and benefits.
- Analysis of adaptation options should address the uncertainty in the timing, magnitude, and nature of the potential impacts of climate change on water resources.

Identification and Evaluation of Adaptive Measures. Countries have applied various methods to identify and evaluate adaptive measures. Analytical methods range from brainstorming sessions with decision makers to the use of multi-criteria decision models and complex simulation models to evaluate the cost-effectiveness of alternative measures under different climate change scenarios. Most countries first evaluate adaptive measures for individual water resource systems (e.g., river basins, reservoirs and lakes, groundwater resources). The results of these system evaluations are then used to prepare adaptation plans and policies at the system, regional, and national levels. All assessment methods tend to follow the same basic steps:

- Characterize vulnerability to the impacts of climate change and contemporary climate variability and identify critical water resource management objectives.
- Select and design alternative adaptive measures.
- Select criteria for evaluating the degree to which the measures satisfy resource management objectives.
- Evaluate alternative measures according to each attribute and under various future scenarios, including climate change. This evaluation can use expert opinion, simulation models, multicriteria decision models, and other tools.
- Present results to key decision makers and affected persons/organizations to develop consensus.

Selection of Adaptive Measures. A wide range of potential adaptive measures can be identified. Issues that must be examined in selecting suitable measures include the following:

- Consider the need for anticipatory adaptation to prepare for potential impacts of climate change, including changes in the frequency and magnitude of extreme events.
- Develop climate change adaptive measures that can be integrated with the development of plans to address future economic and population growth.
- Ensure that adaptive measures are flexible and work under a wide range of possible impacts from climate change and current climatic variability.
- Refine strategies as scientific understanding improves.
- Plan for the likelihood that climate change could have greater impacts on rural areas and require relocation of rural populations.

Identification of Adaptation Priorities. Adaptation priorities vary from country to country, but several measures are generally effective in enhancing resiliency to the potential impacts of climate change:

- Developing comprehensive river basin or lake and reservoir management plans that address climate change along with future growth and other management challenges.
- Implementing water conservation measures.
- Decreasing current pollution of water resources.
- Developing new water supplies.
- Implementing combined use of groundwater and surface water.
- Increasing prices for water use to ensure full cost recovery.
- Increasing the recycling and reuse of wastewater.
- Enhancing capabilities to transfer water resources within or between basins.
- Improving protection against flooding.
- Adding marginal increases in storage capacity.
- Improving drought response planning and systems.
- Increasing public awareness and education about climate change issues.

Integration of Adaptive Measures. Adaptive measures need to be integrated with plans that address future growth and other water resource management challenges. On a practical level, effectively integrating these measures with other water resource management objectives requires that all water resource management plans address the potential impacts of climate change. In addition, assessments of new water resource projects and policies should also address climate change impacts to ensure that new projects or policies will improve resiliency to climate change. Because climate change could alter the frequency and severity of floods, droughts, and other extreme events, the national disaster planning process should also address climate change. Finally, pollution control programs and policies should reflect that climate change may alter future water availability and use.

Regional Cooperation. Opportunities and needs for improved regional cooperation in developing and evaluating water resource adaptive measures have been identified:

- Where river basins extend beyond national boundaries, an effective analysis of adaptation options may require regional studies and development of regional adaptation plans so that upstream/downstream users act together.
- Regional studies of vulnerability and adaptation measures are needed to compare results among countries and help to identify suitable, robust measures for a range of potential impacts and circumstances.
- Developing regional climate models and averaging impact assessments of results can help to improve the reliability of impact assessments, which may help decision makers feel confident in the results and increase their willingness to take action.

Technical Assistance Needs. Technical assistance needed to identify and analyze adaptation strategies includes the following:

- Assist in developing and applying simulation models to evaluate water resource management alternatives.
- Provide technical guidance on methods for selecting suitable climate change scenarios, which will serve as input into the analysis.
- Gather information on the relative merits of different multicriteria decision models that can be used to evaluate adaptation options.
- Give examples of water resource management plans and national plans that address climate change and associated information on the process used to develop and evaluate adaptive measures for these plans.

Future Research Needs. The research that will support development of water resource adaptive measures includes the following:

- Conduct regional studies of vulnerability and adaptation measures.
- Develop regional climate change models and scenarios.
- Continue to substantiate climate change trends and associated changes in water resource variables (e.g., run-off).
- Develop improved techniques for reducing irrigation losses.

- Improve understanding of the adaptation strategies that will help the most vulnerable populations and systems cope with change.
- Enhance the accuracy of water balance models.
- Identify legal measures that can facilitate adaptation.
- Study the effectiveness of measures for resettling populations displaced by climate change impacts and for protecting species most vulnerable to such a disruption.

Summary Assessment. Considerable uncertainty exists concerning the exact timing, magnitude, and nature of the impacts of climate change on water resources. Therefore, it is important to evaluate anticipatory adaptive measures to reduce vulnerability to the potential impacts of climate change on water resources. Countries need to develop adaptive measures consistent with other water resource management objectives. These measures must be sufficiently robust to be effective under a wide range of potential impacts and account for the uncertainty associated with climate change.

Countries are using various adaptive measures, including reducing water resource demand; protecting and enhancing water resource supplies; improving drought and flood response capabilities; increasing water transfer and combined use options; and, in a few cases, marginally increasing storage capacity. It is important to integrate climate change into water resource management plans that address future economic and population growth and other water resource challenges. Integrating water resource adaptation with interrelated resources, such as agriculture and coastal resources, must also be emphasized. Finally, to develop and implement effective adaptive measures, all persons affected should be actively involved and continue to improve public awareness and support.

Water Resources Adaptation Strategy in an Uncertain Environment

Z. Kaczmarek and J. Napiórkowski
Institute of Environmental Protention
Climate Protection Center
Kolektorska 4, Warsaw, Poland

Abstract

Poland has an annual freshwater supply of only 1,500 m^3 per capita, resulting in a scarcity of water in much of the country. A recent impact analysis showed that, for some climate scenarios, the summer run-off from most of Poland's rivers, as well as the amount of soil moisture during the summer, may decrease. At the same time, irrigation water requirements may increase. This combination would increase water deficits in Central Poland. In the framework of the Country Studies Program, the Warta River basin was selected for analysis of possible adaptive measures to cope with adverse effects of climate change. Several alternatives were investigated: (1) reducing economic activities in regions particularly scarce in water, (2) investing in water storage, (3) transferring water among river basins, and (4) establishing a policy aimed at more rational water use. Taking into account the highly uncertain climatic future, the "minimum regret" approach is advocated in formulating a national water-resources strategy, which means that alternative 4 merits priority. Large new investments should be undertaken only when absolutely necessary, that is, when other measures prove insufficient.

Introduction

During an average year, Poland's rainfall and snowfall yield about 197 km^3 of water. Of this amount, 142 km^3 returns to the atmosphere via evaporation, and 55 km^3 feeds the Baltic Sea through the river systems. The spatial distribution of surface run-off is presented in Figure 1. Variations in the run-off from 1951 to 1990 are shown in Figure 2, where the values range between a low of 34.4 km^3 in 1954 and a high of 79.5 km^3 in 1981. An additional 5 km^3 of flow enters Poland from neighboring countries. The annual per capita freshwater supply is about 1,500 m^3, the lowest value in Europe. Because of the inter- and intra-

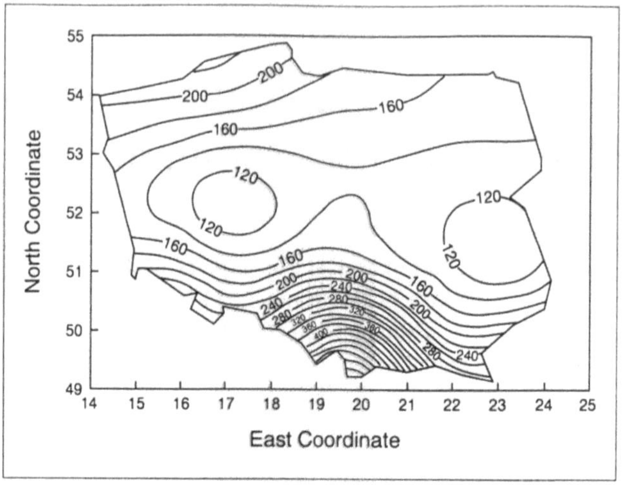

Figure 1. Mean Annual Run-off from Poland (mm yr^{-1})

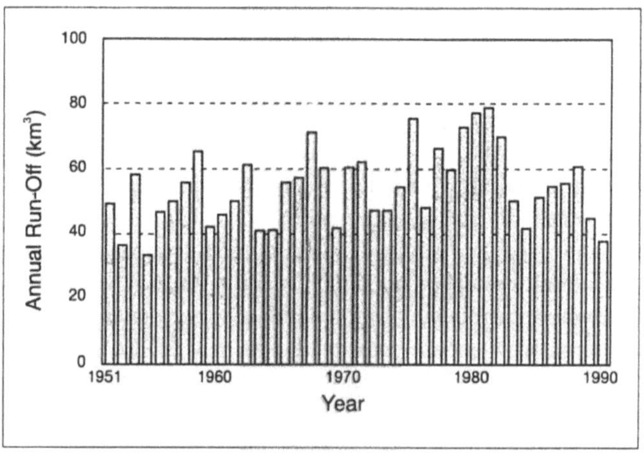

Figure 2. Variability of Annual Run-off from Poland

annual stochastic variability of climatic and hydrologic processes, reliable water resources (available about 95% of the time) equal approximately 22 km^3, but they are unevenly distributed throughout the country. Because of environmental constraints, only 30-40% of these surface-water resources may be effectively used for agricultural, industrial, or residential needs.

Water can become a barrier to social and economic development through several mutually dependent factors:

- Natural water scarcity caused by regional water supply and demand;

- Pollution of rivers, lakes, and groundwater aquifers;
- Technological and economic shortcomings; and
- Institutional impediments and low public awareness.

Only the first two factors are sensitive to climate change. The other two are subject to policy decisions that, if rationally applied, can help to adapt water-resource systems to changing geophysical processes.

Despite natural water scarcity, Poland's economy is water intensive. The water shortages observed in some years in certain regions are deeply rooted, not only in natural scarcity but also in inefficient water use and a high level of water pollution. When a nation's long-term economic future is considered, the issue of climate change cannot be neglected; scenarios of possible trends must be investigated. However, such assessment must be undertaken with the understanding that the main indicators of the water economy projected over the next century will be influenced not only by climate, but also by population processes, economic growth, and technological progress. Many factors can cause water to become a barrier to economic development, and some of these factors are far removed from the water resources themselves.

Possible Changes in Water Supply and Demand

Adaptation strategy must be based on projections of future water supply and demand. Because of the expected global change, such projections should take into account not only demographic and economic processes, but also possible changes of climatic and hydrological processes. In a water-resources impact study by the Institute of Geophysics of the Polish Academy of Sciences (Kaczmarek 1995), several General Circulation Model (GCM)-based climate scenarios were analyzed. Ultimately, the water-resources impact assessment was implemented on the basis of two GCMs (the Geophysical Fluid Dynamics Laboratory [GFDL] R-15 and the Goddard Institute for Space Studies [GISS] GCMs), which were selected because they best reflect two different climate conditions of Poland. For climate changes caused by a doubling of atmospheric concentrations of carbon dioxide (referred to as $2XCO_2$), the GFDL R-15 may be characterized as a "warm-dry" and the GISS as a "warm-wet" scenario. For both cases, data from the model output were interpolated to a grid with a resolution of one degree latitude by one degree longitude. To illustrate possible changes, monthly temperature and precipitation deviations from "historical" values for a station in Central Poland (at coordinates 18.67°E, 52.20°N) are shown in Figures 3 and 4.

To assess the impact of climate on water resources, a model of the hydrological processes is needed. For Poland, a conceptual water-balance model, CLIRUN3, has been applied with lumped input and output variables (Kaczmarek 1993). The model differs from other approaches commonly used

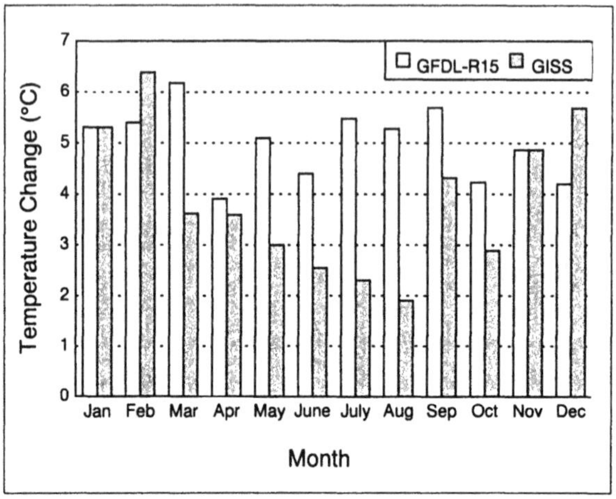

Figure 3. Projected Temperature Change in Central Poland: $2XCO_2$

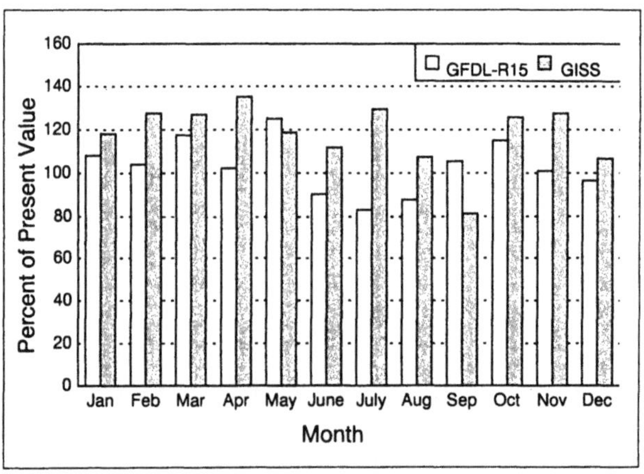

Figure 4. Projected Precipitation Change in Central Poland: $2XCO_2$

in that (1) this model is time-continuous (i.e., the water-balance components vary as continuous functions of time within certain assumed time intervals [e.g., months]); and (2) the stochastic properties of water-balance components are expressed either as a simulated time series or by a set of probabilistic matrices based on stochastic storage theory.

Water-balance components, run-off, evapotranspiration, and storage, were calculated by means of the CLIRUN3 model for 31 river catchments and 60 grid cells (Kaczmarek 1993). The study used climatic and hydrological data measured in 1951-1990 and GFDL R-15 and GISS equilibrium scenarios for $2XCO_2$

conditions. Estimated changes in monthly run-off for one grid box in Central Poland are shown in Figure 5. Possible changes in the surface-water supply estimated for the country as a whole are shown in Table 1. The table presents annual run-off values, as well as values for August, usually the driest month.

Trends in water demands caused by demographic and socioeconomic factors are identified without reference to possible changes in environmental conditions, including climate. The experience of water management agencies in various countries shows that socioeconomic processes influencing water use cannot be accurately predicted for long time periods. In most of the past studies formulating Poland's long-term water strategies, future demands were highly overestimated. Even more difficult to assess were possible implications of climate change on future water requirements.

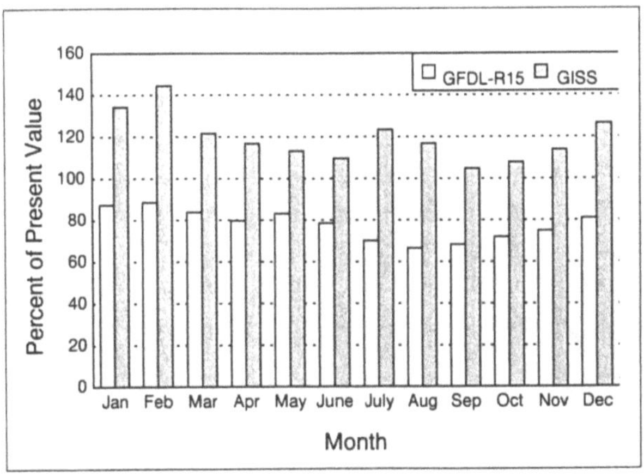

Figure 5. Projected Run-off Change in Central Poland: $2XCO_2$

Table 1. Projected Run-off from Poland

Run-off	Projected Run-off by Year and Model (km^3)				
	1990	GFDL 2020	GISS 2020	GFDL 2050	GISS 2050
Annual — mean	55.2	51.7	58.7	48.3	62.5
Annual — 90%	41.2	38.4	43.6	35.9	46.5
August — mean	3.3	3.0	3.3	2.7	3.2
August — 90%	1.9	1.7	1.9	1.6	1.8

In 1990, water withdrawal in Poland equaled 7.93 km^3, with the following distribution among sectors: domestic use, 2.54 km^3; industry (including water losses in cooling processes), 2.27 km^3; irrigation, 0.60 km^3; other agricultural water use, 1.52 km^3; and all other uses, 1.00 km^3. According to this study's estimates, water demands in mid-2000 may increase by 70% because of nonclimatic factors. Very little information is available on the effects of possible temperature and precipitation changes on water requirements in various sectors of Poland's economy. Preliminary estimates (based mostly on the literature) lead to the conclusion that a temperature increase may have only a moderate impact on industrial and domestic water use. An unknown factor is demand by Poland's agricultural sector, which at present uses a relatively small proportion of the total freshwater withdrawals for irrigation. According to the dry-warm climate scenario, however, the situation may change because the threshold between irrigated and nonirrigated agriculture may be surpassed in most of Poland's lowlands. In such a case, water demand for irrigation may increase substantially.

This study assumed that the area of irrigated agriculture in Poland will increase from the present value of 1.5% to about 4.0% in 2050. The latter figure corresponds to the current level of irrigation in West European countries, where the average air temperature is about 2°C higher than the present average air temperature in Poland. To estimate the unit water requirement for 1 ha of irrigated land, the IRDEM model was developed and applied in various regions. Estimated water demands for 2020 and 2050 are presented in Table 2.

Water Management in an Uncertain Environment

The fundamental problem in responding to possible consequences of global change is deciding what adaptive measures should be undertaken in the face of highly uncertain climatic threats. The choices must be made on the basis of incomplete knowledge of future water supply and demand, and the policy

Table 2. Projected Annual Water Demand in Poland

	Projected Annual Water Demand by Year and Model (km^3)				
Sector	1990	GFDL 2020	GISS 2020	GFDL 2050	GISS 2050
Domestic	2.54	3.25	3.22	3.78	3.71
Industry	2.27	4.09	4.09	5.84	5.84
Agriculture	2.12	3.00	2.77	3.81	3.19
Others	1.00	1.09	1.09	1.12	1.12
Total	7.93	11.43	11.17	14.55	13.86

alternatives must be analyzed with respect to the risks of assuming incorrect future scenarios. During the last century, water-resource management was planned on the assumption that variability of hydrological phenomena is governed by stationary stochastic processes. In the era of global environmental change, this concept has become questionable.

Water systems can be adapted to climate changes by three different approaches. In the first approach, decisions can be postponed until more reliable information on global processes becomes available. The existing water schemes remain unchanged, and new ones are designed in accordance with current procedures. The Commission for Hydrology of the World Meteorological Organization has adopted a "Statement on the Hydrological and Water Resources Impacts of Global Climate Change" (Commission for Hydrology 1988), which reads in part, "... given the added burden of uncertainty about climate change, it is certainly inappropriate at this time to discard available analytical procedures or to engage in expensive alterations to built facilities...." However, this first approach may lead to decisions being made too late to protect water systems from the negative consequences of climate change.

The second approach is the "minimum regret" approach, where decisions are made to solve current problems in the best possible way, while preparing water-resource systems for potential surprises and shocks. Waggoner (1990) describes this policy by saying, "So long as the future remains unsure, and that seems a long time ahead, rational people will make decisions that solve present problems and make water supply robust, resilient, and flexible for any future."

Finally, in the third approach, certain optimality rules are applied to a range of climate and water-resource scenarios. Decisions are made by comparing costs, benefits, losses, and risks for each scenario, partly on the basis of subjective interpretation of the scenarios and the results of the analysis. Some analysts advocate applying Bayesian theory in the decision-making process in an uncertain environment. However, this approach requires prior assignment of probabilities to assumed climate scenarios, which may only reflect different degrees of subjective belief about the accuracy of climatic models.

In Poland's present economic situation, the second policy approach seems to be the most rational. Not much can be done in Poland to reduce natural resource scarcity. Probably the best way of improving the supply/demand balance is implementation of rational demand-management and water-conservation strategies. Appropriate incentives of an economic nature must support these strategies.

Another possibility for coping with the consequences of climate change is to reduce the current variability of run-off by means of increased storage capacity. However, building new reservoirs requires a large investment of capital. In addition, those who favor protection of the environment disagree with those who advocate various kinds of water-resource investments.

If climatic and hydrologic conditions lead to temporal or regional water stress, it is crucial to protect available resources from contamination. The present level of pollution of water resources in Poland is very high, which is the main reason for the possible emergence of water as a barrier to growth. The program of water-quality improvement in Poland should provide for the construction of new wastewater treatment plants and for more effective use of the existing facilities.

Increased social awareness of the country's water problems and the possible negative consequences of changing the environmental conditions is needed. The present lack of perception is not caused by insufficient sensitivity of the population to water issues, but by the lack of belief in the multiplying effect of small, individual undertakings. The notion of "our water" must be given practical meaning, such as in the ongoing process of establishing self-governing regional water authorities in Poland. Water authority would be gradually transferred from the 49 regional administrations to the newly established river basin authorities. This important institutional change has been introduced within the framework of increased decentralized decision making in the entire country.

All of the measures discussed above, demand management, flow regulation, and water quality improvement, should be applied on the basis of rational economic principles, public participation, and more effective institutional arrangements to solve current national water-resource problems. These measures may, at the same time, help to prepare water systems to cope with possible shocks and threats caused by global changes.

Warta River Case Study

Within the framework of Poland's Country Studies Project, the Warta River basin was selected for more detailed analysis. The Warta River basin, located in western Poland, has a catchment area of 53,710 km^2, about 17% of the country's total area. The region is characterized by a moderate climate, a relatively low precipitation of 635 mm yr^{-1}, and an annual temperature of 8.1°C. Mean annual discharge at the mouth is estimated to be 216.7 m^3s^{-1}, which gives an average annual run-off equal to 130 mm or 6.8 km^3, while monthly discharges vary from 73.0 to 729.0 m^3s^{-1}. Spatial distribution of Warta annual run-off is shown in Figure 6.

More than 6.4 million people live in the basin, 33% in four major population areas. Annual per capita freshwater supply is 1,070 m^3, close to the 1,000 m^3 benchmark used as an indicator of water scarcity by The World Bank (Engelman and LeRoy 1993). The Warta basin represents one of the most critical water-resource regions in Poland because of water scarcity and the high level of industrial and agricultural development. Most of the region's agriculture is rainfed, with irrigated arable lands covering only a small percentage of the entire basin. Estimated water supply and demand in the Warta basin for 1990, 2020, and 2050, based on projected future climate conditions, are shown in Tables 3 and 4. Comparison of these values shows that demand will barely be met in 2050.

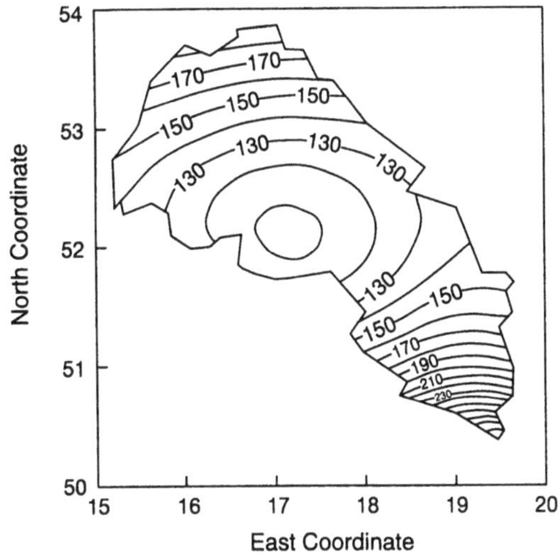

Figure 6. Mean Annual Run-off from the Warta River Catchment (mm yr^{-1})

Table 3. Projected Run-off from the Warta Catchment

	Projected Run-off from Water Catchment by Model and Year (km^3)				
Run-off	1990	GFDL 2020	GISS 2020	GFDL 2050	GISS 2050
Annual (mean)	6.8	6.1	7.0	5.6	7.6
Annual (90%)	4.7	4.2	4.8	3.9	5.3
August (mean)	0.39	0.35	0.37	0.30	0.37
August (90%)	0.21	0.19	0.20	0.16	0.20

The technical infrastructure of the Warta system is very modest. Two reservoirs, Poraj and Jeziorsko, are located along the river; only Jeziorsko has sufficient capacity to affect flow redistribution. The reservoirs together control only 3.2% of the catchment's run-off. Water transfers of limited capacity among various parts of the basin are possible, but their role in the catchment's water management is limited.

Table 4. Projected Annual Water Demands in the Warta Catchment

	Projected Annual Water Demands: Warta Catchment by Model and Year (km^3)				
Sector	1990	GFDL 2020	GISS 2020	GFDL 2050	GISS 2050
Domestic	0.42	0.54	0.53	0.62	0.60
Industry	0.41	0.74	0.74	1.06	1.06
Agriculture	0.36	0.50	0.46	0.63	0.54
Others	0.18	0.20	0.20	0.20	0.20
Total	1.37	1.98	1.93	2.51	2.40

A two-layer optimization technique developed in the framework of the Polish Country Studies (SYMOPT software package) was used to analyze the operation of the Jeziorsko reservoir (Kaczmarek et al. 1995). The reservoir is characterized by a catchment area of 9,063 km^2 (arable land of 1,967 km^2), total storage capacity of 202.8 million m^3, and dead storage of 30.2 million m^3. The optimal storage levels and reservoir outflow were simulated for 40 different hydrologic and water demand monthly time series (each for 1990-2050) for both the GISS and the GFDL climate scenarios. To determine the agricultural water demands, the current irrigation level of 1.5% of the arable land was assumed to increase to 4.0% in 2050. To estimate water requirement per hectare of irrigated land, the IRDEM model was applied for each summer month of all 40 hydrological series. Industrial needs were evaluated according to the expected growth of the gross national product, with some rationing of water use. Domestic water use was assumed to increase proportionally with population growth. In addition, the possibility of water transfer up to 15 m^3s^{-1} from Jeziorsko reservoir to the lower part of the basin was analyzed. A minimum reservoir outflow was assumed to meet hydrobiological criteria ($Q_o = 10.3$ m^3s^{-1}) and ecological criteria (from 25.3 m^3s^{-1} in March-June to 22.8 m^3s^{-1} in July-October).

The results show that, for the GISS scenario, the impact of climate change on the operation of Jeziorsko reservoir may be negligible. For the GFDL scenario series, water deficits may arise after 2020, particularly in 2030-2050. The results of the simulation for a representative year are shown in Figures 7 and 8, and the results generalized for all 40 runs are shown in Figures 9 and 10.

The study shows that the basin's water supply and demand are both sensitive to changes of climatic characteristics and that the region is vulnerable to such changes. In mid-2000, the available freshwater supply may be insufficient to meet requirements in the summer months. Optimal operation of existing reservoirs will not solve the problem in the basin as a whole, although such operation may secure a reliable supply for the upper part.

Burton et al. (1993) listed a number of approaches for coping with negative effects of climate (e.g., prevention of losses, tolerating loss, changing activity or location, etc.). All such actions require comprehensive social and economic analyses and long-term planning.

The list of possible adaptive responses that might be used in the Warta basin to handle future water deficits includes the following:

- Conservation of water by various sectors of the economy;
- Temporary limitation of water use for irrigation in dry years, accompanied by import of food products;
- Improved management of resources through efficient operation of water-resources facilities;
- Development of technical infrastructure (e.g., constructing new storage reservoirs); and
- Transfer of water from other river basins.

The possible role of new water-resources investments in coping with expected water deficits will be investigated in the next phase of the Country Studies Project. Strong opposition exists in Poland to new large-scale hydraulic investments because of the relatively high density of population, lack of lands for additional storage, environmental concerns, and insufficient investment funds. Thus, the most probable approach for adapting to the future climate is water conservation and improved management. An option is to reduce the acreage of irrigated lands and to solve the food supply problem by introducing drought-resistant crops or by importing food. The key recommendation resulting from the Warta study is to undertake an intensive research program on the vulnerability of national agriculture to climate change, with particular emphasis on irrigation strategies. The experience of several European countries also shows that, in the domestic and industrial sectors, water conservation may be an efficient and economically justifiable tool for coping with future water deficits.

Conclusions

It is difficult to formulate definite suggestions until more reliable information on the future climate is available. On the basis of current knowledge, the following conclusions are justified:

- Water managers should be concerned because water supply and demand may be significantly affected as a result of climate change.
- Current-generation climate models do not offer the requisite degree of watershed-specific information on future climate states.
- Design criteria, development plans, operating rules, and water allocation policies must continually be adapted to the newly developed climate scenarios.

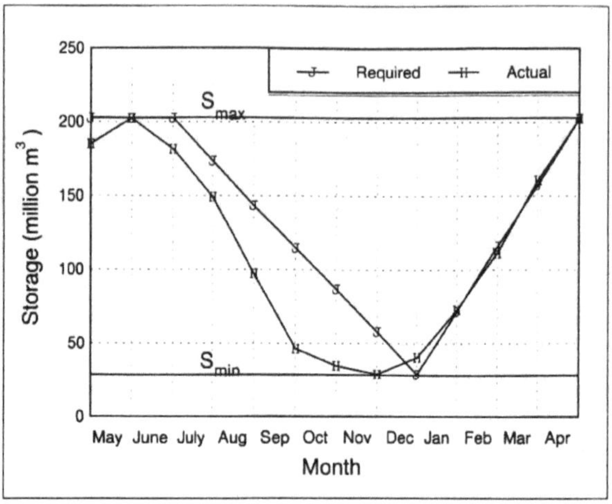

Figure 7. Optimal Storage Path for Jeziorsko Reservoir for a Representative Year (hydrological year begins May 1)

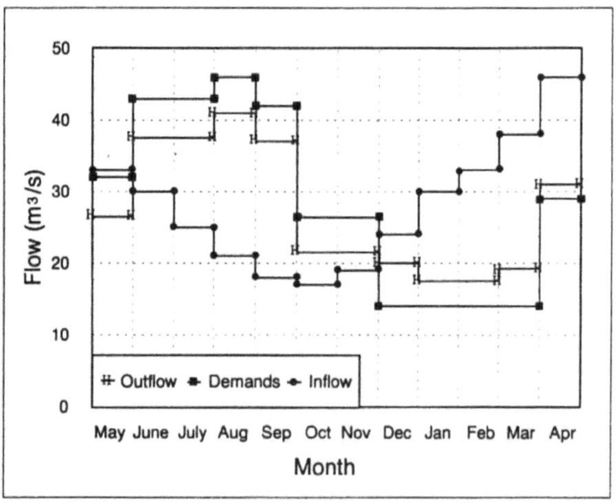

Figure 8. Optimal Outflow Values for Jeziorsko Reservoir for a Representative Year (hydrological year begins May 1)

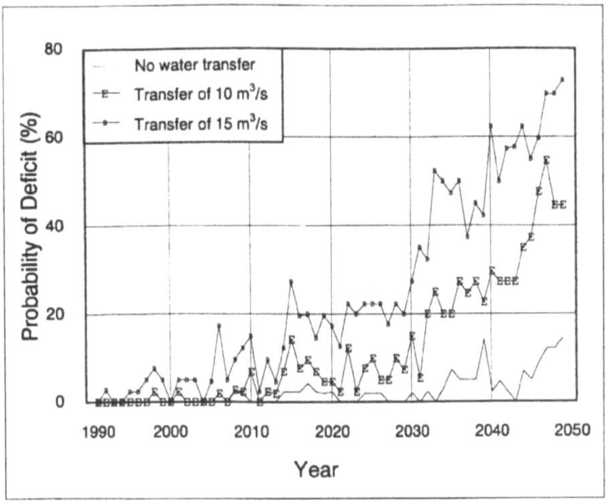

Figure 9. Probability of Water Deficits for Jeziorsko Reservoir: 1991-2050

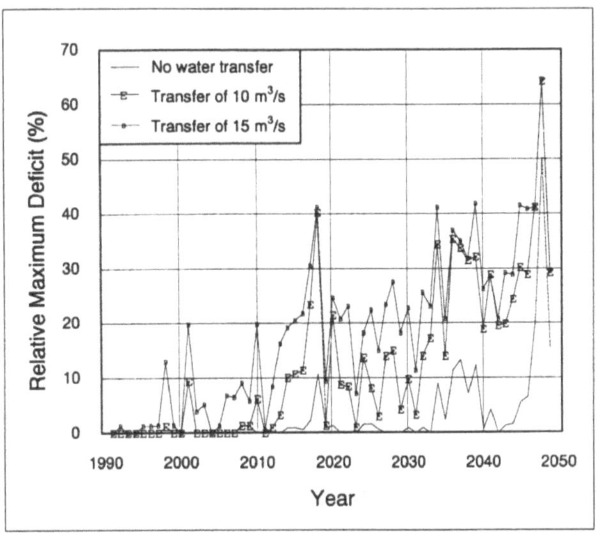

Figure 10. Relative Maximum Deficit (percent of demand) for Jeziorsko Reservoir: 1991-2050

- Vulnerability of water systems to hydrologic change decreases as the level of water-system development and water management increases.
- Water demand, as well as water supply, may be sensitive to climate change, which will affect irrigation-water requirements.
- Primary components for increasing the soundness of water-resources systems under increasing uncertainty due to climate change are improved water-demand management and institutional adaptation.
- Even countries scarce in water may effectively adapt to changed climate conditions, but the cost of adaptation will depend on the extent of the water deficits.
- Lessons drawn from a set of hypothetical case studies should be generalized in the form of guidelines for adaptation strategies.

References

Burton, I., R.W. Kates, and G.F. White, 1993, *The Environment as Hazard*, Gailford Press, New York, N.Y., USA.

Commission for Hydrology, World Meteorological Organization, 1988, *Abridged Final Report of the Eighth Session*, No. 175, Geneva, Switzerland.

Engelman, R., and P. LeRoy, 1993, *Sustaining Water: Population and the Future of Renewable Water Supply,* Population Action International.

Kaczmarek, Z., 1993, "Water Balance Model for Climate Impact Analysis," *Acta Geophysica Polonica* 41(4):423-437.

Kaczmarek, Z., 1995, "National Assessment — Poland," in *Water Management in the Face of Climatic and Hydrologic Uncertainties*, Kluwer Publishing House, Dordrecht, the Netherlands (in press).

Kaczmarek, Z., J. Napiórkowski, and K. Strzepek, 1995, *Climate Change Impact on the Water Supply System in the Warta River Catchment*, International Institute for Applied Systems Analysis, Laxenburg, Austri, in press.

Waggoner, P.E. (ed.), 1990, *Climate Change and U.S. Water Resources*, John Wiley & Sons, New York, N.Y., USA.

Anthropogenic Climate Change and Reduction of Water Resources: Adaptation Issues Related to the Economy in Kazakhstan

V.V. Golubtsov, V.I. Lee, and I.I. Skotselyas
KazNIGMI, Seifullin Pr. 597, Altamy, 480072 Kazakhstan

Abstract

Water resources and their use in the Republic of Kazakhstan are described. The input data and methods for assessing the vulnerability of water resources take into account anthropogenic climate changes. The results of vulnerability and adaptation studies are discussed, and a structure is suggested for adaptive measures that can be applied to water resources under future anticipated climatic changes.

Introduction

Kazakhstan has limited water resources. Estimates show that these resources total 121 billion m^3 (i.e., 24,000 m^3 km^{-2}) (Chebotarev 1978). The plains areas have particularly unfavorable water resources. Some regions of Kazakhstan have relatively large proved reserves of underground water, but the quality does not always meet the requirements of consumers.

Water resources, mainly surface water, are used for irrigating lands and pastures, supplying water to livestock farms, and providing water for the municipal and industrial sectors. Water consumers consist of hydroelectric power stations, fisheries, and water transportation. River run-off is generally used for irrigating lands and supplying water to populated areas. The proportion of run-off used for water management activities varies widely. At times, some water sources are depleted. Water management activity primarily focuses on large rivers, foothills, and plains; however, highlands are also gradually being developed.

Potential climate change caused by greenhouse gas emissions from thermal power stations, vehicles, and other sources can affect the water resources of the entire country. Because water resources are exceptionally important for the development of the Kazakhstani economy, researchers are studying the vulnerability and adaptation of water resources and developing a water management policy to address anticipated anthropogenic climate change. These studies are being conducted within the framework of the Kazakhstani-American project, "Climate Change in Kazakhstan." These issues have never been investigated in Kazakhstan. Investigations are now in the initial stages.

Methodology

To estimate the vulnerability of water resources, scientists used a run-off formation model, developed by KazNIGMI (Golubtsov 1975; Golubtsov et al. 1989). This model presents a river basin as three control reservoirs connected successively and located one above the other. These reservoirs can be identified by the surface, ground soil, and ground run-off formations. The reservoir that characterizes run-off within a hydrographic network is also considered.

The model consists of several unique paradigms. These paradigms describe elementary water balance processes. Such processes include formation of the snow reserve and the water supply to the basin surface; fluctuations in the soil-moisture content; frost-bounding and melting of frozen soil; total evapotranspiration; and formation of surface, ground soil, and ground run-off. These processes also include transforming the run-off into water in-flow to the river-bed system and then to the run-off hydrograph at the closure site. Model parameters are assumed to depend on the latitude and longitude of the locality, the inclination and exposure of the slopes, and the characteristics of the bottom surfaces (e.g., open, forest cover, glacial). The major inputs of the model are the average daily temperature and the daily amount of precipitation measured at meteorological stations.

Anthropogenic climate changes were assumed in accordance with scenarios used in American (Geophysical Fluid Dynamics Laboratory [GFDL] R-30) and Canadian (Canadian Climate Center Model [CCCM]) General Circulation Models (GCMs) for a doubling of CO_2 rates (hereafter referred to as $2XCO_2$). Two intermediate CO_2 concentration levels were considered in the GFDL-T model estimates.

Limitations

Because of the relatively short-term research conducted within the framework of the Kazakhstani-American project, the vulnerability assessments of water resources under anthropogenic climate changes were restricted to two watersheds: the Tobol River basin in the plains and the Uba and Ulba watershed in the mountains. However, Kazakhstan has a great variety of run-off conditions. These studies did not cover watersheds with glacial feed. Therefore, the results of water resources vulnerability assessments cannot be applied to rivers with glacial run-off. Studies of other watersheds will be conducted later.

Results

Vulnerability Assessments of Water Resources

Table 1 lists the calculated values of water resources in the Uba and Ulba watershed under natural and anthropogenic climate change conditions. A $2XCO_2$ (scenario 3) used in the GFDL-T model does not change the volume of water resources. Intermediate CO_2 concentration levels (scenarios 1 and 2) can even increase water resources by 6-12%. The CCCM and GFDL R-30 models predict a 23-29% decrease in water resources.

KazNIGMI climatologists believe that for Kazakhstani conditions, it is advisable to use the GFDL R-30 model to generate climate change scenarios. Thus, it is likely that anthropogenic climate changes could significantly reduce water resources. The most vulnerable water resources would be rivers during plentiful water years. Climate changes could reduce the recurrence of plentiful water years. These conclusions are preliminary, especially when applied to Kazakhstani water resources as a whole because various GCMs produce significantly different results.

Structure of Measures to Adapt to Changes in Water Resources

The issue of using water resources rationally in Kazakhstan becomes even more urgent in view of potential future anthropogenic climate changes. With the

Table 1. Water Resources in the Uba and Ulba Watershed under Natural Conditions and under Anthropogenic Climate Change

GCM	Scenario	Water Resources ($x 10^6 m^3$) Natural	Water Resources ($x 10^6 m^3$) Under Climate Change	Change in Water Resources (%)
GFDL-T	1	8,889	9,974	12
GFDL-T	2	8,889	9,447	6
GFDL-T	3	8,889	8,782	-1
GFDL R-30	3	8,889	6,290	-29
CCCM	3	8,889	6,876	-23

problems associated with climate change now being recognized, the rational use of water resources may be of vital importance later. Therefore, Kazakhstan should accept this strategy (i.e., rational use of water) for its water management policy. Thus, adaptation of water resources to climate change is of exceptional significance, given the recent studies on climate change impacts.

Table 2 gives the structure of evaluation suggested for water resources management: general objectives, and assumed criteria of the assessment. This structure emphasizes two general goals: (1) supporting the development of water management and (2) decreasing the adverse effects of water resources vulnerability. Objective 1 encompasses matters for scientific enquiry, such as environmental protection, optimization of ecosystem development, and assessments of external and internal water resources, which have not been fully exploited. Objective 2 includes minimizing the loss of water resources, minimizing economic and social losses, and enhancing timely decision making. The evaluation criteria used are essentially quantitative.

If the natural water supply is scarce and anticipated anthropogenic climate change decreases water resources, desertification could expand into new territories. Additional ameliorative, agroforest-ameliorative, and agrotechnological measures, such as creation of a sanitary protective zone, may be required. Of even greater concern is the possibility of depleting and polluting water sources,

Table 2. Assessment Structure for Water Resources Management

General Objective	General Assessment Object	Assumed Assessment Criteria
To support the development of water resources management	• Protect the environment • Optimize development of the ecosystem • Reduce greenhouse gas emissions, an attraction of external/internal not fully exploited water resources	• Biovariety, areas • Ecological factors, employment of population • Volume of water, benefits, employment of population
To reduce the effects of vulnerability on water resources	• Minimize losses of water resources • Minimize socioeconomic losses • Enhance decision-making effectiveness of state and economic organizations	• Volume of water, costs • Personal, insurance, social losses • Early and justified forecasting of the length of dangerous periods

especially small rivers (Vladimirov 1991). In particular, economic activity will inevitably be restricted in some areas and, probably, even transferred to other regions. The need to implement measures to cope with floods will remain, since they would probably occur under potential climate change.

The probability of disturbing the biological balance of ecosystems and initiating critical ecological situations increases as water resources decrease. To solve possible ecological problems, it is necessary to understand the following issues:

- Control of plant and animal life,
- Partial migration of the community,
- Chemical and biological water treatment,
- Creation of favorable water and temperature conditions for habitation,
- Reproduction of the most valuable living organisms (e.g., fishes), and
- Water deficit compensation.

Under these circumstances, ecological expertise in new construction projects and enhanced control of water resources should be compulsory.

Additional water resources are also of great importance for supporting the development of a water management plan for land irrigation, pasture irrigation, water delivery to residential areas, and so on. Two sources of such resources could be the most essential: (1) diversion of some of the run-off from the Volga and Siberian Rivers and (2) use of underground water, which has still not gained wide acceptance. Priority should be given to using underground water, since diverting run-off involves high costs and technical difficulties. In addition, it also requires solving complicated intergovernmental problems between the Russian Federation and Kazakhstan, and ecological problems inherent to the Russian Federation. Settling all matters about water management relations with Kyrgyzstan, Uzbekistan, and the People's Republic of China is also important and would be mutually beneficial. Together, these activities could partly reduce the anthropogenic load on surface resources and promote further development of water management approaches.

In determining ways to decrease adverse effects, minimizing the loss of water resources becomes a primary consideration. These losses result from imperfect irrigation systems and failure to observe watering norms, caused by the use of out-of-date equipment in industrial and municipal management. The situation can be considerably improved by instituting water-saving measures, such as reconstructing irrigation systems, improving sowing techniques on irrigated lands, implementing water-free and little-water technologies and recycling systems, imposing strict licensing and reliable control of irrational water use, and introducing sufficiently high water pricing. According to available data, the reconstruction of the irrigation system in the Ili-Balkhash watershed alone would reduce the specific intake by about 30%, which accounts for a water savings of approximately 2 billion m^3 (Bochkov 1989).

As noted, the impacts to water resources can be mitigated by using more underground water and sewage (wastewater recycling). In particular, Zubairov (1989) suggests that using recycled wastewater for irrigation to develop fodder production and cultivate technical agricultural crops is a very promising technique for reducing the impacts of climate change on water resources.

The reduction of water resources places an immediate threat on the carrying capacity of navigable rivers. It could be necessary to dredge, reconstruct piers and moorages, and, probably, replace inland water transport and fishing boats by ships with less draft.

The power sector of the economy will most likely have to revise operating conditions of the hydropower electric stations to shift the weight of power generation to thermal power stations. This change would possibly allow the use of nuclear, solar, and wind energy.

Economic and social losses will be inevitable to adapt economic sectors to changes in water resources. The Republic could be completely or partially deprived of some kinds of produce if they appear to be economically unprofitable. For example, farmers in southern Kazakhstan have recently cultivated rice. Special irrigated tracts of land were created for this purpose. However, this activity disturbed the water balance and adversely affected all of the water management in the area. Naturally, if water resources are reduced, the advisability of rice cultivation will be even more conjectural.

In the future, the lands sown with cereal, fodder, and other crops is expected to be reduced as a consequence of the impacts of anthropogenic climate change on water resources. These losses must be compensated; otherwise, production output must be reduced. It is necessary to evaluate expenditures entailed by the possible migration of people and new construction in newly developed regions. It is important to know which inhabitants will be affected the most, what compensations will be required, and what amounts should be established.

The success of adaptive measures will largely depend on operative actions of managers of different ranks. Conditions necessary to adopt proper and timely decisions are (1) the availability of legislative acts and interstate agreements that regulate water management relations, (2) the reliability of early hydrologic forecasts, (3) the availability of well-supported scientific recommendations and models that allow operative and correct evaluations of different situations, and (4) the readiness for immediate implementation of adopted decisions. Work still needs to be done in these areas.

Many of the above-mentioned problems are unsolved issues that would significantly impact the economic activity associated with water resources. Unfortunately, investigations of some problems are being carried out carelessly: work at others has been suspended as a result of the difficult economic situation in Kazakhstan. The complexity and the number of problems are so great that much time, finance, and effort will be needed to provide economic adaptive measures to counteract the reductions in water resources and changes in the watershed hydrologic regime. To evaluate adaptive measures in detail, it is necessary to create a system that could simulate different situations and then select the most relevant alternatives for managing water resources.

However, priority should be given to measures related to saving water and protecting the environment, including the implementation of water-free and little-water-consuming technologies, the reconstruction of irrigation systems, and the observance of sanitary norms. The water saved can be used to maintain normal ecological conditions, to develop economic activities and, hence, to improve the standard of living. Water-saving techniques can also enable the solution of collateral problems, such as diminishing of "bogging up" and secondary salinization of ground soil, and reducing possible pollution of surface and underground water resources.

Conclusions

The following conclusions resulted from this study:

- According to assessments made by KazNIGMI climatologists, anthropogenic climate change from emissions of greenhouse gases into the atmosphere can reduce water resources in Kazakhstan by 20-30%.
- The most radical adaptive measures are reducing greenhouse gas emissions and diverting a portion of the run-off from the Russian Federation to the Republic of Kazakhstan.
- Priority should be given to water saving and environmental protection measures.
- Comprehensive assessments and management of water resources require the creation of simulation systems.

References

Bochkov, M.I., 1989, "The Rational Use of Water and Land Resources from the Lake Balkhash Basin," *Scientific and Technical Problems of the Development of Natural Resources and the Complex Development of Productive Forces in Pribalkhashye*, plenary report, pp. 62-69, Nauka, Alma-Ata, Kazakhstan (in Russian).

Chebotarev, A.J., 1978, *Hydrological Dictionary*, Hidrometeoizdat, Leningrad, USSR (the former Soviet Union) (in Russian).

Golubtsov, V.V., 1975, "On the Construction of the Mathematical Model for the Run-off Formation in the Mountain Watershed," *Trudy KazNIGMI* 48:3-25.

Golubtsov, V.V., et al., 1989, "The Mathematical Simulation of Formation Processes of Mountain River Runoff in Conditions of Limited Information," *Proceedings of the 5th All-Union Hydrological Congress*, Vol. 6, pp. 374-382 (in Russian).

Vladimirov, A.N., et al., 1991, *Environmental Protection*, Gidrometeoizdat, Leningrad, USSR (the former Soviet Union) (in Russian).

Zubairov, O.Z., 1989, "Prospects of the Sewage Use for Irrigation in Kazakhstan," *Scientific and Technical Problems of the Development of Natural Resources and the Complex Development of Productive Forces in Pribalkhashye*, Section 2, pp. 129-133, Nauka, Alma-Ata, Kazakhstan (in Russian).

Adaptation of Hydropower Generation in Costa Rica and Panama to Climate Change

M. Campos, A. Sánchez, and D. Espinoza
Central American Project on Climate Change
P.O. Box 21-2300
San Jose, Costa Rica

Abstract

Preliminary results from a study of vulnerability and adaptation to climate change in Central America are presented. The main objective of this research is to describe possible measures for adapting to potential impacts of climate change by using the water resources of Costa Rica and Panama as a case study. A methodology was developed that considers the El Niño-Southern Oscillation phenomenon as a climate scenario for Panama and Costa Rica. Adaptive measures dealing with availability, use, distribution, and operation of water resources for hydropower generation in tropical countries serve as a framework for discussing a regional strategy of water resource adaptation to climate change.

Introduction

The latest scientific estimations indicate that changes in the global temperature may affect the intensity and frequency of tropical cyclones in Central America (Brenes and Saborio 1994). Changes in the Intertropical Convergence Zone and the speed and seasonality of the trade winds are also expected. According to the Intergovernmental Panel on Climate Change, climate change may also affect economic activities, such as forestry, biodiversity, water resources, agriculture, human health, and coastal and marine ecosystems (Houghton et al. 1990).

Because Central America is rich in water resources, most countries in this region depend on hydropower for electricity generation. Changes in circulation patterns that affect the seasonality and amount of rainfall would also affect energy generation. Recent experiences with the "El Niño-Southern Oscillation" (ENSO) phenomenon provide evidence of such changes. In Costa Rica and Panama, ENSO manifestation is associated with a decrease in precipitation, especially along the Pacific Coast (Figure 1). The ENSO phenomenon directly affects hydropower sites in Costa Rica, such as the Arenal and Ventanas-Garita dams, where a significant reduction in run-off has been observed (Costa Rica 1989).

Measures for adapting to potential impacts of climate change are described through means of a case study of the water resources of Costa Rica and Panama. The importance of water resources for electricity generation and the benefits this sector generates for the gross domestic products (GDPs) of both countries are studied. Trends in energy production are also studied, and the growing dependence of economic activities on hydropower is evaluated.

The vulnerability and future responses of the electricity generation sector to climate change may be correlated with its response to the ENSO phenomenon. Therefore, this phenomenon is considered as a climate change scenario for future adaptation measures, taking into account possible differences between climate scenarios from General Circulation Models (GCMs) and ENSO scenarios.

Framework Methodology for Climate Change Studies in Central America

The basis for this study is preliminary results from the Central America Project on Climate Change (CAPCC), whose main objective is to estimate the vulnerability of water, agricultural, and coastal resources to a potential climate change. Climate scenarios developed for CAPCC are being reviewed. Seven countries are involved in the project. Each country has organized its own national teams and developed its own studies of the chosen topics. The activities are executed under the guidance, coordination, and cooperation of the Central America Technical Team on Climate Change, a group of experts on each field, assisted by the U.S. Country Studies Program.

The climate change scenarios developed by the CAPCC are based on results from GCMs, and they represent regional-scale climate conditions. This situation presents a problem of scale for watershed planning (Sánchez-Azofeifa 1993). This study uses the basins as the main scale. It uses climate variability as a climate change scenario because of the observed relationship between changes in patterns of electricity generation (thermal and hydraulic) and occurrence of the ENSO phenomenon.

The ENSO phenomenon is an important source of interannual variability in weather and climate around the world. The Southern Oscillation is a global pattern that primarily consists of a seesaw in atmospheric mass involving air exchanges between the Eastern and Western Hemispheres, centered in the tropical and subtropical latitudes. El Niño is an anomalous warming of the eastern tropical Pacific Ocean, but in major "warm events," it extends over much of the tropical Pacific. In such cases, El Niño is manifested as the Southern Oscillation. The ENSO phenomenon results from an interaction between changes in the tropical Pacific Ocean and the global atmosphere. Surface atmospheric winds drive tropical ocean currents and change the sea surface temperatures, which in turn alter the location and strength of atmospheric convection and precipitation in the tropics, and thus change atmospheric heating patterns, atmospheric waves, and winds in the tropics (Trenberth 1990). Figure 1 shows how general changes in major climatic elements due to the ENSO phenomenon may affect Central America.

Case Study: Costa Rica and Panama

Vulnerability studies indicate that economic and population growth in Costa Rica and Panama depend on water resources. Costa Rica has a high capacity for hydropower generation. Total power generation for the 34 main drainage basins is estimated to be 223,000 GWh/yr, and total theoretical energy power is estimated to be 25,450 MW (Costa Rica 1990a). Several small drainage basins have a generation capacity of about 20 MW for a combined production of 1,000 MW.

Energy demand in Costa Rica is strongly correlated with population growth ($r^2 = 0.87$), GDP ($r^2 = 0.85$), number of people with electricity ($r^2 = 0.90$), and urban growth ($r^2 = 0.93$). The rate of energy consumption growth has oscillated since 1966; it has a strong relationship to population growth and also reflects the impacts of climatic variability and economic crises (Figure 2).

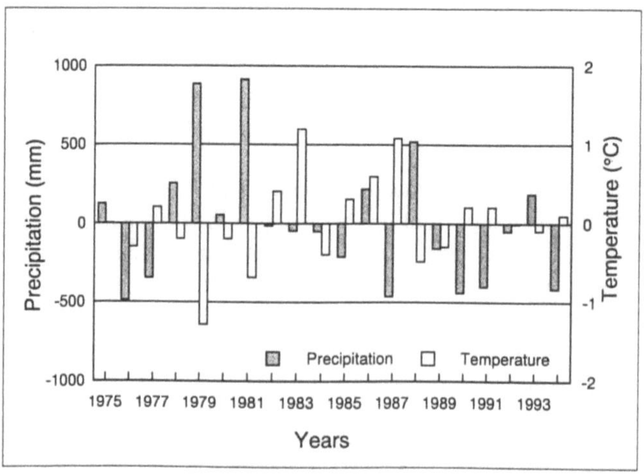

Figure 1. Precipitation and Maximum Temperature Anomalies for the Pacific Coast of Costa Rica

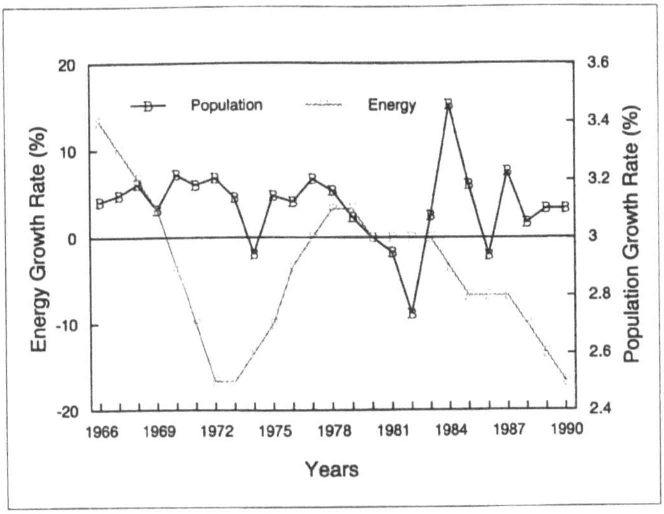

Figure 2. Growth and Population Growth Rates in Costa Rica

Figure 3 shows the evolution of installed capacity in Costa Rica from 1965 to 1989. Most of the increased demand has been supplied by the construction of new hydropower complexes. During 1982-1987, Costa Rica's installed capacity covered 80% of demand, and in 1989, it covered 84%. Current installed capacity covers 98% of total energy demand.

Despite Costa Rica's high capacity for hydropower generation, the combination of population growth, high energy consumption, and the ENSO phenomenon has reduced the gap between supply and demand. As a result, Costa Rica imported 77.8 GWh of electricity in 1986. Energy imports have grown steadily since then, reaching 159.5 GWh in 1987, 190.6 GWh in 1988, and 152.0 GWh in 1989 (Costa Rica 1990b, 1992).

Costa Rica's current policies for meeting energy demand consider new hydropower complexes, geothermal generation, and wind generation as the main resources. Less than 2% of global electricity demand is met by thermal generation, but it is expected to grow as a result of cuts in government plans to expand critical hydropower complexes.

By early 1970, most of the electricity in Panama was generated by burning fossil fuels. Although hydropower was being generated at the Macho Monte Station by the late 1930s, it was not until the mid-1970s, when the Bayano Station began operation, that hydropower equaled thermal generation (Figure 4). Thermal generation has been almost constant since the 1970s; however, hydropower production received a large boost in 1984 when La Fortuna Station was built.

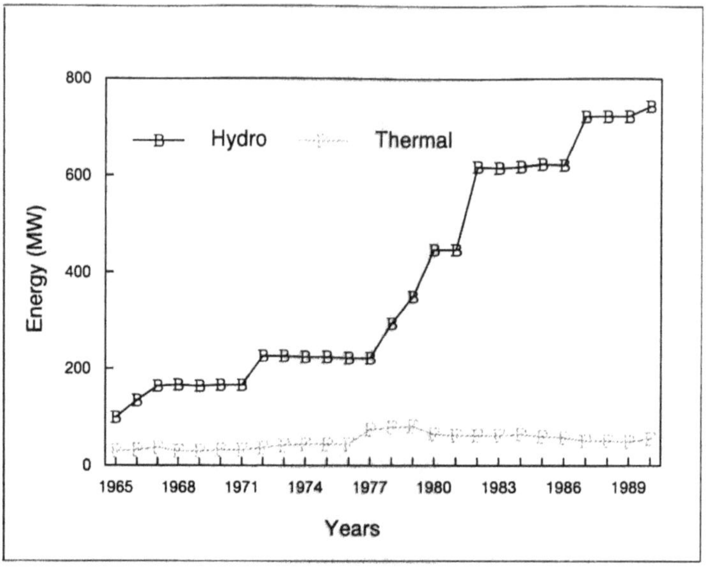

Figure 3. Hydropower and Thermal Installed Capacity in Costa Rica (1965-1990)

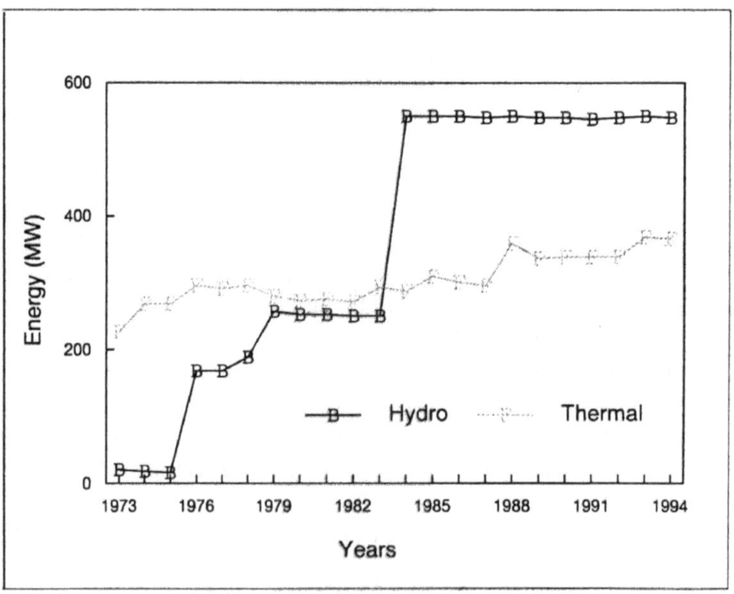

Figure 4. Hydropower and Thermal Installed Capacity in Panama

Energy demand in Panama, as in Costa Rica, is strongly related to growth in the GDP and population. The direct contribution of the electricity sector to Panama's economic growth, though smaller than that of other sectors (3.4% in 1994), represents a large benefit when it is associated with sectors such as industry (9.3% in 1994) and commerce (11.6% in 1994). Thermal and hydropower generation in Panama is driven mainly by demand, but it is also affected by changes in climate, as exemplified by the relationship between the ENSO phenomenon and changes in generation patterns.

Analysis of Results

Figure 5 shows Costa Rica's heavy dependence on hydropower generation. The long-term vision of previous governments, as well as efficient management of demand since 1966, has produced hydropower complexes at critical points of demand. These policies have produced a large storage capability that buffers the effects of climatic variability. Only during the 1977-1978 El Niño did thermal generation increase (Figure 5). As a result of the construction of new hydropower plants, the 1982-1983 and 1986-1987 El Niño anomalies did not affect hydropower generation. However, as Figure 6 shows, during the years that followed El Niño, hydropower generation decreased, and more thermal energy was needed to satisfy demand (Panama 1994). This phenomenon increased Panama's dependence on fossil fuels and, as a result, increased greenhouse gas emissions.

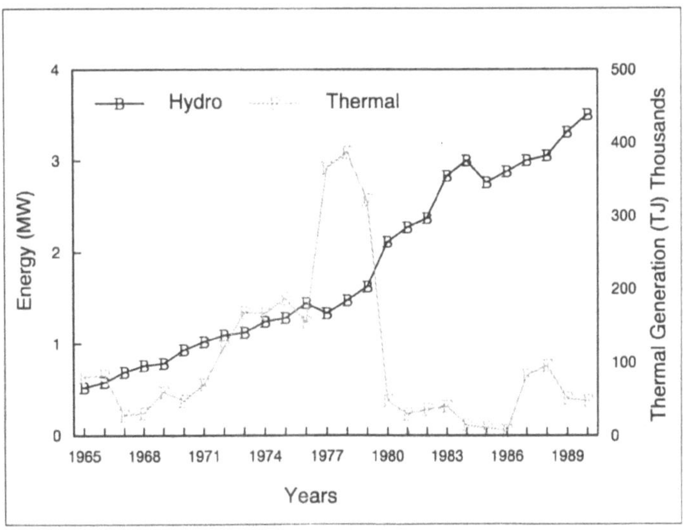

Figure 5. Electricity Generation in Costa Rica by Type

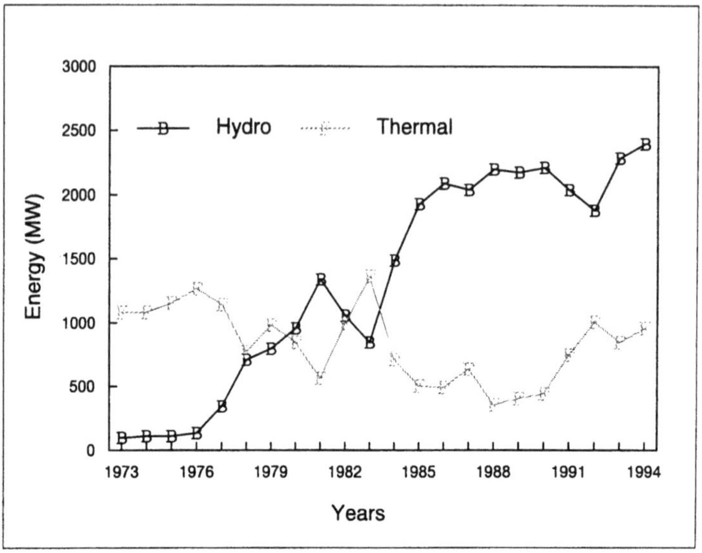

Figure 6. Electricity Generation in Panama by Type

If a climatic phenomenon such as El Niño strongly affects hydropower, a decrease of 15-20% in the total amount of precipitation and changes in its distribution due to climate change may severely affect not only the electricity sector but also associated activities that depend on this service.

Adaptive Measures for Costa Rica and Panama

Conceptual Framework for Adaptive Measures

A strategy for adaptive measures should consider the environment, human society, and social and economic development as a whole, as part of a national strategy for sustainable development. The strategy should be developed around the question posed by Stakhiv (1993): should we adapt incrementally to the explicit signals of climate change, or should we develop an anticipatory strategy to ameliorate the expected adverse impacts of climate change?

Adaptation policies for the water resources of Costa Rica and Panama must consider measures for *long-range planning, demand management, and supply development and management.* Implementation of anticipatory measures is complicated because it is difficult to distinguish between natural variability and long-term climate change. Such measures also deal with both short- and long-term energy plans and should consider the uncertainties in estimates of the effects of climate change.

Long-term planning should consider energy system robustness in terms of the expansion capability of current hydropower plants. Planning goals should consider a "wide soft component" that can be reviewed over time to evaluate the system's response to climate change as more information becomes available. Long-term plans should also consider economic sustainability, ecosystem integrity, social desirability, implementability, and equitability (Klemês 1993).

Demand management and supply development and management are difficult to implement. The general paradigm of price control, legislation, current consumption practices, and allocation of resources by governments is more concerned with satisfying current demand and short-term problems. Despite awareness of the potential problems of climate change, it is difficult to implement measures in developing countries, which are more concerned with what is going to happen in the next five years than in the year 2075 (Linsley and Franzini 1976). As a result, it is almost impossible to implement climate change adaptive measures for long-term scenarios. Past reactions to climatic variability can be an analogy to potential response to climate change in the long term.

Adaptive Measures

Costa Rica and Panama strongly depend on hydropower: a 98% dependency for Costa Rica and 73% for Panama. In both cases, a sound and broad set of measures is needed to confront possible climate change. These measures should serve as an adaptive response and a framework for discussing the availability, use, distribution, and operation of water resources for hydropower.

Costa Rica's Electricity Institute is instituting a nationwide strategy to reduce energy consumption and encourage use of more efficient appliances. A newspaper, radio, and television campaign emphasizes current environmental and climatic stresses near key reservoir sites and their impact on hydropower generation. Incentives for reducing energy consumption are also offered. This first step is probably the most important taken by the government toward energy savings. Under this policy, people learn the value of energy savings and assume responsibility for society.

In Panama, environmental measures aim to improve bill collection and price control. At the same time, a large effort is under way to decrease losses from energy generation, transmission, and distribution, which reached 728 GWh or 21.4% of total energy generated in 1994 (Panama 1994) (Figure 7).

As a result of potential impacts of climate change on water resources in Costa Rica and Panama, construction of new hydropower plans and the use of fossil fuels in thermal generation may increase. Two consequences are expected: a direct impact on the economy and an increase in greenhouse gases due to the use of fossil fuels.

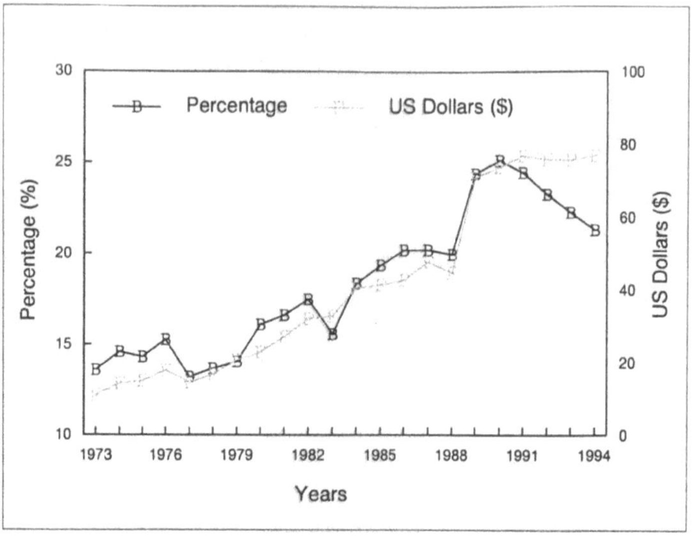

Figure 7. Energy Loss in Panama (1973-1994)

Hydropower is "clean" energy and may be more expensive in the short term. Construction of hydropower plants will directly affect the environment, especially in unexplored areas where little is known about such issues as biodiversity. Hydropower will also directly impact the national economy. Hydropower plants are generally financed by international loans from organizations, such as the Inter-American Development Bank, the International Monetary Fund, or The World Bank. If such projects are not carefully planned, local economies and social systems will be affected by increased long-term external debt.

An increase in the use of fossil fuel for thermal plants presents the highest risk for the global environment. In the short and middle term, an increase in the "petroleum bill" issue will be passed directly to the people as the "thermic factor."

Despite the uncertainties regarding climate change, clean energy is more realistic for the middle and long term. Most current climate change scenarios consider impacts around the year 2050. By that time, most of the current projects will no longer be in operation, and even the ones considered for construction around the year 2010 will be near the end of their life expectancy. Thus, it is necessary to plan new mitigation and adaptive measures that consider sustainable management of the drainage basins in terms of land use as well as a constant evaluation of the soft component of future hydropower plants. As part of this evaluation, state-of-the-art tools, such as remote sensing, digital elevation models, and geographic information systems, will give new insights into natural resource management in the tropics (Sánchez-Azofeifa and Harriss 1994). Also, public awareness of the potential impacts of climate change must be increased through education.

Responses to climate change in Costa Rica and Panama must be considered not only as nationwide plans, but also as part of a regional initiative. For example, Costa Rica's annual electricity deficits are supplied by Honduras, and its surplus is exported to Nicaragua and Panama (Figure 8). Therefore, the construction of new hydropower plants must consider the interdependencies among countries.

Acknowledgments

Funding for this work was provided by the U.S. Country Studies Program, Federal Grant EPA-CX822535-010. Arturo Sánchez was supported by a Fulbright scholarship for Ph.D. studies at the University of New Hampshire. The authors thank the Central America Committee for Hydraulic Resources, Central America Commission on Environment and Development, and Central America Project on Climate Change for their support. Special thanks to J. Smith and the reviewers for their contribution.

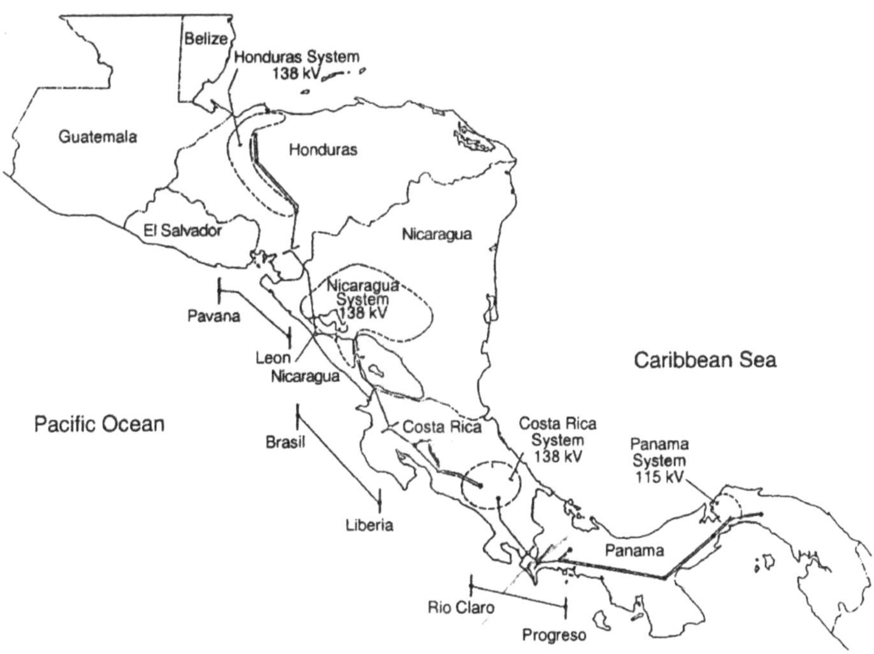

Figure 8. Central American Interconnected System (Source: Panama 1994)

References

Brenes, A., and F. Saborio, 1994, "Changes in the General Circulation and Its Influence on Precipitation Trends in Central America: Costa Rica," *Ambio* 23(1):87-90.

Costa Rica, 1989, *Vulnerabilidad del Sector Energia ante las Amenazas Naturales,* Ministerio de Recursos Naturales, Energia, y Minas, San Jose, Costa Rica (in Spanish).

Costa Rica, 1990a, *Diagnostico del Sector Energia*, Ministerio de Recursos Naturales, Energia, y Minas, San Jose, Costa Rica (in Spanish).

Costa Rica, 1990b, *II Plan Nacional de Energia*, Ministerio de Recursos Naturales, Energia, y Minas, San Jose, Costa Rica (in Spanish).

Costa Rica, 1992, *Memoria Estadistica del Sector Energia de Costa Rica*, Ministerio de Recursos Naturales, Energia, y Minas, San Jose, Costa Rica (in Spanish).

Houghton, J.T., G.J. Jenkins, and J.J. Ephraums (eds.), 1990, *Climate Change, The IPCC Scientific Assessment*, Cambridge University Press, Cambridge, United Kingdom.

Klemês, V., 1993, "Design Implications of Climate Change," in T.M. Ballentine and E.Z. Stakhiv (eds.), *Proceedings 1st International Conference on Climate Change and Water Resources Management*, U.S. Army Corps of Engineers.

Linsley, R.E., and J.B. Franzini, 1976, *Ingenieria de los Recursos Hidraulicos*, Compania Editorial Continental, S.A. Mexico (in Spanish); also published by McGraw-Hill Book Company, N.Y., USA, 1984.

Panama, 1994, *Historian Electric National*, Institute de Recursos Hidraulicos y Electrificacion, Ciudad de Panama, Panama (in Spanish).

Sánchez-Azofeifa, G.A., 1993, "Hydrology, Energy and Sustainable Development in Costa Rica," in *International Workshop on Climate Variability, Global Change and Their Impacts in Latin America and the Caribbean*, National Meteorologic Institute, San Jose, Costa Rica.

Sánchez-Azofeifa, G.A., and R. Harriss, 1994, "Remote Sensing of Watershed Characteristics in Costa Rica," *International Journal of Water Resources Development* 102:117-130.

Stakhiv, E.Z., 1993, "Water Resources Planning and Management under Climate Uncertainty," in T.M. Ballentine and E.Z. Stakhiv (eds.), *Proceedings 1st International Conference on Climate Change and Water Resources Management*, U.S. Army Corps of Engineers.

Trenberth, K.E., 1990, "Atmospheric Circulation Changes and Relationships with Surface Temperature and Precipitation," in *Observed Climate Variations and Changes: Contributions in Support of Section 7 of 1990 IPCC Scientific Assessment*, Boulder, Colo., USA.

Managing Water Resources for Climate Change Adaptation

Eugene Z. Stakhiv
Institute for Water Resources
U.S. Army Corps of Engineers
Alexandria, Virginia 22315-3868, USA

Abstract

Water resources managers have historically led efforts to respond to natural climate variability, shifts in population and resultant demands, and changes in public preferences that influence the value and uses of water. Therefore, the potentially serious effects of climate change (both floods and droughts) forecast by the Intergovernmental Panel on Climate Change and the General Circulation Models are not expected to result in similar adverse consequences for water resources or management, even in the semi-arid areas of the United States. This conclusion is based on a comprehensive study conducted by the U.S. Army Corps of Engineers on 10 separate river basins and municipal water supply systems in the United States. No direct linear correlation was found between hydrologic response sensitivity and water management vulnerability; the effects on water resources will depend on the ability of agencies and institutions to address contemporary water resources problems.

Introduction

The Intergovernmental Panel on Climate Change (IPCC) Response Strategies Working Group (RSWG) (IPCC 1990) listed near-term strategies or adaptive actions for water resources management that should be implemented whether or not the climate changes. These strategies included the following:

- Determining the flexibility and vulnerability of current hydrologic and water management systems;
- Improving systemwide operation;
- Enhancing scientific measurement, monitoring, knowledge, and forecasting;
- Implementing water conservation measures;

- Addressing the escalating demand for water through proper pricing;
- Establishing institutional mechanisms to ensure that water is directed to where it is most productive;
- Modifying agricultural irrigation practices;
- Implementing design modifications and changes in operation;
- Improving flood plain management, and warning and evacuation systems; and
- Protecting the quality of estuarine areas and adjacent groundwater that is susceptible to saltwater intrusion from potential sea-level rise.

In addition, the IPCC RSWG defined a series of principles to guide contemporary resources managers in evaluating, selecting, and implementing appropriate adaptive response strategies. These strategies should be undertaken when they

- Are beneficial for other reasons and justifiable under current evaluation criteria;
- Are efficient and cost-effective;
- Serve multiple social, economic, and environmental purposes;
- Are adaptable to changing circumstances and technological innovation;
- Are compatible with the concept of sustainable development; and
- Are technically feasible and implementable.

The list of adaptive actions and principles recommended by RSWG to respond to potential global warming is familiar to most water resources managers in the United States and other developed countries. It is virtually the same strategy proposed following the two National Water Assessments conducted by the U.S. Water Resources Council (1968, 1978); in the groundbreaking, comprehensive, policy-oriented report of the National Water Commission (1973); by President Carter's Water Policy Initiatives Task Force (1978); and in countless subsequent studies conducted by commissions, task forces, and working groups. The list mirrors the actions, reforms, and initiatives included in the "National Action Plan for Global Climate Change" (U.S. Department of State 1992), prepared by the United States to fulfill its obligations to the United Nations Framework Convention on Climate Change.

The subsequent "U.S. Climate Action Report" (1994) outlines adaptive strategies for water resources. The activities chosen by the federal government as focus areas to facilitate adaptation consist of five complementary approaches:

- Improving demand management through conservation and market-oriented pricing;
- Improving supply management (by improving coordination and joint management of groundwater and surface water supplies and of reservoirs and reservoir systems);
- Facilitating water marketing and related types of water transfers;
- Improving planning for floods and droughts; and
- Promoting the use of new analytical tools that enable more efficient operations.

What the "U.S. Climate Action Report" does not discuss in detail is that, for each approach suggested, federal and state water resources agencies have already undertaken a long list of actions and efforts that fulfill the objectives of the original IPCC RSWG and both U.S. national action plans for water resources adaptation. Moreover, little difference is found among the various documents in terms of the adaptive measures proposed. It would be difficult to develop a set of substantive actions that could be undertaken by water resources managers to respond to climate change, *other* than those listed by the IPCC, that have not already been implemented or are in the process of being implemented by federal water agencies. Federal water agencies are responding quite well, albeit deliberately, to the various proposed policy reforms that have gained widespread acceptance and have been promulgated as agency regulations, planning procedures, design criteria, and operating rules. The deliberate nature of these agencies' responses simply reflects slowly changing social priorities and the difficulty of mobilizing the political process to respond in a timely manner. The most recent example of this political process at work is the range of concrete proposals and policy responses to the devastating floods of the Upper Mississippi River basin in 1993 (U.S. Interagency Floodplain Management Review Committee 1993).

The reality is that many of the adaptive response strategies formulated by the IPCC (1990), the U.S. National Academy of Sciences (1992), and other water policy commissions have actually been *derived* from actions and conventional practices undertaken by various of federal, state, local, and private agencies responsible for water management in the United States and the European Community. Despite the literature's suggestion regarding climate change, these adaptive measures represent conventional practices rather than possibilities for dealing with climate change. However, the measures are being introduced as a diverse set of normal business and sound engineering practices driven largely by changes in demands for water and conflicts among water uses and sectors

The solutions are dispersed in myriad legislative authorities, conventional design practices, planning procedures, operating rules, cost-sharing policies, and life-cycle project management approaches. What is missing is a coherent strategy that reaches beyond the federal water management sector and explicitly integrates the potential effects of climate change. The "U.S. Climate Action Report" (1994) attempts to develop that coherent strategy, but does not explicitly address factoring climate uncertainty into conventional operating decisions, design standards, and project analysis.

Water Resources Managment under Climate Change Uncertainty

The real question, then, is not *whether* nations are capable of adapting to future climate uncertainty, but whether they can conduct meaningful analyses of the potential effects of climate change, given the poorly developed state of the General Circulation Models (GCMs), and use that information to make decisions today about the need for future adaptive measures. Is the available information adequate for making important water management decisions? Rogers (1992) believes that water managers and planners do not need to be concerned with changes that could occur beyond 30 or 40 years in the future. He argues that, in the future, uncertainties in water demands and supplies will make little difference in scheduling and sizing water management systems.

Society is continuously adapting, incrementally, to a variety of changes in diverse ways (Goklany 1992). The polices, procedures, technologies, and practices being carefully integrated into the current water management philosophy take into account both contemporary climate variability and rapidly changing public preferences that influence the value of water. The manner in which water resources management has evolved makes it responsive to changes in information, technology, and public preferences. These factors ensure that water resources managers will respond effectively to changes that may be required to adapt to the projected rate of climate change. The key policy debate is whether society should *anticipate* or *incrementally adapt* to the potential change. The choice of an anticipatory path requires a profound investment and behavioral changes that cannot be justified by climate change analyses completed to date.

The second fundamental issue is not *whether* adaptation should occur — it is already an integral part of water resources management — but when and how adaptation will occur. Those questions can be responsibly answered only by considering the costs and ultimate benefits of the adaptive measures, the risks and uncertainties inherent in any strategic planning, and the availability of innovative technologies that can be implemented through increased investment in research and development. Many of today's problems result from misallocation, waste, and incorrect pricing of water. These problems, which are being slowly but steadily rectified in all aspects of water management, should not be confused with an inability to adapt technically feasible solutions to changing conditions. The availability, relative effectiveness, and technical implementability of virtually all water management options are very well known because water resources managers have been conducting economic and financial analyses of their projects and systems for nearly 50 years. Of course, the ultimate success of any water resources management system depends on the capacity of the individual country to adapt the wide range of existing management resources to match its needs.

This failure to differentiate between the technical feasibility of various adaptive measures and the relative capacity to implement well-known, accepted, and relatively conventional water management practices has led to considerable confusion in the debate about the relative susceptibility of societies to the socioeconomic consequences of climate change. Developing nations in water-rich areas may have more difficulty implementing conventional water supply measures to meet the ever-increasing water demand caused by rapid population growth than countries in arid or semi-arid areas. In his study of the water resources needs of the Middle East, Rogers (1994) notes that in most countries that have a water delivery system in place for irrigation, a reduction of 10% of the water currently used for agriculture would meet the increasing demands of cities and industries through the year 2025. In countries where water delivery systems are poorly developed because of the abundance of natural surface water supplies, susceptibility to water resource problems may increase because of water pollution, increased demands, and climate change. Again, the problems would have little to do with climate change or variability, but everything to do with management and the relative importance attributed to efficient use of resources.

It would be useful to differentiate hydrologic *sensitivity* to climate change from water management *vulnerability* and societal *susceptibility* to economic disruptions and dislocation as a consequence of climate change. Hashimoto et al. (1982a,b) introduced a taxonomy to account for the risk and uncertainty inherent in water resources system performance evaluation. The following five definitions expand upon the key components of more traditional engineering reliability analysis (i.e., they focus on the sensitivity of parameters and decision variables to uncertainty, including some aspects of strategic uncertainty):

- *Robustness* — describes the overall performance of a water resource project under the uncertainty of future demand forecasts, complementing traditional benefit-cost analysis. This extends Fiering's (1976) definition from sensitivity of system design parameters to variability in future events to sensitivity of total costs to variability in forecasts.
- *Reliability* — a measure of how often a system is likely to fail.
- *Resiliency* — how quickly a system recovers from failure (floods, droughts).
- *Vulnerability* — how severe the consequences of failure may be.
- *Brittleness* — the capacity of "optimal" solutions to accommodate unforeseen circumstances related to an uncertain future.

Water management vulnerability, then, is a function of hydrologic sensitivity (as input to the managed system) and the relative performance (robustness) of a water management system as it affects societal susceptibility. Without the capacity to successfully manage its water, society becomes increasingly susceptible both to population-driven increases in water demands and to climate change variability. In many cases, improving the capacity of developing nations to implement sound water management practices may be a larger barrier than the potentially drastic changes in water availability caused by climate change (World Bank 1993).

Adaptive Measures

This section discusses the relative success of the adaptive measures outlined in the "National Action Plan for Global Climate Change" (U.S. Department of State 1992). The objective is to examine the feasibility and effectiveness of such measures in addressing current climate variability and their potential for meeting the challenges associated with future climate change. The discussion clearly shows that many agencies are routinely implementing the five initiatives specified in the U.S. National Action Plan to solve contemporary water resources problems. Current water management practices are considered to be sufficient to address future climate change and variability.

The evaluation assumes that the technical and financial capacity exists to implement these solutions; conclusions are drawn entirely from examples in the United States. It is important to remember that many of the problems in water resources management faced by nations today would exist *even without* the threat of climate change. Most countries would have to implement sound water resources management measures and strategies merely to meet the needs of increasing populations and greater concentration of those populations in urban areas. Demand-driven water management actions will invariably precede adaptation driven by climate change, based on our best information regarding the rate of population growth versus that of a doubling of carbon dioxide concentration ($2xCO_2$) (World Bank 1993). If societies cannot meet the challenges associated with population growth and the resultant increases in agricultural, industrial, and municipal water demand, they will not be able to cope with the consequences of climate change. On the other hand, if countries confront the need to manage the foreseeable consequences of growth and development, the same water resources management actions will serve as a platform for adaptation to climate change.

Another issue that comes up repeatedly in climate change impact studies is the failure of researchers to differentiate risk, uncertainty, and residual risks in assessing future vulnerability. Societies are already exposed to many natural hazards, which present a known risk as well as a considerable degree of variability and uncertainty (i.e., the timing and incidence of the hazards). Water resources management can offer a high degree of reliability in the delivery of services and significantly reduce the known risks. However, residual risks will always surface because the cost to ensure a risk-free society makes such an undertaking impossible. Therefore, the definition of vulnerability is inherently linked to the degree of residual risk a community or society is willing to bear and to the amount of resources that society is willing to mobilize to reduce each additional increment of risk and uncertainty. The degree of risk-bearing, or vulnerability, is, paradoxically, a choice to be made by society. There is no fixed criterion for social risk-bearing; different societies and communities make different choices about the degree of flood protection or water supply reliability that they are willing to pay for.

Demand Management

Since publication of the National Water Commission report in 1973, supply-side planning by U.S. federal water agencies has shifted fundamentally to demand-side management and more effective operation of existing systems. The reason for this shift is that water resources managers have responded to changes in the demands for water and in public preferences regarding the relative value of water resources for different purposes. On a national scale, the combination of countless small regulatory, legal, economic, and technological actions has resulted in a major shift in water-use trends. As late as 1975, researchers — even the highly regarded U.S. National Water Commission — were forecasting water-use values three and four times higher than actual water use. Credible contemporary estimates of water use for 1980, 1985, and 1990 (Solley et al. 1993) are superimposed in Figure 1, which also shows the range of water-use forecasts developed by many respected and influential U.S. commissions. Water quality regulations, which forced increased industrial recycling, account for the largest proportion of the reduction in water use. This reduction was accomplished despite a doubling of manufacturing activity between 1970 and 1980 (Foxworthy and Moody 1985).

Effective demand management requires a careful balancing of economic, efficiency, equity, and environmental considerations (Frederick 1993). It also requires increased flexibility on the part of the institutions that manage and

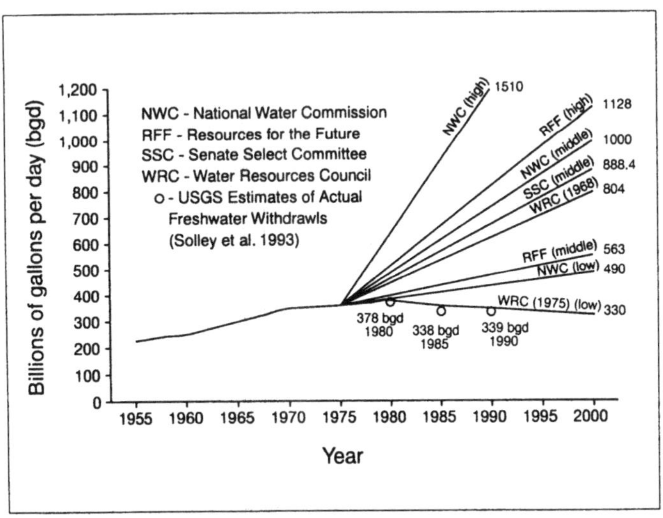

Figure 1. Historic and Projected Freshwater Withdrawals (1955-2000)

allocate water. Many of these changes are now taking place. The California drought of 1987-1992 is the most recent and best documented example of how the adaptive process works. Although not an efficient process, it seems to be the only way that complex democratic societies can resolve sometimes controversial resource management issues. California's previous drought (1976-1977) resulted in many reforms in operating and regulating the state's extensive water management system and in implementing a few significant institutional and water demand management measures. In fact, although population grew significantly (+27%) in the decade between the two droughts (1980-1990) and is projected to grow another 40% by 2010 (U.S. Army Institute for Water Resources [IWR] 1994a), total water use in California has actually declined. While the amount of water used in the cities increased from 14% (in 1980) to 16% (in 1985) of total water use in California, water used for irrigation declined from 84% (in 1980) to 81% (in 1985). So total water use dropped by 2% in that five-year period and is not expected to reach 1980 levels again until the year 2010. During the droughts, especially in 1991 and 1992, agricultural water use declined by about 20-30%.

The 1976-1977 drought prompted significant reforms in water supply management and, more important, a considerable reduction in the demand for water. The recent and more intense drought led the state to study and adopt an even broader set of water management actions. These included a range of strategic (long-term), tactical (short-term), and contingency (emergency) measures that were available but never implemented for a variety of legal, institutional, or technical reasons. Long-standing legislative constraints on the operational flexibility of California's Central Valley Project were removed by Congress, which also reallocated water for in-stream uses to protect aquatic habitat. The California State Water Control Board published a Water Rights Decision requiring stricter adherence to water quality standards during droughts, again to protect aquatic ecology, especially in the Sacramento-San Joaquin Delta. Price and water-use rationing were used in conjunction in urban areas, resulting in a successful program of urban water conservation during the drought. These controls will have lasting effects in the future (U.S. Army IWR 1994a).

The drought forced water resources agencies and professionals to seek creative solutions to water shortage problems, showing that market forces can be used effectively to reallocate restricted water supplies. In 1991, California established the Drought Emergency Water Bank at the height of an accumulation of six years of water deficits. More water was offered for sale than buyers were willing to purchase, even at a reasonable price of \$125 1,233 m^{-3}. By comparison, the cost of developing new water supply sources today ranges by a factor of three. Many more administrative, institutional, and legislative changes were made at all levels of government. These will form the basis for a new water management ethic until the next major perturbation stresses the system. The key point is that the measures implemented by California are the same measures offered for adaptation to climate change. As the California experience demonstrated, society is continuously adapting to the combined forces of climate variability and shifting demands. Gleick et al. (1995) drew similar conclusions in their analysis of California's water needs: "...the changes necessary to achieve a sustainable water future for California can be brought about by encouraging and guiding positive trends that are already underway."

Efforts like those in California are under way in many river basins and urban areas, with comparable results. The Boston metropolitan area offers another example of the focus on demand management that is yielding dramatic results. Increases in the price of water resulted in a 25% decrease in water use in the greater Boston metropolitan area from 1988 to 1992 (Figure 2). It is expected that water use will continue to decline in the City of Boston, mainly because of the national plumbing code standards passed as part of the Energy Policy Act of 1992 and additional price increases (Figure 3). All of these changes and forecasts were made possible through the development and application of innovative water demand forecasting models and simulation models that were reported to Congress and developed as part of the U.S. Army IWR (1994b) "National Study of Water Management during Drought."

Improved Supply and System Management

The U.S. Army Corps of Engineers conducted two recent systemwide operation review studies for the Columbia and Missouri Rivers (U.S. Army Hydrologic Engineering Center 1991a,b). Both studies were prompted by changing water

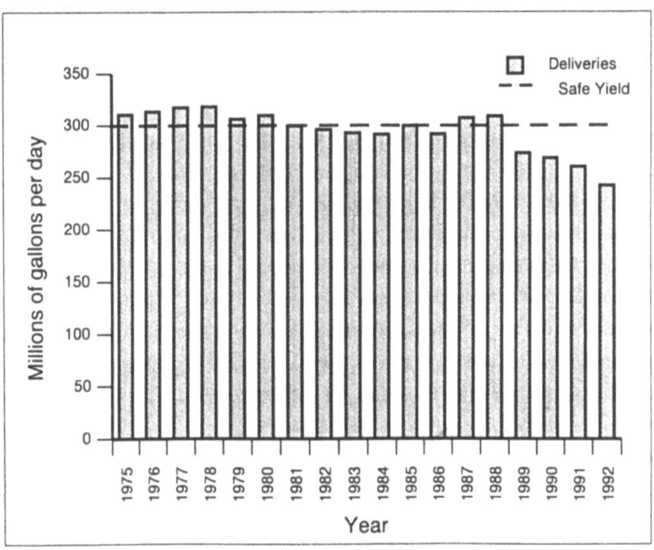

Figure 2. Water Use in the Boston Area (1975-1992)

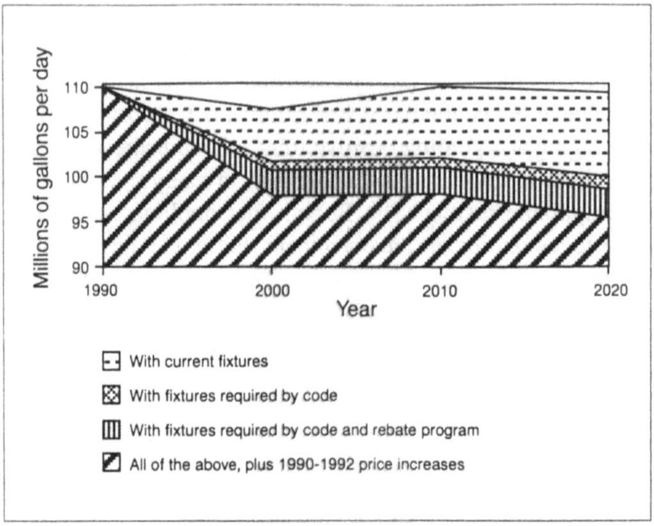

Figure 3. Water Use Forecasts for Boston under Four Scenarios

uses and increasing demands that were not foreseen during the design of the systems, and both were accentuated by increasing conflicts among users that were exacerbated by recent experiences in climate extremes, floods and droughts.

In the U.S. Columbia basin, increased irrigation demands compete with the need for hydropower and the need to enhance the dwindling anadromous fishery. In the Missouri basin, the recent three-year drought prompted a review of the region's existing Master Water Control Manual in light of the large increase in recreation, the importance of the Missouri River in augmenting flow to the Mississippi River for commercial navigation, and the explicit requirement that water regulation criteria for endangered and threatened species be included in the manual.

Both systems analysis operating studies introduced new social, economic, and hydrologic data and analytical tools. The most important of these tools was a mathematical programming optimization model, called a network-flow model. Considerable study was required to quantify both the supply and the competing demands on each of the systems and to develop economic benefit functions for each of the authorized water uses and for incidental water management purposes (e.g., white-water rafting, in-stream flow needs for aquatic ecosystem maintenance). Upgrading the hydrologic analysis and conducting numerous flow sensitivity tests led to a refined understanding of the groups of reservoirs as a system and the operating flexibility of that system under a broader range of climate changes and unforeseen demands.

This study combined analytical steps, such as risk analyses, economic optimization, sensitivity analyses, risk-cost trade-offs, factoring in environmental constraints, and testing countless permutations of single-reservoir and multiple-reservoir allocation and operating schemes under historically likely combinations of droughts and floods. This type of analysis should lead to improved robustness and resiliency of water management systems. These comprehensive systems approaches are the practical expressions of what a generation of theoretical water systems analysts have advocated for sensible management, both for current needs and for conditions of future uncertainty.

Water Marketing and Transfers

The recent California experience in water marketing is a successful real-life example of such a mechanism. Water marketing and flexible, voluntary water transfers are particularly useful tools for making water available for more economically efficient uses. Many administrative problems need to be resolved, and the important issue of water markets and transfers for in-stream aquatic habitat has yet to be resolved satisfactorily. The mechanisms require a system of transferable property rights, well-functioning markets, and flexible institutions (Wahl 1989). This system has not been established in many developing countries or countries with rapidly changing economies.

The basic principle underlying water marketing is that water needs to be viewed as an economic commodity. If a specific cost, passed directly to the consumer, is attached to a good, both the demand for and the supply of that good will change. Many countries, particularly those in arid and semi-arid areas, subsidize water development and delivery, so there is little incentive to conserve. Typical fees for water in the arid belt of Middle Eastern countries (e.g., Qatar, Morocco, Tunisia, Kuwait, Algeria) are currently less than or equivalent to typical rates for U.S. municipal supply (\$0.1-1.4 m^3). These fees are clearly subsidized and do not reflect the true recovery costs of water in these arid areas, where 80% of the water supply is used for irrigation (Rogers 1994).

Improved Planning

Countless water resources planning efforts are under way in the U.S. by federal, state, and local agencies. These agencies are involved in a continuous and overlapping series of river basin, watershed, urban metropolitan area, and water utility planning efforts in virtually every area of the United States that has had a recent flood, drought, or water shortage caused by increased demand. The recently completed "National Study of Water Management during Drought" (U.S. Army IWR 1994b) documented many such efforts and outlined a systematic, uniform approach to planning for water-related events.

One large-scale effort, undertaken in the aftermath of the devastating floods of the Upper Mississippi River basin, is described in a U.S. Federal Interagency

Floodplain Management Review Committee report to the Clinton Administration (1994). That report, which included recommendations that will affect flood management policies of all federal agencies for the entire nation, is a comprehensive review of all of the overlapping policies and practices. Its recommendations are being implemented through various legislative and administrative changes and executive orders. The effort stands as a classic example of the catalyzing effect that large natural hazard events have on the formulation and implementation of adaptive response mechanisms that improve efficiency and reduce vulnerability.

Countless other watershed studies that focus on restoring and protecting wetlands; combining sewer overflow; improving water quality; restoring ecosystems; developing water supplies; rehabilitating hydropower, navigation, and flood control; and restoring anadromous fish are under way in various regions of the United States. Changes in operations, policies, institutional criteria, administrative mechanisms, and technologies are being proposed and implemented in these watersheds. These adaptive measures will form the basis for subsequent improvements in the performance of those systems.

Analytical Tools

The past three decades have witnessed an explosion of computer models and analytical methods designed for use in water resources planning and management. The scale and complexity of water resources problems analyzed with the aid of computer models vary greatly. Complex models design urban storm-water detention basins in watershed areas that measure less than 1.6 km^2. Other models optimize major multiple-reservoir, multipurpose (e.g., flood control, hydropower, water supply, recreation, navigation) systems that regulate the water resources of river basins covering thousands of square miles.

In 1982, the U.S. Office of Technology Assessment (U.S. Congress OTA 1982) completed a comprehensive assessment of the use of mathematical models in water resources planning and management. Numerous experts from agencies, universities, and consulting firms participated in the study. The findings were broad in scope and included the following observations:

- Models capable of analyzing many pressing water resource issues are available and have significant potential for increasing the accuracy and effectiveness of information available to managers, decision makers, and scientists.
- Models have significantly expanded researchers' ability to understand and manage their water resources.
- Models have the potential to provide even greater benefits for making decisions about water resources in the future.
- Models are used in most areas of water resources planning and management.

Since 1982, when the OTA report was published, the use of water management models has exploded with the advent of the microcomputer. The model user community has grown dramatically, particularly local public agencies, private consulting firms, and other nonfederal users. Most of the water management models include user-friendly executable (ready-to-run) versions for desktop computers. Essentially everyone in the U.S. water management community now has convenient access to the computer hardware needed to run the available software.

A comprehensive guide to water management models is available from and in use by the U.S. water resources community through the U.S. Army IWR (1994b). The guide covers every significant hydrologic, reservoir system optimization, and water demand forecasting model developed by federal agencies and academia. Many of these models have been used to predict hydrologic perturbations and assess water management vulnerability. Consultants routinely use these models in most studies in developing nations; these studies are financed by international donor agencies.

Case Study Appraisals

Given that the U.S. has both the technical capacity and the economic and institutional resources to implement a wide range of both adaptive and anticipatory management measures, an appraisal of water resources management under climate uncertainty in the U.S. inevitably focuses on system performance. The issue is not whether the agencies and institutions can adapt to climate change, but identification of the most cost-effective solutions for improving the resilience and reliability of water resources management systems and minimizing social susceptibility to residual risks. Future appraisals must focus on better defining the ranges of risk and uncertainty inherent in basic climate change information input into the hydrologic and water management models.

The IPCC, for example, proposed several standard CO_2 emissions scenarios for analysis, along with three GCMs that use the $2xCO_2$ rates and provide temperature and precipitation outputs. The three GCMs are the *transient* versions of an earlier generation of *steady-state* models. The newer transient models generally show smaller temperature increases and lower reductions in precipitation. Many areas of the world show dramatic increases in precipitation and consequent run-off, according to the new models. Considerable difference between the transient and steady-state models was observed in each of the case studies examined.

Great Lakes

The Great Lakes of North America represent a useful system to study and test the hydrologic sensitivity of climate change, along with the relative effectiveness of adaptive water management measures and their impacts on societal susceptibility. The reason is that the Great Lakes region represents a macrocosm not only for testing how agencies and institutions could deal with hydrologic and water management changes using conventional approaches, but also for evaluating the relative effectiveness of nontraditional adaptation mechanisms. Also, the Great Lakes undergo periodic fluctuations of lake levels of 1-2 m, simulating the physical effects of sea-level rise and consequent social, economic, and environmental impacts.

In 1933, the International Joint Commission (IJC) completed a comprehensive, detailed six-year study of water management options that addressed contemporary water resources problems of the Great Lakes (IJC 1993). A rigorous and extensive planning process undertaken to analyze the range of problems that exist today, as well as emerging issues, was of critical importance in this effort. Researchers examined the implications of current and future water supply needs, navigation requirements, recreation and hydropower demands, flood damage, shore erosion control, water quality, wetlands, and fisheries issues in great detail. Numerous models, data inventories, and economic benefit-cost analyses were conducted to help to evaluate socially optimal solutions. The result was the development of numerous alternative solutions (adaptive management options) to cope with the wide range of problems.

Climate change considerations were discussed, but not explicitly brought into the evaluation and decision-making phase. This phase presented an opportunity to use the excellent sources of information developed for the conventional water resources problem-solving approach and ask the same decision makers involved in the original IJC study which alternatives they would select if (1) they knew with certainty that a particular climate change scenario would occur and (2) each of the GCM scenarios could occur with equal probability.

The modeling framework used for decision making is described in Chao et al. (1994) and Hobbs and Chao (1995). The overall results of the Great Lakes study can be summarized as follows:

- The Geophysics Fluid Dynamics Laboratory (GFDL) transient scenario is the most benign in terms of Great Lakes impacts and requires no additional adaptive measures for addressing the range of climate change impacts forecast.
- The United Kingdom Meteorological Office (UKMO) transient model predicted moderate adverse effects, especially during extended drought periods reflecting historic low lake levels. The effects were felt mostly in the economic sectors of hydropower production and periodic disruptions of navigation, especially in Lake Erie, which is the shallowest lake among the Great Lakes. Introducing a water control structure be-

tween Lake Erie and Lake Ontario would alleviate most of the problems, but could not be justified on a benefit-cost basis.
- The Max-Planck Institute (MPI) transient model predicted the most severe and widespread impacts in all sectors, especially after 2030. Water control structures that stabilize lake levels by controlling flow between lakes would ameliorate most of the negative economic, environmental, and social effects. Furthermore, the cost of these structures, which would stabilize wetland resources, fisheries, and cold-water habitat, would be economically justified.

In summary, the cadre of 20 key decision makers involved in the original IJC study, who helped to formulate, evaluate, and select the preferred solutions, under a conventional planning paradigm with steady-state climate assumptions, concluded the following:
- Climate change modeling is too uncertain to be used in making management decisions about real water resources.
- The gradual nature of climate change allows adequate time for a deliberate adaptive approach.
- Current conventional management measures are more than suitable for serving the potential range of adaptive responses suggested by the GCM outcomes.
- Sensitivity analysis of GCM outputs should be conducted only within a realistic framework of trade-offs among alternatives, with full recognition of economic opportunity costs.
- It is well within the capacity of the jurisdictions and municipalities studied to adapt to projected climate changes in the Great Lakes area, especially because the rate of change provides adequate time to adapt incrementally and cost-effectively.

Similar conclusions were derived for several other studies using a comparable methodology and a consistent set of IPCC and GCM scenarios. Only the real-time decision making and trade-offs component was omitted because the case studies were selected on the basis of relatively innovative water management strategies already in place. The objective was to test the performance of those systems and implicit water management strategies under a variety of IPCC/GCM scenarios.

Although an increasing number of studies have examined potential changes in real water resource supply systems, few have considered the feasibility or costs of adapting to change. Fewer still have used the latest IPCC (1992a) transient GCM scenarios.

One such study addressed the water management supply-and-demand impacts of climate change on a municipal system, the Washington, D.C., metropolitan area (Steiner et al. 1995). The study involved developing water demand forecasts for the metropolitan area, which includes Washington, D.C., and 11 surrounding counties. The forecasts were made on the basis of future population projections and various assumptions about future water demands in the three major water- use sectors (i.e., residential, commercial, industrial) and a sector that represents water that is unaccounted for, primarily leakage. Forecasts of water use were made by using a sophisticated desegregated water demand forecasting model for six climate change scenarios, including both steady-state and three transient GCMs based on the latest IPCC (1992a) scenarios for summer, winter, and average annual use.

The results of the analysis showed that all the scenarios for municipal water demand are very close to the stationary climate case in terms of both summer use and average annual water demand. The largest variation in summer use was only 5.6%, despite a wide range of GCM results for the region. During droughts, restrictions on outdoor watering, which represents the largest water use in the suburban metropolitan area, are typically imposed. In many cases, the results are likely to be quite different because of the importance of agricultural water use. In most countries where considerable agricultural irrigation occurs, that sector of water use accounts for 75-95% of all water withdrawals (Postel 1992).

Tacoma/Green River Water Supply

The Green River, located in the state of Washington, provides water for the Tacoma metropolitan area (near Seattle). Lettenmaier et al. (1995a) examined the sensitivity of the Tacoma water supply system to global warming in detail, with emphasis on water supply, in-stream flows, and flood impacts. The climate change scenarios used in this study were based on the following:

- Results from transient climate change experiments preformed with coupled ocean-atmosphere GCMs for the 1995 IPCC assessment. The GMCs were transferred to the local level by means of a perturbation approach (scaling by fixed monthly ratios for precipitation and fixed monthly shifts for temperature).
- A regional circulation pattern (RCP) method, which classifies regional circulation patterns based on gridded sea-level pressure over the eastern North Pacific, and corresponding sequences of gauge precipitation and temperature consistent with the regional circulation classes. (The specific methods are described in Lettenmaier et al. [1995b].)

The simulation showed that the hydrologic effects of all of the climate change scenarios evaluated (IPCC and RCP) would be similar in character to those found in other studies of snowmelt-dominated streams; the spring snowmelt run-off peak would shift toward a rainfall-dominated peak in the winter. However, despite the substantial changes in the characteristics of the reservoir inflows simulated under the altered climate scenarios, the simulations showed that the *performance* of the water supply system would only be affected slightly. Rather, the greatest impacts of the altered climate scenarios on the Green River system would be on in-stream flows and flooding.

Savannah River System

The Savannah River system is a multipurpose system of five major reservoirs operated by the U.S. Army Corps of Engineers and the Duke Power Company primarily for hydropower, recreation, and flood protection (Lettenmaier et al. 1995c). The climate change scenarios and basic approaches to this analysis were identical to those used in the Tacoma/Green River study.

Hydrologically, changes in precipitation and, to a lesser extent, potential evapotranspiration would be the most important factors affecting the hydrology of the Savannah River system under the climate change scenarios investigated. Simulated annual average reservoir inflows were higher for two of the three sets of IPCC transient climate change scenarios (GFDL and UKMO) and slightly lower for the third (MPI) scenario, compared with historical averages. The RCP $2XCO_2$ simulations are much drier, and summer flows especially are much lower.

In terms of the three major economic uses of the reservoir system, the net effect of the climate change scenarios was positive for two of the three (GFDL and UKMO) and nearly neutral for the third (MPI). The largest average annual change, +$18.8 million, occurred for the GFDL scenarios in Decade 5. Changes for the UKMO scenarios were about one-half as large, and for the MPI scenarios, the net changes were close to zero.

Investigation of the environmental performance of the system showed that, for the GFDL and UKMO scenarios, system performance in terms of spawning constraints (on the basis of May-June lake-level variations) would be about the same as for the historical simulation. For the drier MPI scenarios, there would be a slight increase in the fraction of violations of the spawning constraint. The average minimum flow at Augusta would be relatively unchanged for the GFDL and MPI scenarios; the average minimum flow would be about 5% lower for the most extreme MPI scenario. Reservoir system performance, as measured by refill characteristics, would be only slightly changed under all of the altered climate scenarios. The robustness of the system performance in response to changes in the hydrology is primarily due to the large amount of conservation storage in the system and the relatively conservative policies presently used to operate the system.

Boston/Massachusetts Water Resources Authority

The Massachusetts Water Resources Authority (MWRA) system provides water for the City of Boston and 45 surrounding communities. The National Weather Service River Forecast System soil moisture accounting and snowmelt models simulated daily flows (aggregated to monthly) of the Ware River at Colbrook, Massachusetts, and the Connecticut River at Montague, Massachusetts. The MWRA system supply (to Quabbin and Wachussett reservoirs) was indexed to the Ware River at Colbrook, while low-flow reservoir release requirements were indexed to the Connecticut River at Montague. The climate change scenarios used in this study are based on results from transient climate change experiments performed with coupled ocean-atmosphere GCMs for the 1995 IPCC assessment; the results are transferred to the local level by means of a perturbation approach, as described in Lettenmaier et al. (1995d).

Hydrologically, the response of the MWRA system inflows to climate change is somewhat similar to that observed in other studies of streams in temperate climates. For the warmer climate scenarios, the historical run-off peak in the Connecticut and Ware Rivers, which occurs in early summer, would shift to late winter because of reduced snow accumulation. However, because the capacity of the MWRA system reservoir is large relative to the demand, reservoir system performance is more sensitive to long-term changes in the mean and statistics of annual stream flows than to seasonal changes. In fact, system performance sensitivity was found to be greatest for the MPI scenarios, which have much smaller temperature changes than the GFDL and UKMO runs, but which have the largest precipitation changes (precipitation decreases primarily from late spring through early fall).

The main conclusion of this study was that the MWRA system is relatively insensitive to climate change, as long as system demand remains at or near its current level. MWRA system demand has decreased by more than 20% since 1987, primarily because of system loss reduction, rate increases, and a regional economic downturn. By using current demand, the only effect of any of the climate change scenarios would be to slightly increase MWRA revenues (because of increased demand). No system failures would occur.

A slightly higher demand (1990 level) would still show no system failures under any of the 10 scenarios associated with two of the climate model simulations (GFDL and UKMO). For the MPI model scenarios, the system would be unable to meet full system demand in a few months. However, the deficiencies would be minor and could be entirely mitigated by drought management.

The results of this study contrast sharply with those of a similar study of the MWRA system by Kirshen and Fennessey (1993), who found that system safe yield would decrease by as much as 40% for some climate scenarios. Although the previous study used a baseline demand considerably higher than present, the difference in outcomes appears to be related primarily to the use of current generation GCM results in Lettenmaier et al. (1995d) (1995 IPCC transient scenarios). The most recent scenarios differ from those used earlier in that (1) they are transient, rather than steady state, and account for thermal lags (especially caused by ocean coupling); (2) global warming is predicted on the basis of assumptions that result in lower cumulative emissions than the $2XCO_2$ runs used in most previous studies; and (3) the higher resolution of the more recent results has resolved some model deficiencies, notably undersimulation of summer rainfall along the east coast of North America.

Conclusions

Water resources management, as practiced in the United States, has been demonstrated to be relatively robust and resilient to the range of anticipated effects of climate change. However, this finding does not mean that serious adverse effects will not occur in regions other than those studied. The case studies merely show that the current approaches to water management are virtually the same as those suggested for adaptation to climate change and that they have been demonstrated to be cost-effective and reliable responses. Researchers acknowledge that *residual risks* of system failure and associated adverse social, economic, and environmental consequences will still be associated with water management. No system is currently designed to be fail-safe, and it is not expected that the planning and design strategy will change in the future. If anything, greater reliance on nonstructural measures (e.g., flood warning and evacuation, water conservation) increases societal susceptibility to extreme natural hazards.

A crucial component of an analysis that focuses on the responsiveness and performance of existing water management adaptation options is the assumption that agencies and institutions have the capacity to implement such measures. Conventional water management approaches have proved to be cost-effective and reliable in ameliorating the impacts of natural hydrologic hazards caused by contemporary climate variability. These same management measures can serve as the basis for climate change adaptation *only if* the individual countries have the capacity and capability to implement and manage those solutions. Water resources management can decrease susceptibility to climate change once a socially acceptable level of residual risk is determined. Clearly, if those solutions cannot be implemented because of financial constraints, institutional failure, or technical deficiencies, societal susceptibility will increase. In that event, however, normal trends in population growth and increases in water demands will force those countries to face the same water resources issues well before climate change emerges as a serious problem.

References

Chao, P., B. Hobbs, and E.Z. Stakhiv, 1994, "Evaluating Climate Change Impacts on the Management of the Great Lakes of North America," in L. Duckstein and E. Parent (eds.), *Engineering Risk and Reliability in a Changing Environment*, Proceedings of NATO ASI, Deauville, France, Kluwer Academic Publishers, Dordrecht, the Netherlands.

Fiering, M.B., 1976, "Reservoir Planning and Operation," in H.W. Shen (ed.), *Stochastic Approaches to Water Resources*, Vol. 2, Chap. 17, Water Resources Publishers, Fort Collins, Colo., USA.

Foxworthy, B.L., and D.W. Moody, 1985, "National Perspective on Surface Water Resources," in *National Water Summary 1985 — Hydrologic Events and Surface Water Resources*, Paper 2300, U.S. Geological Survey, Water Supply, Reston, Va., USA.

Frederick, K.D., 1993, *Balancing Water Demands with Supplies — The Role of Management in a World of Increasing Scarcity*, World Bank Technical Paper No. 189, The World Bank, Washington, D.C., USA.

Gleick, P., P. Loh, S. Gomez, and J. Morrison, 1995, *California Water 2020 — A Sustainable Vision*, Pacific Institute, Oakland, Calif., USA.

Goklany, I., 1992, "Adaptation and Climate Change," draft paper presented at the Annual Meeting of the American Association for the Advancement of Science (Feb. 6-11).

Hashimoto, T., J.R. Stedinger, and D.P. Loucks, 1982a, "Reliability, Resiliency and Vulnerability Criteria for Water Resource Systems Performance Evaluation," *Water Resources Research* 18(1)14-20.

Hashimoto, T., D.P. Loucks, and J.R. Stedinger, 1982b., "Robustness of Water Resources Systems," *Water Resources Research* 18(1)21-26.

Hobbs, B., and P. Chao, 1995, *Evaluation and Decision Making under Climate Uncertainty — Resource Management of the Great Lakes*, draft report submitted to the U.S. Army Institute for Water Resources, Fort Belvoir, Va., USA.

IJC, 1993, *Methods of Alleviating the Adverse Consequences of Fluctuating Water Levels in the Great Lakes — St. Lawrence River Basin*, report by the International Joint Commission to the governments of Canada and the United States, Washington, D.C., USA.

IPCC, 1990, *Climate Change — The IPCC Response Strategies*, report prepared by the Response Strategies Working Group, Intergovernmental Panel on Climate Change, Island Press, Washington, D.C., USA.

Kirshen, P.H., and N.M. Fennessey, 1993, *Potential Impacts of Climate Change upon the Water Supply of the Boston Metropolitan Area*, report 68-W2-0018, submitted to the U.S. Environmental Protection Agency, Washington, D.C., USA.

Lettenmaier, D.P., S. Fisher, R.N. Palmer, S.P. Millard, and J.P. Hughes, 1995a, *Water Management Implications of Global Warming: The Tacoma Water Supply System*, report prepared for the U.S. Army Institute for Water Resources, Fort Belvoir, Va., USA.

Lettenmaier, D.P., G. McCabe, and E.Z. Stakhiv, 1995b, "Global Change: Effect on Hydrologic Cycle," in *Handbook of Water Resources*, L. Mays (ed.) (in press).

Lettenmaier, D.P., A.W. Wood, R.N. Palmer, S.P. Millard, J.P. Hughes, and S. Fisher, 1995c, *Water Management Implications of Global Warming: The Savannah River System*, report prepared for the U.S. Army Institute for Water Resources, Fort Belvoir, Va., USA.

Lettenmaier, D.P., A.E. Kerzur, R.N. Palmer, and S. Fisher, 1995d, *Water Management Implications of Global Warming: The Boston Water Supply System*, report prepared for the U.S. Army Institute for Water Resources, Fort Belvoir, Va., USA.

National Water Commission, 1973, *Water Policies for the Future: Final Report to the President and to the Congress of the United States*, U.S. Government Printing Office, Washington, D.C., USA.

OTA, 1982, *Use of Models for Water Resources Management, Planning and Policy*, Office of Technology Assessment, U.S. Government Printing Office, Washington, D.C., USA.

Postel, S., 1992, *Last Oasis — Facing Water Scarcity*, W.W. Norton and Company, New York, N.Y., USA.

President Carter's Water Policy Initiatives, 1978, Report to Congress, Washington, D.C., USA (May 12).

President Carter's Water Policy Directives, 1978, Directive to the Federal Agencies to Implement Water Policy, Washington, D.C., USA (July 12).

Rogers, P., 1992, "What Managers and Planners Need to Know About Climate Change and Water Resources Management," in T. Ballentine and E.Z. Stakhiv (eds.), *Climate Change and Water Resources Management*, Proceedings of the U.S. Interagency Conference, U.S. Army Institute for Water Resources, Fort Belvoir, Va., USA.

Rogers, P., 1994, "Water Resources Management Needs in the Arab World," in P. Rogers and P. Lydon (eds.), *Water in the Arab World: Perspectives and Prognoses*,Harvard University Press, Cambridge, Mass., USA.

Solley, W.B., R.R. Pierce, and H.A. Perlman, 1993, *Estimated Use of Water in the United States in 1990*, U.S. Geological Survey Circular 1081, Reston, Va., USA.

Steiner, R., N. Ehrlich, J.J. Boland, S. Choudhury, W. Teitz, and S. McCusker, 1995, *Water Resources Management in the Potomac River Basin under Climate Uncertainty*, draft report prepared by the Interstate Commission on the Potomac River Basin, Maryland, for the U.S. Army Institute for Water Resources.

U.S. Army Hydrologic Engineering Center, 1991a, *Missouri River System Analysis Model — Phase I*, PR-15, Davis, Calif., USA.

U.S. Army Hydrologic Engineering Center, 1991b, *Columbia River System Analysis Model — Phase I*, PR-16, Davis, Calif., USA.

U.S. Army IWR, 1994a, *Lessons Learned from the California Drought (1987-1992) — Executive Summary*, IWR 94-NDS-6, Institute for Water Resources, Fort Belvoir, Va., USA.

U.S. Army IWR, 1994b, *National Study of Water Management during Drought*, report to the U.S. Congress, IWR 94-NDS-12, Institute for Water Resources, Fort Belvoir, Va., USA.

U.S. Climate Action Report, 1994, *Submission of the United States of America under the United Nations Framework Convention on Climate Change*, U.S. Government Printing Office, Washington, D.C., USA.

U.S. Department of State, 1992, *National Action Plan for Global Climate Change*, Bureau of Oceans and International Environmental and Scientific Affairs, U.S. Government Printing Office, Washington, D.C., USA.

U.S. Interagency Floodplain Management Review Committee, 1994, *Sharing the Challenge: Floodplain Management into the 21st Century*, report to the Administration Floodplain Management Task Force, U.S. Government Printing Office, Washington, D.C., USA.

U.S. National Academy of Sciences, 1992, *Policy Implications of Greenhouse Warming — Mitigation, Adaptation, and the Science Base*, National Academy Press, Washington, D.C., USA.

U.S. Water Resources Council, 1968, T*he Nation's Water Resources — The First National Assessment*, U.S. Government Printing Office, Washington, D.C., USA.

U.S. Water Resources Council, 1978, *The Nation's Water Resources — 1975-2000: Second National Water Assessment*, Summary Report, U.S. Government Printing Office, Washington, D.C., USA.

Wahl, R.W., 1989, "Markets for Federal Water: Subsidies, Property Rights, and the Bureau of Reclamation," in *Resources for the Future*, The Johns Hopkins University Press, Baltimore, Md., USA.

World Bank, 1993, *Water Resources Management*, Policy Paper, Washington, D.C., USA.

Potential Effects of Sea-Level Rise in the Pearl River Delta Area: Preliminary Study Results and a Comprehensive Adaptation Strategy

H. Yang, Division of Marine Policy
China Institute for Marine Development Strategy
No. 8 dahuisi Haidianqu,
Beijing 100081, People's Republic of China

Abstract

Predictions show that the sea level in the Pearl River Delta region, in south-central Guangdong, People's Republic of China, will rise by 40-60 cm by the end of 2050. A preliminary assessment was made to determine the vulnerability of Pearl River Delta coastal areas to sea-level rise. The many potential effects are analyzed, and a comprehensive strategy is presented for preventing or mitigating those effects by applying practical adaptive measures. In addition, the total cost of implementing the recommended measures is evaluated. The cost would be about 0.12% of the annual local gross domestic product (GDP), or 3.5% of the annual local government expenditure anticipated during the next 110 years. If these measures are not adopted, the average annual economic loss for the next 110 years will be 121.9 billion RMB yuan, which amounts to 54% of the annual GDP of this deltaic coastal area.

Introduction

The coastal region (1.27 million km^2) of the People's Republic of China (hereafter referred to as China) includes Tianjin and Shanghai D.C.; Liaoning, Hebei, Shandong, Jiangsu, Zhejiang, Fujian, Taiwan, Guangdong, Hainan provinces; and Guangxi Zhuangzu Zizhiqu. This region accounts for 13.24% of the total land area of the country, yet it supports 40.2% of China's population and contributes 55% of the country's gross agricultural and industrial output (Yang and Tian 1994). The narrow 40-50-km-wide portion of this region along the coastline (called the "coastal zone" in China's Climate Change Country Study [CCCS]) includes 44 coastal cities with prefecture status, 35 coastal cities with county status, and 111 coastal counties or districts from two D.C. and

9 provinces (not including Taiwan, Hong Kong, and Macao). Although this coastal zone (0.278 million km^2) makes up 2.9% of area of the country, its population constitutes 13.43% of the total, which makes the population density 4.7 times the average for all of China. The region's total social output value is 28.8% of the total, and the output value per unit is 9.9 times the average for the country. This socially and economically important coastal zone is the main area that would be affected by a sea-level rise due to global climate change.

Because most areas within the coastal zone are very sensitive to changes in sea level, several institutions and organizations have recently begun to study the potential effects of sea-level rise and the policies needed to deal with those effects. For example, since 1994, the Country Studies Program has supported the Study of the Impacts of Climate Change and an Accelerated Sea-Level Rise on Nature and the Social Economy in China's Coastal Zone and Vulnerability and Adaptation Assessments, sponsored by the State Science and Technology Commission (SSTC 1994).

The coastal zones can be listed on the basis of their vulnerability to a rise in sea level and their present socioeconomic importance to China as follows: Changjiang (Yangtze River) Delta, including the North Jiangsu coastal plain and coastal area of Hangzhou Bay; Zhujiang (Pearl River) Delta, Huanghe (Yellow River) Delta, and the coastal areas of Laizhou and Bohai bays; the east coast of Guangdong Province and west coast of Guangdong Province; the coastal area from East Zhejiang Province to the north of Fujian Province; the Lower Liaohe Delta plain; the coastal area of Guangxi; and the southeast coasts of Hainan Province. China's other coastal zones are of secondary importance.

The Pearl River Delta region is one of the three largest delta areas in China, where the economy, urbanization, and industrialization have developed very quickly. It lies in south-central Guangdong Province and is an alluvial plain formed by the lower West, North, and East Rivers of the Pearl River water system. According to the Zhujiang Water Conservancy Committee (ZWCC 1988), the Pearl River Delta area consists of 8 cities and 17 counties (or districts), with a total area of 26,820 km^2, including 421.26 km^2 for the Kowloon Peninsula of Hong Kong and 18 km^2 for the Macao Peninsula (Table 1). The river estuarine area amounts to only 16,351 km^2 and consists of 17 counties (cities or districts) in a network of waterways, according to the Ministry of Water Conservancy (MWC 1992). On the basis of analyses and surveys by Li et al. (1994), involving 347 land topographic maps with a scale of 1:10,000, 6,186.10 km^2 of the total deltaic area consists of low-lying plains with elevations lower than 5.0 m (Yellow Sea Datums). These areas are most sensitive to an accelerated sea-level rise. The total deltaic area studied here consists of 31,164 km^2, and the area of affected cities and counties is 18,726 km^2 (Table 1).

Table 1. Pearl River (Zhujiang) Delta Area Administration

Plan Basis	Total Area (km^2)	Administration Areas
Integrated Delta Use Plan of Zhujiang River Basin (ZWCC 1988)	26,820	Guangzhou, Huaxian, Conghua, Zencheng, Fanyu; Shenzhen, Baoan; Zhuhai, Doumen; Foshan, Sanshui, Nanhai, Shunde, Gaoming; Jiangmen, Xinhui, Taishan, Enping, Kaiping, Heshan; Zhongshan, Dongguan; Qingyuan; Huizhou-Longmen; Yangjiang; Hong Kong, and Macao.
Development Plan of China's Estuary (MWC 1992)	16,351	Guangzhou, Fanyu; Shanzhen, Baoan; Zhuhai, Doumen; Foshan, Sanshui, Nanhai, Shunde, Gaoming; Jiangmen, Xinhui, Taishan, Heshan; Zhongshan; Dongguan.
China Climate Change Country Study Program	31,164 18,726[a]	Guangzhou,[a] Huaxian, Conghua, Zengcheng, Fanyu;[a] Shenzhen,[a] Baoan;[a] Zhuhai,[a] Doumen,[a] Foshan,[a] Sanshui,[a] Nanhai,[a] Shunde,[a] Gaoming;[a] Jiangmen,[a] Xinhui,[a] Taishan,[a] Enping, Kaiping, Heshan; Zaoqing, Gaoyao, Sihui, Guangning, Xinxing; Huizhou, Huiyang, Boluo; Yangjiang;[a] Dongguon;[a] Zhongshan;[a] Qingyuan.

[a] Areas very easily affected.

Determining the Effects of a Sea-Level Rise on the Pearl River Delta Area

Sea-Level Trends

Many factors contribute to relative sea-level fluctuations. This impact and vulnerability assessment is based on a sea-level rise due to global climate change, local vertical crustal movements, ground subsidence, and rising estuary levels.

Preliminary estimates of the contribution from global climate warming, including the unexplained global mean sea-level rise observed over the last century, range from 25 to 80 cm by 2100, with the best estimate being 50 cm. Even the lower estimates represent rates two to four times the rate experienced during the last 100 years. However, in accordance with the Intergovernmental Panel on Climate Change (IPCC) Common Methodology for assessing vulnerability to sea-level rise, the arbitrary values of 30, 65, and 100 cm have been used for analyzing the Pearl River Delta area.

Local vertical crustal movements vary according to location. The average annual downward movement is 1-2 mm in the plains, and the average upward movement of 1 mm yr^{-1} occurs in hilly areas in the Pearl River Delta, according to monitoring by the State Seismological Bureau and its Guangdong subbureau. For this study, 1 mm was used as the annual rate of vertical movement.

The united embankment and gate-dam project, a large-scale activity, dramatically changed the river and channel levels in the Pearl River Delta just after its completion. This type of river-level change has been described as "rising in the beginning and then falling for about 10 years after the rising period," according to data from the Zhujiang (Pearl River) Water Conservancy Committee. As a result, the long-term impact of these changes was not examined because the effects will balance out over time. Ground subsidence is not considered because it is small in this area.

Taking the above factors into consideration, it is predicted that the sea level in the Pearl River Delta region will rise 15-30 cm and 40-60 cm by 2010 and 2050, respectively (Chinese Academy of Sciences [CAS] 1994). Because of vertical crustal movement variations, the rate at which the sea rises will differ in various parts of the delta.

Impact on the Pearl River Delta Region

The sea level will rise slowly and continuously and will have a long-term cumulative effect that will threaten flourishing socioeconomic conditions, as well as natural habitats and local environments in the Pearl River Delta region. The anticipated effects are submergence of many lowland areas, aggravated storm surges and floods accompanied by a decreased capacity for flood control and storm-surge prevention along rivers and sea coasts, increased saltwater encroachment into estuarine streams, increased silting of river beds, irrigation and drainage difficulties, increased investment in infrastructure for towns and rural areas, and other related environmental and socioeconomic problems.

The preliminary results of a study of the Pearl River Delta area by the CCCS and investigations by the Department of Earth Science, CAS, are discussed in the following sections.

Inundation of Low Lands

Many lowland areas will be inundated. The CCCS research group used a 200- × 200-m digital topographic network to compute the probable inundation area along the coastline and its ratio to the total land area. The group considered three levels of potential sea-level rise as coastal tidal gauges: the mean high-water level of the spring tide, the highest high-water level on record, and the 100-yr high tide level (30, 65, and 100 cm). All calculations were performed for two scenarios: no defensive installations and the present number of defensive installations (Table 2).

Increase of Storm Surges and Floods

Damage caused by storm surges and floods will increase. The Pearl River Delta plain is characterized by a network of criss-crossing rivers and low elevations. One-quarter of this area is cultivated land lower than 0.4 m, according to the Pearl River Datums Level (PDL); one-half is lower than 0.9 m. From ancient times to the present, dikes and seawalls have protected this land along the rivers and seashore. By the end of 1991, 952 dikes and seawalls had been constructed in this area, totaling 6,333 km in length. They surround 7 million mu (1 hectare = 15 mu) of cultivated land, covering more than 50% of the total area. They also protect 7.88 million inhabitants — 43% of the general population in the Pearl River Delta region.

The flood and sea-tide prevention standard of these embankments, dikes, and seawalls is sufficient for about a 10-20-yr frequency flood event. In general, some larger or more important dikes can protect for more than a 20-year flood frequency. Only the most important dikes were designed according to the 50-100-yr flood and tide standard. If the sea rises by 0.4-0.6 m, the sea tide in the Pearl River Delta region will cause storm surges that currently occur at a frequency of 50 or 100 years to possibly occur every 10 years (He 1994). That is, the risk of storm surges will increase 5 to 10 times.

Intense Saltwater Intrusions

Saltwater intrusions upstream will intensify. A rise in sea level will cause the tidal boundary to move up, bringing saltwater farther upstream. The saltwater boundary will generally move more than 3 km with a sea-level rise of 0.4-0.6 m in the Pearl River Delta region. In the future, saltwater intruding even to Guangzhou City in a low-water period will create difficulties for coastal inhabitants. This effect could be offset or enhanced by changes in run-off in the Pearl

Table 2. Probable Inundated Area in the Pearl River Delta Region

	Sea-Level Rise					
	30 cm		65 cm		100 cm	
Scenario	Area (km^2)	Ratio (%)	Area (km^2)	Ratio (%)	Area (km^2)	Ratio (%)
Without protection						
Mean high water of spring tide	2,190	7	3,744	12	4,282	14
Highest high water on record	5,546	18	5,967	19	6,543	21
With protection						
	1,153	4	3,453	11	6,520	20
100-yr high tide level	1,719	6	2,875	9	7,823	25

River, but deviations in runoff of the Pearl River have been minor over the last 40 years.

Increased Difficulty with Irrigation and Drainage

Irrigation and drainage difficulties will increase. Sea-level rise and saltwater intrusion upstream will cause the original natural irrigation and drainage system in the Pearl River Delta region to lose its effectiveness. Because the present 20-yr flood frequency will decrease to an approximate 5-yr frequency, the cost of irrigation and drainage will increase by 15-20%.

Threatened Infrastructures

Towns and important infrastructures will be threatened. Most of the main cities and towns of the Pearl River Delta area are located in low-lying areas near rivers or the sea. For example, 4.25 km^2 of Guangzhou is lower than 1.40 m, 13.01 km^2 is lower than 2.40 m, and 27.22 km^2 is lower than 3.40 m (PDL). If the sea level rises 0.7 m, the high-water level (without storm surge) in Guangzhou will be 3.20 m (PDL), which means that 40% of the total area would be inundated, and the economic loss for one event could reach 20 billion RMB yuan (price index in 1990: 1 U.S. dollar = 4.3 yuan).

The elevations of Shenzhen, Zhuhai, Zhongshan, Foshan, Jiangmen, Dongguan, Fanyu, Shunde, and so on, are only 0.5-2.0 m (PDL). Numerous airports, sea and river ports, and other infrastructures distributed along the coast would be affected. Engineering evaluations suggest that the area should be raised and that newly developed zones should be located in areas of higher elevation. The softening of building foundations will also require more investment.

Increased Water Pollution

A rising sea will cause a large amount of tidal water to travel upstream in the river delta, thereby blocking the irrigation and drainage system in the deltaic embankment areas. As a result, it will be difficult to drain wastewater from cities and towns, which will increase pollution of the river network and embankment area.

Adapting to a Rising Sea Level: Strategic Framework

Many opportunities exist to take relatively inexpensive precautions now in the Pearl River Delta areas to prevent and mitigate many of the unfavorable effects of a sea-level rise on the socioeconomic development of this area. Such measures would protect inhabitants, environment, resources, industries, commerce, and national security and stability. However, this strategy must be comprehensive and will involve complex engineering.

Strategic projects can serve as models to be adopted by the local and national governments to prevent and mitigate the unfavorable effects of a sea-level rise. They include strategic principles, integrated priorities and essential activities, research and implementation components, and domestic and international cooperation (Figure 1). The implementation of integrated coastal zone management (ICZM) should be recognized because, even if the sea level does not rise or the amount of the rise is not as high as has been estimated, enhancing coastal zone management is still a very urgent activity that has benefits for the present situation.

Determining the Most Promising Measures for Implementation

Dikes and Seawalls

Improving the design standard of river dikes and seawalls must consider factors related to a rise in sea level. Global warming and sea-level rise could increase the intensity and frequency of storm-surge disasters that threaten united embankment and urban flood and storm-surge prevention in the Pearl River Delta area. This area has 774 river dikes, with a total length of 4,132 km, and 178 seawalls, with a total length of 2,201 km. Among them are four dikes (North River Great Dike, Jiaosang United Embankment, Zhongshun Great Embankment, and Jiangxin United Embankment) that protect more than 300,000 mu; 28 dikes and embankments that protect 50,000-300,000 mu; and 59 dikes that protect 10,000-50,000 mu.

According to the Socioeconomic Plan for the Pearl River Delta area, several dikes and embankments are being strengthened and reconstructed according to increased design standards. The repaired North River Great Dike was reconstructed according to a 50-yr flood standard. Foshan Great Dike was reconstructed by using a 50-yr flood and storm-surge standard. Placement of key dike and embankment repair and reconstruction projects in the Eighth Five-Year Plan (1991-1995) means that the Jiaosang United Embankment, Zhongshun Great Embankment, Jiangxin United Embankment, and Jingfeng United Embankment will also meet the 50-yr flood and storm-surge standard when repairs and reconstruction are completed by the end of 1995. However, the dikes and embankments that greatly need to be strengthened in accordance with elevated design standards are in Fanyu, Doumen, Zhongshan, and Zhuhai cities (counties). These are the Fanshun United Embankment, Jiaodong Embankment, Wanqinsha Embankment, Baijiao, and Chikan and Zhongzhu United Embankments. The designs should follow a 20-50-yr flood and storm-surge standard.

In addition, the 1,737 dams with a net width of more than 3 m and a total net width of 13,317 m should be repaired and reconstructed at the same time that the

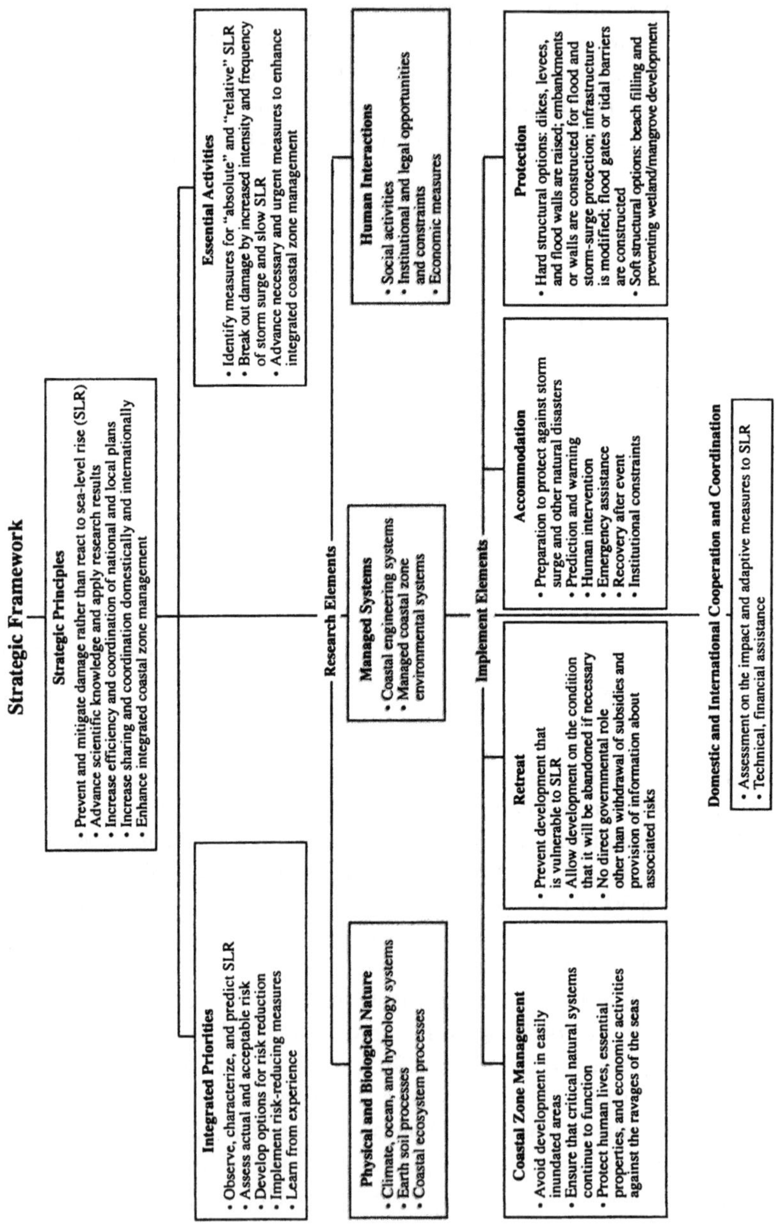

Figure 1. Strategic Framework for Adapting to the Pearl River Delta Sea-Level Rise

united embankments in the Pearl River Delta are reconstructed. The engineering involved will be extremely extensive, and the investment will be considerable.

Upgrades in Design Levels

The design levels of municipal engineering, communication, and other important infrastructures could be upgraded. The designs of irrigation and drainage systems, highways, and other important engineering infrastructures in the coastal cities of Guangzhou, Shenzhen, Zhuhai, Foshan, Jiangmen, Zhongshan, and Dongguan should be studied and revised by the provincial government to facilitate urban irrigation and drainage and to mitigate damage to buildings and structures in low-lying areas. All completed important engineering projects, such as sea or river ports, airports, power stations, iron and steel plants, oil and chemical plants, and so on, should be reinforced with protective devices. The construction designs in the new economic development regions should take projected sea-level rise into account.

Adjustment of Socioeconomic Development Plans

The formulation and implementation of local socioeconomic development plans should be adjusted to provide for the rising sea level and the increased intensity of storm-surge disasters. Implementing the flood and storm-surge prevention projects outlined in "The Pearl River Comprehensive Utilities Plan" approved by the State Council of China should accommodate sea-level rise as a first step. All urban construction in low-lying areas near or along rivers and seacoasts and all development plans that involve important industries, communications, energy, and agriculture should also consider sea-level rise and the possibly increased intensity and frequency of storm-surge disasters.

Developing Long-Term Adaptive Measures

Integrated Coastal Zone Management

Integrated coastal zone development and management are the most important parts of the socioeconomic development plan for the Pearl River Delta area. Knowledge of the destructiveness of human activities to regional resources and environments has existed since the 1980s. Additional knowledge about the potential damage to be caused by a global climate change and an accelerated rise in sea level has increased the concern about the rapid loss of various natural and man-made structures on the coasts. Therefore, coastal zone management needs to add the goal of coordinating sustainable development of the coastal economy in a way that accounts for sea-level rise, increased coastal disasters, and other environmental problems in those areas. A "more sensible stage" of development that coordinates society with nature is promoted by the implementation of coastal zone management strategies.

Reinforcement of the Monitoring Network

The basic task of research in understanding sea-level rise and its effects involves setting up and enhancing the monitoring network, improving the accuracy of observation, and accumulating sufficient observational data to provide a scientific foundation for strategic decisions regarding early warning, defense, and planning. Tides, vertical crustal movements, surface subsidence, coastal erosion, river bed and bay deposits, ground softening and subsidence due to sea embankments, and other important examples of engineering should be monitored.

Long-Term Plans for Agriculture and Aquaculture

A long-term plan should be implemented to transform agriculture and animal husbandry into aquaculture. The probable inundation of 57.5 km^2/yr corresponds to a sea-level rise of 65 cm by 2100, assuming that the defensive measures on the rivers and seacoasts remain at the 1990 level. If agriculture and animal husbandry in a 30-km^2 area are transformed into fish farming, the gross domestic product (GDP) balance will more or less keep up (corresponding to an absence of a sea-level rise).

Education and Public Awareness

The education and dissemination of information to the public about global warming and sea-level rise, as well as the regional socioeconomic effects and adaptive measures, are critical to the implementation of coastal zone management goals and sustainable development.

It is important that government officials at all levels become aware of the need to support the visionary coastal zone management activities. Some educational materials could also be incorporated into school textbooks.

Estimating the Cost of Implementing Adaptive Measures

The estimated additional input, calculated by a preliminary study by the CCCS, needed to implement the above adaptive measures totals 29.85 billion RMB yuan for the next 110 years. The average annual input is 270 million yuan,

Table 3. Estimated Cost for Implementing Adaptive Measures (in million RMB yuan by 1990 standing price index)

Adaptive Measures	Cost/Year	Price/km	Total Cost
Strengthen and reconstruct dikes and seawalls (total length 4,132 km for 774 river dikes and 2,201 km for 178 seawalls)		2.85	18,059
Raise the design standard	30.0		3,300
Additional cost for preventing flood and storm surge disasters	10.0		1,100
Additional cost for monitoring network	10.0		1,100
Transform agriculture and animal husbandry to aquaculture	30.0		3,300
Scientific, education, and other "soft" structural measures			3,000
Total			29,850

covering 0.12% of the GDP and 3.5% of local government expenditure (Table 3) of the Pearl River Delta area. If these measures are not adopted, the average annual economic loss for the next 110 years will be 121.9 billion RMB yuan, which accounts for 54% of the annual GDP.

Conclusions

This case study of the Pearl River Delta area discusses many effects of an accelerated sea-level rise due to global warming and highlights the importance of considering local adaptive measures that seem to be practical within the comprehensive strategy suggested here. On the basis of the calculated total implementation costs of these measures, local policymakers should spend about 0.12% of the annual local GDP, or 3.5% of local government expenditures anticipated for the sustainable socioeconomic development of this deltaic area.

Acknowledgments

This study is funded by the results of many research works on this area. I am grateful to Prof. Academician Su Jilan, the Director of the Second Institute of Oceanography, who edited the report on Impacts of Sea-Level Rise on China's Delta Area and Adaptation Strategy to provide relevant information to this study. I would also like to thank J. Huang, L.H. Primo, and J. Smith for their reviews and the English version and many other comments that much improved an earlier draft of this paper. All opinions expressed in this paper are purely the author's.

References

Chinese Academy of Sciences, 1994, "Impact of SLR on Economic Development of Pearl River Delta and Adaptation Measures (February 15, 1993)," in *Impacts of Sea-Level Rise on China Delta's Area and Adaptation Strategy*, pp. 5-7, Advisory Report by Academicians of the Department of Earth Sciences, Chinese Academy of Sciences, Science Press, Beijing, People's Republic of China.

He, H.G., 1994, "The Possible Influence of Storm Surge on the Zhujiang Delta Area Due to Sea-Level Rise," in *Impacts of Sea-Level Rise on China Delta's Area and Adaptation Strategy*, pp. 174-180, Advisory Report by Academicians of Chinese Academy of Sciences, Science Press, Beijing, People's Republic of China.

Li, P.R., et al., 1994, "Essential Features of Zhujiang Delta and Effects of Sea-Level Rise," in *Impacts of Sea-Level Rise on China Delta's Area and Adaptation Strategy*, pp. 315-324, Advisory Report by Academicians of Chinese Academy of Sciences, Science Press, Beijing, People's Republic of China.

Ministry of Water Conservancy (MWC), 1992, *Development Plan of China's Estuary*, Beijing, People's Republic of China.

State Science and Technology Commission of China, 1994, *Proposal on China Country Study with U.S. Support to Address Climate Change*, Beijing, p. 58, People's Republic of China.

Yang, H.T., and S.Zh. Tian, 1994, "Sea-Level Rise and Coastal Zone Management," in *Impacts of Sea-Level Rise on China Delta's Area and Adaptation Strategy*, pp. 303-310, Advisory Report by Academicians of Chinese Academy of Sciences, Science Press, Beijing, People's Republic of China (in Chinese, with English abstracts).

Zhujiang Water Conservancy Committee, 1988, *Report of the Integrated Delta Use Plan of the Zhujiang River Basin*, Ministry of Water Conservancy, Beijing, People's Republic of China.

4

COASTAL RESOURCES

Rapporteur's Statement

Frank Rijsberman
Resource Analysis, Zuiderstraat 110
2611 SJ Delft, the Netherlands

The impacts of sea-level rise on coastal resources is one of the most pressing issues related to climate change. The major issues addressed in the following pages are briefly discussed in the paragraphs below.

Adaptation Assessment Methods. A good starting point for adaptation assessment for coastal resources is *The Common Methodology to Assess Vulnerability to Sea Level Rise*, developed and published in a number of evolving versions by the Coastal Zone Management (CZM) subgroup of the Intergovernmental Panel on Climate Change (IPCC). However, this methodology has a number of limitations, including the following:

- The Common Methodology concentrates on the impacts for a country as a whole or for specified regions. The methodology should be expanded to account for the distribution of costs and benefits of climate change impacts as well as adaptation strategies of different social groups or interested persons and organizations.
- The Common Methodology is perceived as a top-down approach. This approach needs to be complemented with bottom-up (e.g., participatory planning) approaches to enable different interested parties to become involved and to account for traditional or community-based management practices, such as those used in the small Pacific island nations.
- Regional development factors (e.g., policies of neighboring countries) are not included explicitly in the Common Methodology. In adaptation assessments, regional development factors can be very relevant and should be taken into account explicitly as scenario variables.
- The Common Methodology focuses particularly on flooding and flood risks as the main impact category of climate change and sea-level rise. Other climate change impacts on the coastal zone should also be studied.

Two stages are involved in preparing for climate change adaptation. Stage 1, the analysis stage, identifies and evaluates adaptation strategies. Stage 2, the consensus-building stage, communicates the results of the analysis to interested parties and supports the negotiation process to achieve socially acceptable strategies with sufficient backing for their implementation. The Common Methodology addresses the analysis stage of the preparation for climate change adaptation (vulnerability and adaptation assessment), but additional methodologies need to be developed to address the consensus-building stage.

Adaptive Measures. The IPCC CZM identified three major adaptation strategies for coastal zones: *retreat, accommodate*, and *protect*. In the retreat strategy, inhabitants of the coastal zone are resettled, and existing structures are abandoned. New developments must be set back specific distances from the shore, as appropriate. In accommodate strategy, vulnerable areas continue to be occupied. A greater degree of flooding is accepted, and risks are controlled by accommodation measures, such as early warning systems. Development strategies and land use are adjusted (e.g., convert farms to fish ponds). In the protect strategy, vulnerable areas — especially population centers, economic activities, and natural resources — are defended through flood protection and associated measures.

The most widely considered adaptive measures in the coastal zone relate to coastal engineering measures for flood protection and coastal erosion control, such as dikes and levees, abutments, revetments, groynes, breakwaters, seawalls, and beach nourishment schemes.

Many other adaptive measures need to be considered in selecting the most suitable adaptation strategy:

- Creating services to improve or change agricultural practices to be more flood-resistant or less flood-damage sensitive (for example, the floating rice cultivars used in Bangladesh).
- Negotiating regional water-sharing arrangements.
- Providing efficient mechanisms for disaster management.
- Developing desalination techniques.
- Planting mangrove belts to provide flood protection.
- Planting salt-tolerant varieties of vegetation.
- Improving drainage facilities.
- Establishing setback policies for new developments.
- Developing food insurance schemes.
- Devising flood early warning systems.

It is important to consider the traditional knowledge and expertise of the people living in vulnerable areas. They often have evolved adaptive measures to sometimes extreme climatic conditions over centuries. Such measures can also be used as components in strategies to adapt to climate change.

Integration of Adaptive Measures. As the IPCC CZM recommended and subsequently adopted in Agenda 21, the measures to adapt to climate change in the coastal zone should be incorporated into the framework of national CZM plans. Such plans, based on integrated CZM concepts, should address present-day stresses on the coastal zone system as well as long-term concerns that result from climate change.

Regional Cooperation. The small island nations threatened by sea-level rise are well organized in the international community through the Association of Small Island States. Other low-lying nations (e.g., those with deltas and low, open coastlines) should also consider organizing themselves and cooperating in this area. In countries as physically similar and socially diverse as Bangladesh and the Netherlands, local people have developed ways to cope with flood risk and take advantage of opportunities offered by the system, such as aquaculture, horticulture, and adapted agricultural activities.

Technical Assistance Needs. Many countries have undertaken a first assessment of vulnerability to sea-level rise and have identified some adaptive measures; however, most countries have not yet drawn up CZM plans. Technical assistance, focused on capacity building for integrated CZM, is a priority. Transfer of technology related to coastal morphology and engineering, as well as technology related to a range of "soft" adaptive measures, can contribute significantly to the capacity of vulnerable coastal nations to adapt to changing climate conditions. For a number of countries, the Stage 1 enabling activities (preparation for adaptation), which can be financed by the Global Environment Facility under the U.N. Framework Convention on Climate Change, will provide opportunities for technical assistance related to coastal zones.

Future Research Needs. Future research needs in the coastal zone relate to the following:

- An improved understanding of the dynamics and functioning of coastal ecosystems, such as wetlands, mangroves, and coral reefs, and the impacts of climate change on these ecosystems, as well as their adaptation capacities.
- Development of effective institutional arrangements to manage coastal zone systems in a sustainable manner.
- Bottom-up CZM approaches that account for traditional and community-based management systems and enable interested parties to participate in the planning and management process.

Summary Assessment. Coastal resources is probably the sector where adaptation to climate change is best accepted and most developed. Efforts to analyze impacts and adaptation to sea-level rise have been widespread. Small island nations are particularly vulnerable to sea-level rise and require international support to implement adaptation strategies. More attention should be given to impacts of climate change other than sea-level rise on the coastal zone, such as changes related to freshwater resources in the coastal zone.

The Common Methodology is a good starting point for making adaptation assessments, although this methodology has a number of important limitations. Specifically, the Common Methodology should be expanded to address the distribution of climate change impacts, and costs and benefits of adaptation, over social impact groups or interested parties.

Integrated coastal zone management (ICZM) concepts urgently need to be implemented. These concepts can put long-term concerns, such as sea-level rise, in a common framework with present-day pressures on the coastal zone (such as erosion, overfishing, pollution, and habitat destruction) as well as structure the analysis of adaptation strategies for different interested groups.

The Common Methodology and ICZM concepts, which are sometimes perceived as top-down approaches, need to be implemented with participation from interested parties. Such groups can be involved through participatory planning and appraisal techniques. The capacity of countries to deal with future rises in sea level and related climate change impacts can be increased significantly by strengthening CZM institutions.

Sea-Level Rise along the Lima Coastal Zone, Perú, as a Result of Global Warming: Environmental Impacts and Mitigation Measures

N. Teves
Federico Villarreal National University
G. Laos
Direction of Hydrography and Navigation, Marine Department
S. Carrasco
Peruvian Marine Institute
C. San Roman
Federico Villarreal National University
L. Pizarro and G. Cardenas
Direction of Hydrography and Navigation, Marine Department
A. Romero
Federico Villarreal National University
Universidad Nacional de Engenieria AV, Tupac Amaru S/N
Apartado 1301
Lima, Perú

Abstract

If the sea-level rises, the beaches along the Lima Coastal Zone, Perú, will be submerged. The effects of two scenarios proposed by the Intergovernmental Panel on Climate Change, a 0.30-m and a 1.00-m rise in sea level, are considered, and adaptation measures are recommended. In recent years, marine erosion has greatly affected La Punta, requiring that certain areas be protected with groynes and breakwaters. The sea has retreated, and the narrow beaches have widened. These measures are good examples of how to protect a coastline should the sea rise as a result of global warming or other reasons.

Introduction

At a conference on the effects of climate changes in the Southeast Pacific held in Santiago, Chile in 1992, the South Pacific Permanent Commission (CPPS 1994) concluded that the region will probably be affected by sea-level rise as a result of global climate changes. The conference resolved to (1) improve communication of sea level and temperature measurements to appropriate institutions and (2) consult related studies about the effects of global climate changes when planning a response to the anticipated sea-level rise.

In 1993, with support from CPPS, a Peruvian group initiated a case study to analyze climate-related data (CPPS 1994). To date, preliminary information has been gathered. When completed, the data analysis may serve as a model for a coastal vulnerability study.

Perú has participated in the U.S. Country Studies Program, a large multinational program, since November 1993. The program studies the primary areas that could be affected by sea-level rise and considers the following scenarios proposed by the Intergovernmental Panel on Climate Change (IPCC): a 0.30-m and a 1.00-m rise in sea level.

The study area for this report, the Lima Coastal Zone, lies midway along the Peruvian coast, between Ancón-Ventanilla and Pucusana (Figure 1). This coastal area, known as Costa Verde ("Green Coast"), has sand and gravel beaches at the foot of cliffs. The town of La Punta at the northern end of Costa Verde is 1.5 m above sea level. Two million tourists visit the area in the summer, drawn by the beaches and new summer resorts south of Lima.

This paper has four objectives: (1) to establish a baseline of physical, biological, and socioeconomic factors for the study area; (2) to determine areas with potential for flooding on the basis of marine dynamics evaluations; (3) to assess the impacts of anticipated flooding; and (4) to propose mitigation measures.

Environmental Baseline

Existing data helped to establish physical, biological, and socioeconomic baselines as described in the following sections.

Physical Environment

Geology/Geomorphology

Figure 2 is a geological and geomorphological map of the region. The study area (between Ancón and Pucusana) is characterized by deposits that are the product of alluvial, marine, aeolian, and mass erosion processes. These deposits are interrupted by isolated coastal massifs and hills of Mesozoic volcanic rocks and Cretaceous sedimentary rocks. To the east, the western flank of the Andes Mountains forms part of a granodioritic coastal batholith. Geomorphological study of the lower reaches of the Chillon, Rimac, and Lurin Rivers revealed the presence of alluvial deposits of ancient and Holocene age.

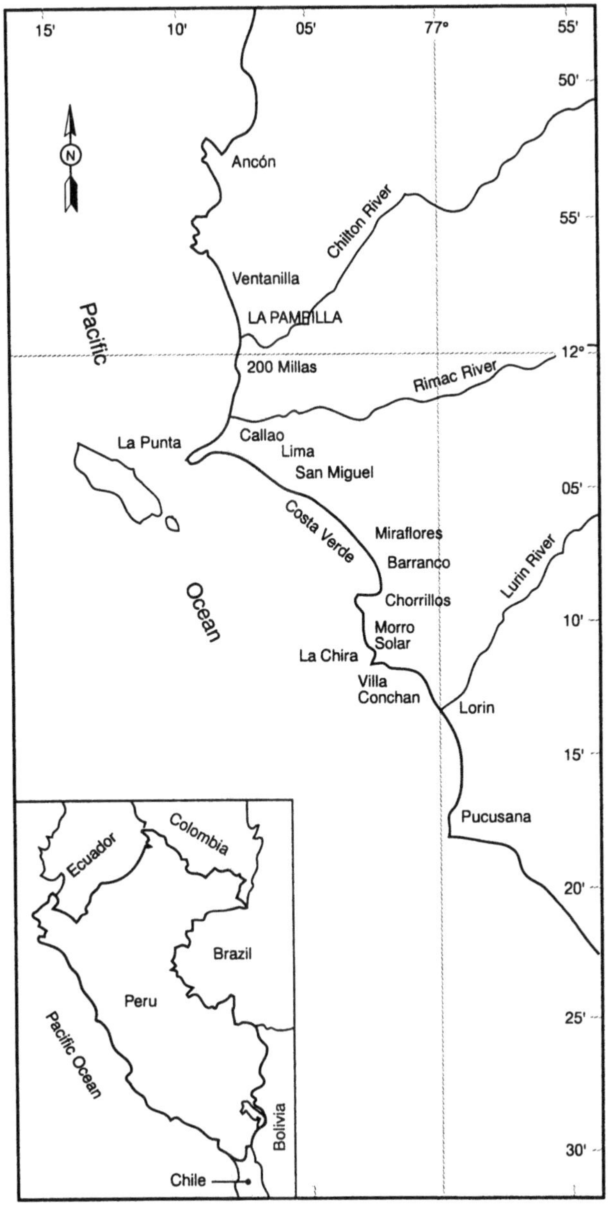

Figure 1. Map of Metropolitan Lima and Callao

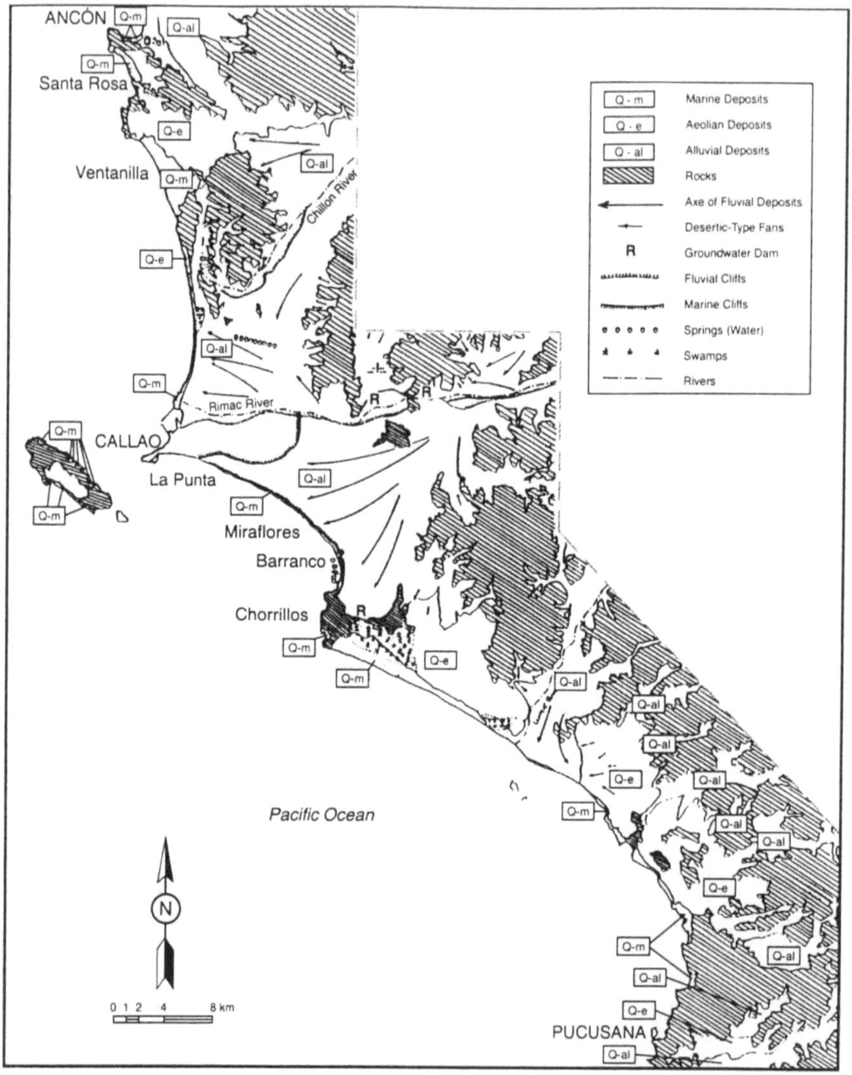

Figure 2. Geological and Geomorphological Map of Metropolitan Lima and Callao

Erosion and Sedimentation

In the north, Santa Rosa and Ventanilla beaches receive increasing amounts of sediment from the Chillon and Rimac Rivers (Teves and Gagliano 1970). The port of Callao also receives some of the sediment transported by the Rimac River, which makes it necessary to dredge every four or five years. La Punta is the result of marine erosion, jutting out into Lima's coastal area, protected by the San Lorenzo and Fronton islands (Teves 1967).

The Costa Verde beaches between La Punta and La Herradura are characterized by deposits of smooth pebbles and sands. The coastline between La Punta-Callao and Miraflores consists mainly of pebbles with small areas of sand. At the center of the bay, storm waves have destroyed the roadway, which was built on fill. Toward the south, the sandy beaches continually grow because of the construction of piers. The beaches generally have two or three berms; beach "sinuosities" occur in the higher parts and "beach cusps" in the lower parts.

Morro Solar is a rocky massif with high cliffs formed by strong marine erosion as a result of wave refraction. Wave refraction, a deformation of waves that concentrates its action primarily in the La Herradura and La Chira beaches, grown noticeably in recent years. To the south, in the Villa and Conchan areas, a beach ridge protects the low-lying land that once constituted a lagoon. Several beaches are growing toward the south, such as Arica, Punta Hermosa, Punta Negra, San Bartolo, Santa María, and Naplo.

Bathymetry

In general, isobaths follow the coastline. A 10-m isobath is parallel to the coastline at an approximately 1.5-km distance.

Meteorology and Climatology

The study area is affected by the Semipermanent Anticyclone of the South Pacific (ASPS). This high-pressure center is located in the Southeast Pacific. Its position is semifixed, with slow seasonal displacements. The ASPS determines the stability or instability of the air masses, which run vertically because of significant phenomena, such as turbulence and advection. Winds are from the southeast to the west along the entire coast.

Temperature isotherms parallel the coastline. Air temperatures vary from 13.6 to 28°C, decreasing closer to the Pacific Ocean because of the thermoregulator effect, which reduces the range of variation in temperature. Figure 3 shows monthly average air temperatures at Callao (Senamhi 1984-1994).

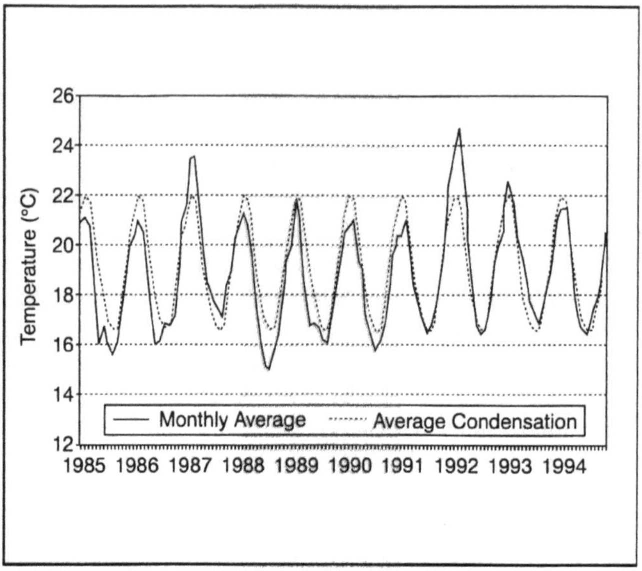

Figure 3. Air Temperature at Callao (1985-1994)

Sea Currents and Hydrochemical Aspects

Sea Currents

The port of Callao is classified as an upwelling center, that is, a small area along the sea near the coast (e.g., Ilo, Pisco, Callao, and Chimbote) where elevated cold water mixes with subtropical (oceanic) water (upwelling). Seasonal variations are influenced by winds and water masses that form the circulation regimen of the area. The general pattern of currents in the upwelling area is to the northeast; this pattern differs from the shore system (Guillen 1976). Direct measurements were taken no farther than 0.5 nautical miles from the coastline. In Miraflores Bay, the surface current at low tide was 0.36 knots north-south, and the subsurface current was 0.28 knots south-southeast. In Callao Bay, velocities varied between 0.12 and 0.64 knots, and directions were northeast, south, and southeast.

Hydrochemical Aspects

The average surface temperature of the Peruvian Sea fluctuates from 17 to 25°C; the lower temperatures are near the coast, and the higher ones are offshore. The surface temperature isotherms follow the coastline, while the subsurface ones are transversal.

Table 1 summarizes the hydrochemical parameters for Miraflores and Callao bays. The sea-surface temperature of coastal water in Callao Bay varies from 15.1 to 17.9°C (Figure 4). (During the El Niño phenomenon, the maximum temperature was 24.2°C.) Salinity varies from 34.60 to 35.09%. Dissolved oxygen varies from 1.5 mg L^{-1} in upwelling areas to 6 mg L^{-1} in areas of intense photosynthesis. In April 1994, the sea-surface temperature north of Miraflores Bay averaged 22.6°C, and the temperature at the bottom of the sea was 18.85°C. The surface salinity in Miraflores Bay varied from 34.54 to 35.11% for 1988-1994 (Table 1).

Biological Environment

The phytoplankton distribution in the Callao area fluctuates seasonally with the upwelling process. Other local influences, such as the Rimac River and the residual unloadings of domestic wastes (as at Callao Bay), also control phytoplankton production, especially in the area of Mar Brava, where there is a concentration of nutrients from waste sources.

The main types of phytoplankton found in the Callao area are cold-water diatoms, such as *Skeletonema costatum, Lithodesmium undulatum, Thalassionema nitzschiodes*, and *Leptocylindrus danicus*, which are present most of the year. The *Dinoflagellata* are not significantly represented, except as *Gymnodimium* and *Prorocentrum micans*, which generally occur in large quantities during "red tides." The highest concentration of the microscopic algae is in the euphotic layer (0-30 m).

Table 1. Hydrochemical Parameters for Miraflores and Callao Bays

	Miraflores Bay		Callao Bay	
Parameter	0 m	10 m	0 m	10 m
Temperature (°C)	18-22.7	18.85	18.8-21.5	17
Salinity (%)	34.54-35.11	34.79-35.11	34.8-35.13	34.88-35.13
Dissolved oxygen (mg/L)	1.9-6.1	1.9-6.1	1.23-9.19	1.23-9.19
Nitrates (mg/L)	0.13-0.28	0.01-0.24	0.02-0.46	0.02-0.30
Nitrites (mg/L)	3.5-8.4	4.6-9.8	0.5-10.5	3.9-10.15
Phosphates (µg atm/L)	1-4	1-4	1.91-3.43	1.91-3.43
Silicates (µg atm/L)	1-30	1-30	1-30	1-30

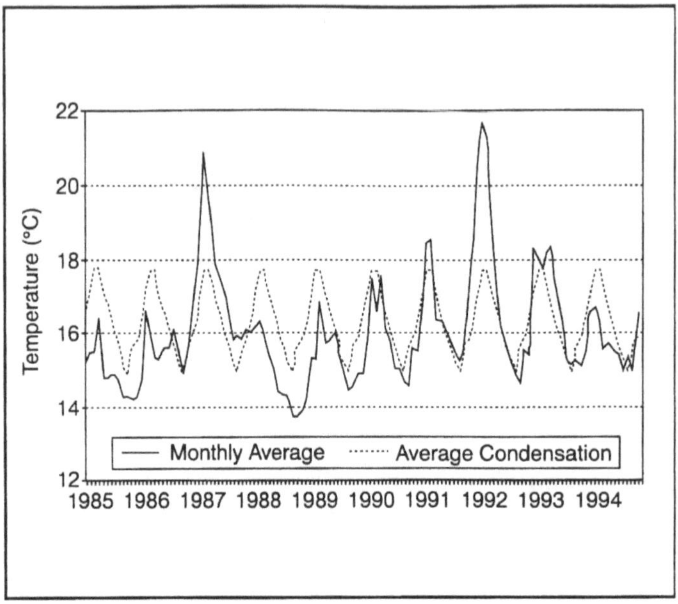

Figure 4. Sea Temperature at Callao (1985-1994)

The phytoplankton distribution on the surface of Callao Bay, within 100 km of the shore, is similar to that of *Chlorophylla*, reaching values above 0.5 g m^3 day^{-1}. Great variability occurs in the vertical distribution, with an average of 4 g m^3 day^{-1} 80 km from shore in summer and fall, decreasing as the distance from the coast increases. The primary phytoplankton production rates on the Peruvian coast depend on the origin of the upwelling water and on the photosynthetic efficiency of the species. Growth rates of 0.9 doubling time have been reported for the Callao area.

Small amounts of zooplankton are recorded in the fall and winter, with high values in spring and maximum values in summer. The surface zooplankton studies carried out between 1986 and 1991 in Callao Bay found larval phases of holoplankton and copepod (calanoid and harpacticoid) organisms common to the coastal areas (Santander and Guzman 1986; Ochos and Gomez 1988). Areas adjacent to river mouths have few zooplankton, possibly because of a lack of food, the presence of pollutants, or an intolerance for the mixed waters.

The following organisms are also found in the Callao Bay area: (1) birds, *Sula variegata*, *Larus dominicanus*, *Pelecames thagus*, and *Phalacrocorax olivaceus*; (2) fishes, *Engraulis ringens*, *Sarda chilensis*, *Odontesthes regia*, *Trachurus picturatus*, *Muil cephalus*, etc.; (3) crustacea, *Tegula atra*, *Mytilus chorus*, *Emerita analoga*, and *Petrolisthes sp.*; and (4) macroalgae, *Ulva papenfussi*, *Gigartina chamissoi*, *Gracilaria lemanneformis*, *Polysiphonia paniculata*, and *Polysiphonia confusa*.

Socioeconomic Aspects

Water Quality

High levels of water pollution are found in Callao and Lima, but the pollution originates from different sources (domestic, industrial, mining-metallurgical, and oil hydrocarbon wastes). The marine pollution in Callao and Lima includes high concentrations of copper, iron, and lead, which have their sources in waters from the Rimac River. In addition, 93% of domestic waste for coastal Peruvian cities is unloaded in this area (Conopuma et al. 1986; Conopuma 1987; Jacinto 1990).

Coastal Area Uses

The coast, with its bays, estuaries, and lagoons, is intensely used by people. In addition to being one of the most important fishing systems of the world, Lima's ports and bays are used as transport centers. The most common uses of the coast are recreation (walks, landscapes, fishing, etc.); education (scientific research, ecological sampling); tourism; transportation; aquaculture; shelter for ships (with natural protection from swells and winds); construction of new housing; industry; and waste unloading.

Lima and Callao contain approximately 50 urban planning districts in which an estimated 6.5 million people live (about 30% of Perú's population). These areas are projected to have 14 million inhabitants by the year 2000.

Fishing

The port of Callao, which is located at 12°03'S and 77°09'W (Figure 1), is in the central zone of the Peruvian upwelling ecosystem. It is an important disembarkation point for the industrial fishing fleet that operates in this area, as well as for smaller fishing fleets, which supply much of the country's population with fish.

The fishing industry catches around 180,000 Mg of fish per year in this area, equivalent to 4% of the national fishing industry. The capture volume of fish for human consumption in this area amounts to 15,000 Mg annually, or 17% of the total capture for the coastline. Callao Bay is the site of fish powder plants, crude oil plants, and fish canning plants.

Potential Flood Areas

Calculation Methods

To delimit the areas of maximum flooding should the sea level increase in the next century, the following parameters of marine dynamics were considered:

(1) observed maximum high tide, (2) average upper maximum high tide (during one month), (3) heights of normal waves, (4) heights of storm waves, and (5) increases of 1.00 and 0.30 m in mean sea level as the result of global warming.

The wave analyses were performed by using data collected in Ventanilla, where the heights of significant waves and normal waves have been measured for more than a year. Wave heights and periods were calculated in deep waters. With this information and the bathymetry obtained from Miraflores Bay, the wave pattern was projected orthogonally toward the coastline by using refraction charts. The wave heights were calculated by using different coefficients.

Results

Assuming an increase of 1.00 m above the present sea level (the maximum scenario proposed by the IPCC), the height of maximum flooding would be 3.40 m in northern La Punta (Callao Bay) and 3.66 m in southern La Punta (Miraflores Bay) (Table 2).

The observed maximum high tide (the highest tide during one month from historic records) and the upper maximum high tide (the average of highest tides each day in a month) were 0.76 and 0.40 m, respectively. The data reviewed were from 1942 to present, were obtained by a marigraph installed in La Punta-Callao, and are valid for Miraflores and Callao bays (DIHIDRO 1942-1992).

Table 2. Scenarios for Sea-Level Rise (m)

Location	Present Condition	0.30-m Sea-Level Rise Rise	0.100-m Sea-Level Rise
Miraflores Bay			
Ordinary condition (High tide level)	1.24	1.54	2.24
Extreme event (Storm surge)	2.66	2.96	3.66
Callao Bay			
Ordinary condition (High tide level)	1.13	1.43	2.13
Extreme event (Storm surge)	2.40	2.70	3.40

The potential areas of flooding and the anticipated impacts are described in the following sections on the basis of the topographic evaluation carried out in Ventanilla, Calla-La Punta, Costa Verde (San Miguel-Barranco and Chorrillos), La Chira, and Conchan-Pucusana. Table 3 lists the sizes of the potentially flooded areas and of the affected populations.

Ventanilla

The potential area of flooding in Ventanilla beach is 7.21 km^2, 61% of the total area of the beach that is less than 5.0 m above sea level. The sandy beach is mainly used for recreation in the summer. The services area is 5.0 m above sea level and contains a parking lot, medical facility, public bathrooms, and footpaths. Next to this infrastructure is an area below sea level that contains agricultural areas and a coastal ecosystem, where a small population of webfooted birds of different ages has been observed. The socioeconomic impacts are related to the seasonal recreational use of Ventanilla Beach, where 400,000 patrons swim and sunbathe on the broad, sandy white beach.

Callao-La Punta

Between Ventanilla Beach and the Port Callao are "marginal" (low-income) human settlements and several industries. This area is in danger of flooding, which would affect approximately 15,000 people.

Table 3. Areas between Ventanilla and Pucusana Bay Potentially Flooded by a 1-m Sea-Level Rise

Coastal Area	Potentially Flooded Area (km^2)	Affected Population[a]
Ventanilla Beach	7.21	400,000
La Punta	1.28	15,000
Costa Verde	2.61	2,000,000
La Chira Beach	1.08	3,000
Conchan-Pucusana	11.51	500,000

[a] Most are summer tourists.

Although an increase in the sea level at the port of Callao will reduce the need for dredging and therefore reduce maintenance costs, the potential negative impacts could encompass the entire infrastructure of services. La Punta is one of the most critical areas. The affected area would be 1.28 km^2, 71% of the area that is less than 5.0 m above sea level. Because the maximum height above sea level in La Punta is 3.4 m, it is possible that the entire urban area of La Punta could be affected (e.g., the beach resort of Cantolao, the Navy School, and the breakwaters and piers).

Costa Verde

Approximately 2 million people would be affected by the projected flooding of the San Miguel-Barranco and Chorrillos areas. The area of potential flooding at San Miguel-Barranco is 2.06 km^2, which is 58.2% of the area that is less than 5.0 m above sea level. The main service infrastructure that could be affected is the Costa Verde Roadway, which relieves traffic congestion between the northern and southern areas. This roadway also connects the different beaches, piers, sporting areas, and restaurants. In addition, an area of San Miguel receives solid wastes.

The Chorrillos area is made up of the Los Yuyos, Sombrilla, Agua Dulce, and Pescadores beaches. The potential flooding area is 0.55 km^2, corresponding to 37% of the area that is less than 5.0 m above sea level (which is the height where the cliffs begin). This area is of social significance because it provides a recreational area for the majority of Lima's inhabitants during the summer. The affected infrastructure consists of piers, public bathrooms, medical centers, sports areas, restaurants, and a dock for recreational fishing. Next to Chorrillos Beach is a private resort; all of its buildings would be affected by a rise in sea level.

La Chira Beach

The area of potential flooding at La Chira Beach is 1.08 km^2, corresponding to 49% of the area that is less than 5.0 m above sea level. Approximately 3,000 people would be affected by this flooding. The beach houses a water treatment plant; solid wastes and residual waters are unloaded through the La Chira main sewage channel.

Conchan-Pucusana

The area of potential flooding in Conchan-Pucusana totals 11.51 km^2, corresponding to 59.9% of the area that is less than 5 m above sea level. Area beaches serve as recreational sites for more than 500,000 inhabitants. The area also has urban sites, private clubs, a fish powder plant, and the Pucusana fish distribution center.

Response Strategies

Present Situation

Since 1940, various sea defense structures have protected the bases of the cliffs in Costa Verde and Miraflores Bay, and recreational areas and beaches have been created. The first groyne, more than 200 m long, was constructed on Agua Dulce Beach, toward the southern end of the bay, at right angles to the shore. A vast amount of sand accumulated in a short time. As a result of this success, several main groynes and shorter intermediate groynes were constructed at intervals. The result was the growth of sandy beaches more than 100 m wide on the southern sides of the groynes in both the Barranco and Miraflores areas.

In the 1960s, SOGREAT (a French institution) studied the hydrodynamics in Miraflores Bay. They used miniature models and proposed the sea defenses for La Punta-Callao. In the early 1970s, two Peruvian consulting companies, Corporacion Peruana de Ingeniería and Aramburu Menchaca Asociados, studied the coastal engineering in Miraflores Bay and proposed constructing groynes at 250-m intervals. They also studied the stability of the cliffs to establish geotechnical zones on which to build hotels, clubs, houses, or gardens, according to the quality of the terrain.

Marine erosion is extensive north of Miraflores Bay, mainly in the southern side of La Punta, known as Mar Brava (Rough Sea). Two long breakwaters have been constructed at oblique angles to the shore, where the higher waves break on the seaward face. A sand ridge closed the opening between the two breakwaters, stopping the water flow between them, and resulting in sedimentation and subsequent decomposition of mud and organic material accompanied by a putrid smell. Openings were made in the breakwaters to improve water circulation, but these were not as successful as anticipated. The breakwaters have been weakened by the two openings, and storm waves in recent years have partially loosened the igneous rocky blocks. The height of the northern breakwater was recently raised to 5 m, and the seaside of the breakwater was covered with a layer of concrete to protect it from the "piston effect." A plan to construct hotels and clubs was abandoned. About 30,000 m^2 of new terrain was gained inside the breakwaters, and buildings have been constructed there.

Toward the south, the Callao coastline was protected by three groynes at right angles, with inadequate results. The positioning of the groynes was modified into a T shape with effective results, protecting the coast and forming semicircular beaches. All of the coastal beaches in La Punta and Callao are stony (mainly 60- to 120-mm particles), but inside the groynes in the T shape, the sedimentation is of coarse sand (1- to 2-mm particles) and small granules (2 to 20 mm).

South of Mar Brava in San Miguel, storm waves have eroded the coastline, flooded the recently constructed road, and destroyed nearly 50% of it. Seawalls of igneous rocks are now being erected in the reconstruction of the two-way road.

In Callao Bay, Cantolao Beach is growing. The port of Callao is protected by two oblique breakwaters; the streets are flooded only during storm waves. The Ferroles area has fishing powder plants with piers for unloading material. The port of Callao must be dredged every five years because of heavy sedimentation discharged by the Rimac River. This coastal zone is growing rapidly, mainly in the Ventanilla and Santa Rosa beach areas. South of Lima, the beaches have grown in Conchan, Arica, San Bartolo, and Santa María. This region is the site of some clubs and summer resorts.

Proposed Adaptation Measures

Adaptation measures to protect the coast from marine erosion have been derived from IPCC recommendations, past experience, and the results obtained in this study. The IPCC Working Group of Response Strategies suggested three mitigation alternatives: retreat, accommodate, and protect. In the case of La Punta, most of the urban area is at 1.50 m above sea level (IPCC 1992). La Punta is the site of many buildings, houses, and clubs, and the Navy School with its expansive infrastructure. The city has 6,000 inhabitants; therefore, in a preliminary evaluation, protection would be the best alternative. South of La Punta, the two breakwaters could be elevated to protect against normal and storm waves. Recently, part of the northern breakwater was raised and covered by a layer of concrete; the results await evaluation. The levee along Cantolao Beach, north of La Punta, could also be raised. Erosion is extensive between Dos de Mayo School and the Maggiolo shipyard; construction of a groyne parallel to the coastline is recommended. Other responses may include:

- The port of Callao has sedimentation originating from the Rimac River. An increase in sea level would favor the entry of the ships into the port, requiring that the heights of the two breakwaters and the piers be raised.
- Beaches in the Ferroles area would be flooded with an increase in sea level. The construction of seawalls is recommended to protect the fish powder plants.
- North of Callao Bay, Ventanilla and Santa Rosa bays have large sandy beaches created by the continued accretion of sands supplied by the Rimac River. Groynes will need to be constructed there.
- Costa Verde is a long beach of more than 14 km. It could be protected by elongating the groynes for sand accretion and constructing longitudinal seawalls like the ones being constructed for the San Miguel Roadway. The foot of the cliff is 5.0 m above sea level.

South of Lima, the beaches of Conchan, Arica, Punta Negra, San Bartolo, and Santa María, are growing. Conchan Beach has a longitudinal ridge around an ancient filled lagoon. Some summer resorts are being constructed there, which would be in danger with an increase in sea level during storm waves. The beaches would be covered and eroded by marine waters by an increase in sea level, affecting more than 500,000 tourists during the summer. The construction or elongation of groynes and seawalls is recommended to encourage sand accretion on the upper part of the beach ridge and to protect the urban area.

Evaluation of Adaptation Measures

The measures that have the greatest chance of being implemented are the construction or the elongation of groynes and the construction of seawalls. Protecting La Punta is a priority, but protection measures will likely be complex. Perú's Environmental Office, which reports to the Department of the Presidence, coordinates measures to protect the environment. This Office is charged with recommending laws and rules to protect areas against potential flooding and approving the best measures to accommodate the affected population. The National Fund for Lodging (FONAVI) is currently responsible for moving persons affected by flooding.

Conclusions

The following conclusions were reached as a result of this study:

- Lima's coastline suffers from erosion and accretion, the latter due mainly to the construction of groynes and piers.
- The best way to protect the area in danger of flooding from a rise of sea level appears to be the construction of breakwaters, groynes, levees, and seawalls.
- La Punta, with a population of 6,000, is also in danger, but the solutions in this area are complex.
- Coordination with national governmental offices and local governments is going well.

References

Conopuma, C., 1987, *Inventario y Caracterización de la Contaminación Marina por Fuentes Domésticas, Industriales, Agrícolas y Mineras en el Area Costera de Lima Metropolitana*, 1986-87, Informe de CONPACSE, Callao, Perú.

Conopuma, C., G. Sanchez, and M. Echegaray, 1986, "Inventario de fuentes de contaminación en el área de Lima Metropolitana a nivel industrial," *Doc. Seminario Taller*, PNU-MA/CPPS/ECO/CEPIS, Lima, Perú.

Dirección de Hidrografía y Navegación (DIHIDRO), 1942-1992, Estadística de Mareas de la Estación Mareográfica del Callao, Callao, Perú.

Guillen, O., 1976, "El Sistema de la Corriente Peruana, Parte I, Aspectos Físicos," *Informe de Pesca* No. 185 *FAO,* Rome, Italy.

Intergovernmental Panel on Climate Change (IPCC), 1990, *Strategies for Adaptation to Sea Level Rise*, R. Misdorp et al. (eds),Response Strategies Working Group, Coastal Zone Management Subgroup, Ministry of Transport, Public Work and Water Management, The Hague, the Netherlands.

Intergovernmental Panel on Climate Change (IPPC), 1992, *Global Climate Change and the Rising Challenge of the Sea*, Report of the Coastal Zone Management Subgroup, Response Strategies Working Group, Rijkswaterstatt, the Netherlands.

Jacinto, M., 1990, *Calidad del Medio Marino en Areas Costeras Peruanas*, Informe de CONPACSE, Callao, Perú.

National Institute of Statistics and Information (INEI), 1993, *Población Nominalmente Censada por Departamentos, Provincias y Distritos y por Sexo,* Resultados preliminares del Censo Nacional de 1993, Lima, Perú.

Ochoa, N., and O. Gomez, 1988, "Variación Espacio-Temporal del Fitoplancton frente a Callao, Perú 1986," in H. Salzwedel and A. Landa (eds.), *Inst. Mar Perú. Bol. Vol. Ext.* pp. 51-58, Callao, Perú.

Santander, H., and S. Guzman, 1986, *Informe sobre el Zooplancton en la Prospección frente al Puerto del Callao*, Prospección de Mareas.

Senamhi, 1984-1994, *Informes Trimestrales de las Condiciones Meteorológicas en el Litoral Peruano*, Comisión ERFEN-SENAMHI, Lima, Perú.

South Pacific Permanent Commission (CPPS), 1994, "Socioeconomic Implications of the Increase of Sea Level in Metropolitan Lima and Callao as a Consequence of Global Warming," CCPS/PNUMA, Lima, Perú.

Teves, N., 1967, "Erosión Marina y Formación de Canturrales en el Litoral de La Punta-Callao," *Centro Pesq. Facultad de Oceanog. y Pesq.*, UNFV Pub. No. 10:1-12, Frederico Villarreal National University, Lima, Perú.

Teves, N., and S. Gagliano, 1970, "Geología Litoral y Submarina de la zona comprendida entre el Río Rimac y la Playa Santa Rosa Ancón, Lima," *I Cong. Latin. Geol. Anales* V:37-54, Lima, Perú.

Response Strategies and Adaptive Measures to Potential Sea-Level Rise in The Gambia

Bubu P. Jallow
Department of Water Resources
7, Marina Parade
Banjul, The Gambia

Abstract

The Gambian coastal zone has 70 km of open ocean coast and about 200 km of sheltered coast dominated by extensive mangrove systems and mud flats. About 20 km of the coastline is developed. Other areas are largely underdeveloped, except for fish landing sites and cold storage infrastructure for processing and storing fish and shrimp. Socioeconomic, land-use change, and ecological data were used to determine the nation's vulnerability in economic terms to a rise in sea level at a coastline area. Banjul (the capital city) is expected to be overtaken by the sea, displacing about 42,000 inhabitants. Land loss translates to US$217 million, not including loss of infrastructure. The effects of sea-level rise on the nation's coastal zones are examined and adaptive measures are suggested that can be taken now and in the future to protect resources essential to the economy.

Introduction

The Gambia extends from Buniadu Point and the Karenti Bolong in the north to the mouth of the Allehein River in the south. The Gambian coastal zone consists of 70 km of open coastline and about 200 km of sheltered coastline. About 20 km of this coastline is significantly developed, including Banjul (the capital city), Bakau and Cape St. Mary, Fajara, and the tourism development area. Thirteen hotels and resorts have been built in this area.

The coastal zone contributes significantly to the nation's economy. From October to May, more than 100,000 tourists visit The Gambia, and the tourist industry employs about 7,000 persons. In 1990-1991, the government collected about 48 million Gambian Dalasis (D) (US$5.3 million) in direct and indirect taxes from tourism. This amount makes up about 10% of the government's revenue (United Nations Environmental Programme [UNEP] 1992). Other sectors with important economic activities in the coastal zone include fisheries and agriculture.

The U.S. Country Studies Program has provided data sets to characterize the coastal zone, assess vulnerable areas, and determine the infrastructure and people at risk from the effects of climate change. From 2000 to 2050, most land loss will be caused by inundation. This loss is estimated to be about 92.32×10^6 m^2 for a 1-m sea-level rise, 45.89×10^6 m^2 for a 0.5-m sea-level rise, and 4.96×10^6 m^2 for a 0.2-m sea-level rise. Most of this area will be wetlands and mangrove (about 66,900 ha) systems, which are important fish spawning and wildlife habitats. From 2051 to 2100, most land loss will be caused by direct erosion. Bruun Rule calculations show that this loss could vary from 0.068 m/yr in erosion-resistant cliffs to 8.80 m/yr in other areas.

Socioeconomic, land-use change, and ecological data were used to determine the nation's vulnerability in economic terms to sea-level rise for one coastline area. Land loss translates to US$217 million, not including the loss of infrastructure. The potential climate changes will exacerbate erosion. Hence, it is necessary to respond now and identify adaptive measures. Responses include (1) managing available sand to let it flow to sand-starved areas around the cemeteries and Banjul and (2) rehabilitating the groyne systems. Anticipatory adaptive measures include construction of revetments, seawalls/bulkheads, and breakwater systems to protect economically and culturally important areas. In the longer term, building regulations, urban growth planning, wetland preservation and mitigation, and development of a Coastal Zone Management Plan should be instituted.

Coastal Characteristics

On the basis of the geology, physical structures, socioeconomic data, and a review of the literature, the study area was divided into nine units (Figure 1 and Table 1), each exhibiting one or a combination of coastal types. Some of the units in the study are based on the UNEP/Oceans and Coastal Areas Programme Activity Centre report (Quelence 1988).

Impacts of a 1-m Sea-Level Rise on The Gambian Coastal Zone

Following an aerial videotape-assisted vulnerability analysis (AVVA) to assess the impacts to the coastline, an assessment was performed to determine the vulnerability of the Gambian coastal zone to a 1-m sea-level rise, and potential impacts were identified (Tables 2 and 3). Details of the AVVA technique and its

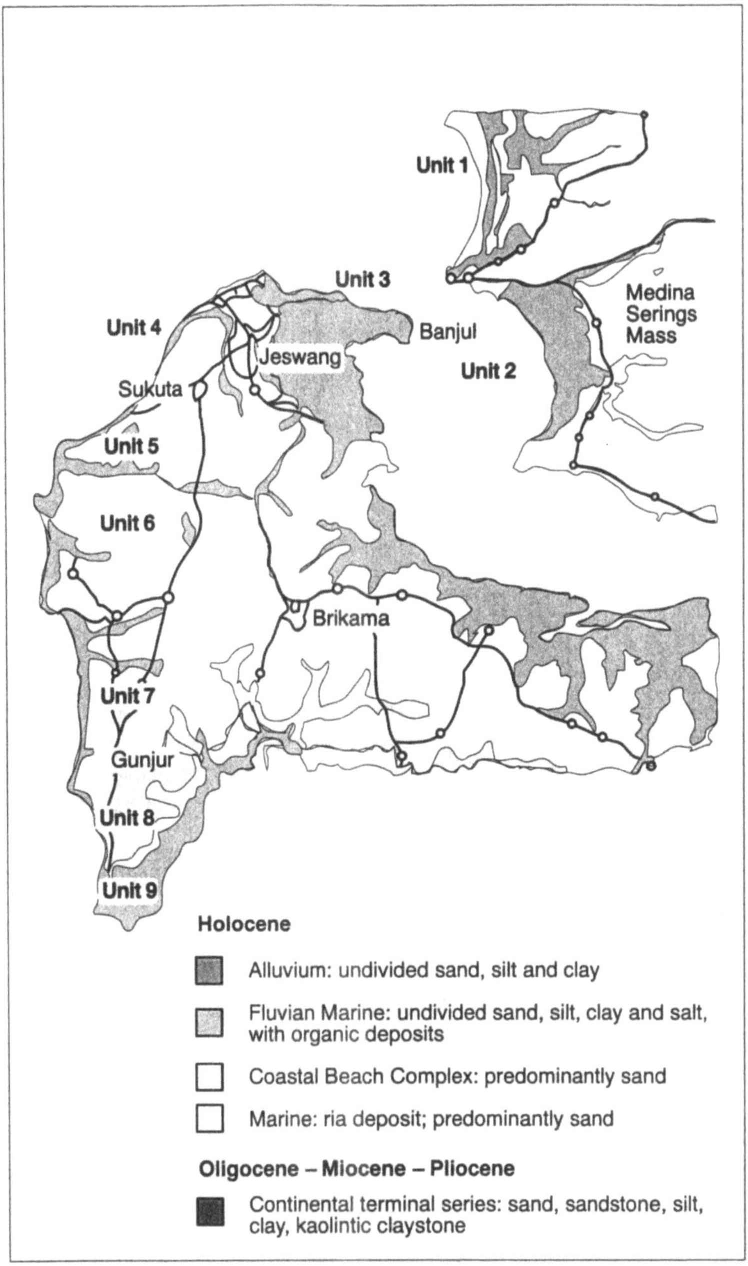

Figure 1. Geological Map of the Coastal Zone of The Gambia

Table 1. Names of Units in the Gambia Study Area

Unit	Name
1	Buniadu Point/Kerenti Bolong to Barra Point
2	The Gambia River Estuary
3	Banjul Point to Cape St. Mary
4	Cape St. Mary to Fajara
5	Fajara to Bald Cape
6	Bald Cape to Solifor Point
7	Solifor Point to Benchmark KM125
8	Benchmark KM125 to Kartong Point
9	Kartong Point to Allehein River

applicability can be found in Leatherman et al. (1995). The losses and impacts were then expressed in monetary terms and extrapolated with time. Adaptive measures and response strategies were assessed, costs projected, and the most viable option selected. The methodology for estimating land loss and response options is detailed in Nicholls et al. (1995).

Estimates of Population at Risk and Cost of Loss

A comprehensive evaluation of the economic consequences of a 1-m rise in sea level in The Gambia was attempted for Unit 3 (Banjul to Cape St. Mary) of the coastal zone. If the sea were to rise 1 m, the entire city of Banjul would be covered by water in the next 50-60 years because most of the city lies below 1 m in sea level. The mangrove systems on St. Mary's Island, Kombo St. Mary, and the strand plains on the north bank from Barra to Buniadu Point would be lost. The entire population of Banjul (42,000 inhabitants) would be displaced. Currently, these inhabitants live in eastern Bakau and Cape St. Mary, the swampy parts of Old Jeswang, Kanifing Industrial Estate, Eboe Town, Tallinding Kunjang, Fagikunda, and Abuko.

A preliminary analysis of the data from the Department of Lands on the value of land and sample properties between Banjul and Kololi Beach Hotel

Table 2. Area of Land to be Lost in the Gambia Because of Various Sea-Level Rise Scenarios[a]

Coastal Unit	Sea-Level Rise Scenarios ($m\ 100\ yr^{-1}$)			
	0.2	0.5	1.0	2.0
Senegal Border (north) to Barra Point	562,500	14,062,500	28,125,000	56,250,000
Gambia River Estuary	937,500	23,437,500	46,875,000	95,750,000
Banjul Point to Cape St. Mary	1,560,000		7,800,000	17,630,000
Cape St. Mary to Fajara	0	0	0	0
Fajara to Bald Cape	10,000	25,000	50,000	200,000
Bald Cape to Solifor Point	406,000	1,015,000	2,030,000	2,890,000
Solifor Point to Benchmark KM125	268,000	670,000	1,340,000	1,810,000
Benchmark KM125 to Kartong Point	34,000	25,000	170,000	170,000
Kartong Point to Senegal Border (south)	1,186,000	2,695,000	5,930,000	8,850,000
Total land loss	4,964,000	45,890,000	92,320,000	181,550,000

[a] Totals may not equal because of rounding.

Table 3. Application of the Bruun Rule to Determine Land Retreat and Erosion Rate in The Gambia

	Depth of Closure (m)	Overfill Ratio	Active Profile Width		Dune or Cliff Height (m)	Scenario	Retreat $R = G$ $(LB + h)S$
			Map (cm)	Actual (m)			
Coastal Unit	h	G	L	L	B	S	R
Senegal Border (north) to Barra Point	5.9	1.0	3.5	2,610.5	2.0	1.0	330.4
Gambia River Estuary	NA[a]	NA	NA	NA	NA	NA	NA
Banjul Point to Cape St. Mary	5.9	1.0	3.0	2,284.2	2.7	1.0	275.6
Cape St. Mary to Fajara	5.9	1.0	1.6	1,232.5	6.1	1.0	102.2
Fajara to Bald Cape	5.9	1.0	2.8	2,080.8	2.7	1.0	264.1
Bald Cape to Solifor Point	5.9	1.0	8.6	6,475.5	1.8	1.0	839.2
Solifor Point to Benchmark KM125	5.9	1.0	6.8	5,096.3	3.0	1.0	597.2
Benchmark KM125 to Kartong Point	NA	NA	NA	NA	NA	NA	NA
Kartong Point to Senegal Border (south)	NA	NA	NA	NA	NA	NA	NA

[a] NA = not assessed.

suggests that the amount of land lost due to sea-level rise would be worth D1,950 billion (US$217 million).

Response Strategies and Adaptation Policies

Response Strategies

If sea-level rise were to occur as projected in this study, and considering the present economy of The Gambia, the most appropriate response strategy would be to protect the important economic areas (i.e., Units 3 and 5 of the coastal zone). Land loss in these areas would represent about 8.5% of the total land to be lost if the sea level rose 1 m. However, in economic terms, this land would represent the greatest loss to the country.

Banjul, Cemeteries, Wadner Beach, Sarro, and the Cape Point Hotel Complex

In Unit 3, it will be necessary to protect the entire coastline of Banjul, the shoreline area from the cemeteries to the Wadner Beach Hotel, the infrastructure at Sarro (the former Gambia Produce Marketing Board facilities), and the hotel complex at Cape Point. The following sections describe shoreline hardening and stabilization techniques that could protect this area.

Innovative Sand Management Approach

This study established that the area from the Utility Holding Corporation water tanks to Banjul Point has been sand-starved by a growing recurved sand spit. As a result, significant erosion has been observed in this area in the last five years. An innovative approach to solve this problem in the near future is to tie the end of the sand spit to the main land area around the Palm Grove/Utility Holding Corporation water tanks so that it becomes a sand feed to the area in the east.

This approach is likely to introduce some problems in the future. The huge sand supply to the cemeteries and Banjul will first lead to the loss of the expanding lagoon to the west of the Wadner Beach and Palm Grove hotels. It will also lead to large unwanted deposits of sand at the Banjul-Barra Ferry Terminal and The Gambia Ports Authority facilities. To protect these facilities from the huge unwanted sand, a groyne system must be used.

Groynes

Constructing groynes is a major technique for stabilizing beaches. In this technique, a rigid structure is built at an angle from the shore to protect the shore from erosion. In The Gambia, trunks of rhun palm trees (Figure 2) are jetted into the beach and tied together with timber. (The existing groynes in the area

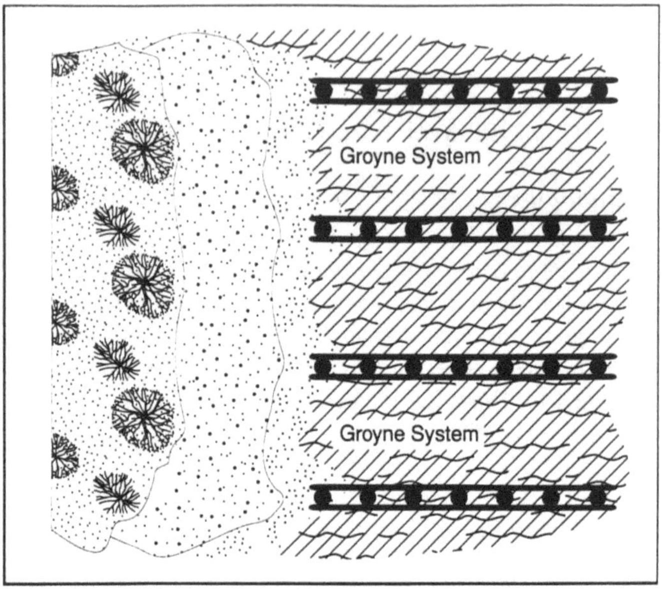

Figure 2. Plan View of a Groyne System

between the Wadner Beach and Palm Grove hotels to Banjul Point were built in the late 1950s by the government of The Gambia, possibly through an aid program.)

Groynes are easy to build, are fairly cost-effective, and locally hold the beach or capture more sand in the long sediment transport (LST) system. They do not "create" new sand, but they do help to redistribute the sand along the beach. To be more effective in distributing sand and stabilizing the beach, the groynes should be filled with sand, so that the net LST system is continuous. When the groynes are not filled with sand, the sand is trapped in the upwind side of the groyne, while downwind more sand erodes.

Repairing these groynes will help in transporting sand to the cemeteries and Banjul and in stabilizing the beach in this area. To reduce the flow of sand to The Gambia Port and Ferry Terminal, a long, high terminal groyne should be constructed at Banjul Point between State House and Albert Market. This terminal groyne should not be filled with sand to avoid halting or reducing transport of longshore sediment.

Revetments

The innovative sand management approach and the groyne system of shoreline stabilization techniques are useful short-term solutions, especially with the erosion in the area. Figure 3 shows a cross section of a revetment. A revetment

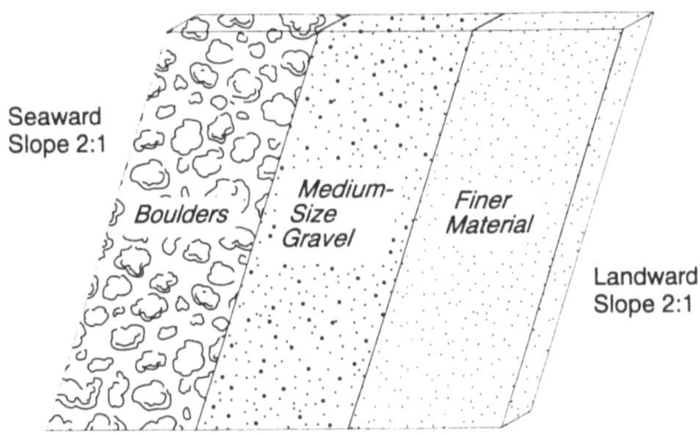

Figure 3. Cross Section of a Revetment

consists of a layer of large boulders on the seaward side and a layer of filter screening material on the landward side. These two layers are separated by another layer of medium-size gravel used to protect the screening material. The voids structure of the outer layer on the seaward side dissipates wave energy. The filter screening material on the landward side allows water to pass through.

Generally, the structure is built with a slope of 2:1 and must be designed with sea level, subsidence, and wave climatology in mind. When properly constructed, a revetment accumulates sand at its base and thus helps reduce erosion and stabilize the beach.

No data are available on wave climatology. Therefore, the following assumptions are made to assess the breaking wave height: (1) during storms, a change in atmospheric pressure may lead to a 0.2-m rise in the water level; (2) prevailing winds may lead to a sea-level rise of about 0.5 m; (3) waves may contribute another 0.5 m; and (4) tidal variation may contribute to a rise of 0.75 m. Hence, the breaking wave height H_b can be estimated as:

$$\begin{aligned} H_b &= 0.78\,(0.2 + 0.5 + 0.5 + 0.75) \\ &= 0.78 \times 1.95 \text{ m} \\ &= 1.5210 \text{ m}. \end{aligned}$$

With this breaking wave height, it is safer to build a 2-m-high revetment to avoid overtopping. When a wave breaks over the top of a revetment, it digs a hole behind the filter, which could cause the revetment to collapse in the future.

Seawall/Bulkhead

Although the area discussed is exposed to wave action, the waves are relatively small. Thus, it may be necessary to build a low-cost seawall or a bulkhead (Figure 4). The seawall has a slope of 1:2, a 2-m beam, and a height above water of 1.2 times the number used in the sea-level rise scenario.

National Partnership Enterprise Cold Storage Plant to Gambia High School and Areas Bordering the Mangrove Systems

To protect these areas, it is sufficient to use dikes composed of about 1.5-2 m of sand on which vegetation can be grown.

Kololi Beach Hotel to Kairaba Hotel

With a 1-m sea-level rise, the shoreline is expected to retreat about 264 m in the next 100 years. This number translates to an erosion rate of about 2.64 m/yr. The area is not in immediate danger, but will need to be protected by building a revetment, seawall, or breakwater (Figure 5).

Adaptation Responses

Public Awareness

The public should be informed about the risk of living in coastal and lowland areas that could be affected by sea-level rise through saltwater intrusion, flooding, erosion, and so on. Timely public education about erosion, sea-level rise, and risks of flooding could be a cost-effective way to reduce future expenditures.

Physical planning and building control measures and regulations should be instituted and implemented. The lands, physical planning, and building control institutions should avoid allocating land likely to be flooded, such as the dried-up streams on the Kombo Peninsula, which flooded during the rainy seasons in the late 1980s.

Increase in the Height of Coastal Infrastructure and Urban Growth Planning

When the construction of coastal infrastructure, such as roads, fishery landings, and curing plants, is approved, the authorities and owners of these infrastructures should ensure that plans include marginal increases in the height of the structures. These increases would account for sea-level rise and other related phenomena. Large facilities or those that pose significant hazards when flooded should be built on sites away from sensitive lands (i.e., toward less vulnerable areas).

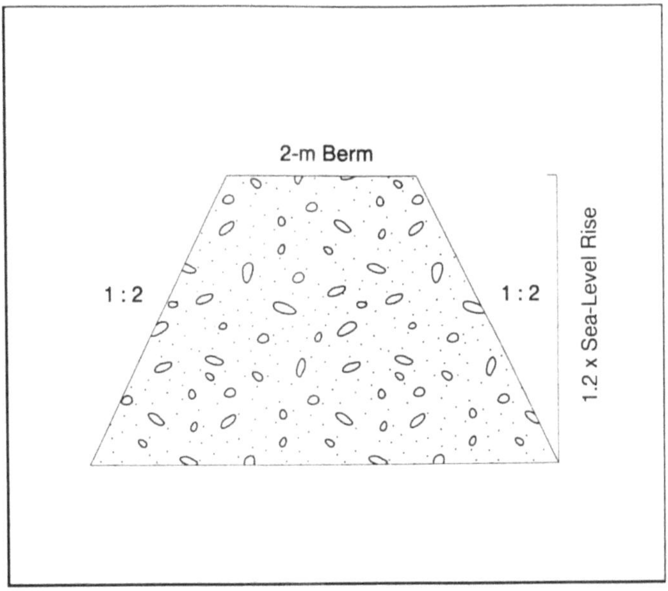

Figure 4. Low-Cost Seawall/Bulkhead

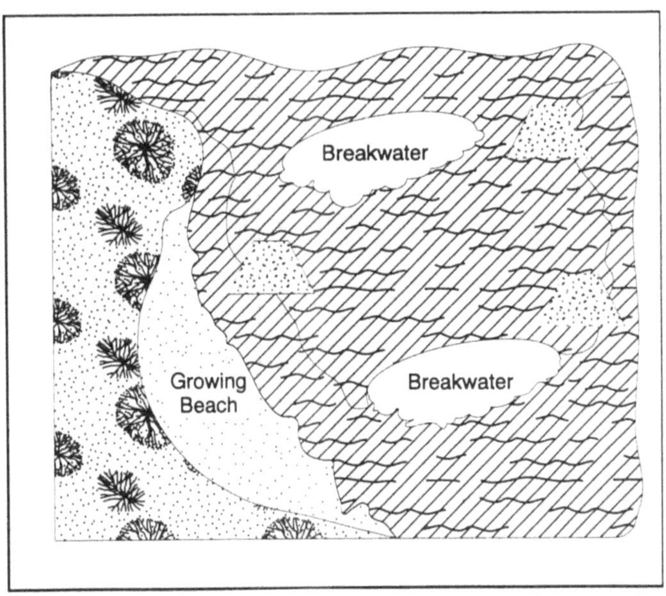

Figure 5. Plan View of a Breakwater

People located in high-risk areas should be offered incentives to relocate. Policies should be instituted that allow high-risk areas to be used for natural preserves or for low-value use. Marginal increases in the height of the infrastructure during construction and redirection of growth away from sensitive lands are relatively inexpensive options for reducing the impacts of sea-level rise and risks of flooding. Policies that may lead to relocation from high-risk areas could reduce the need and cost of disaster relief in the future.

Wetland Preservation and Mitigation

The estuary of the Gambia River contains useful wetlands and mangrove systems. The mangrove systems on Kombo St. Mary are important breeding grounds for various aquatic species. Efforts should be made to protect these areas by declaring them "protected wetlands." This designation would discourage exploitation of resources in these wetlands.

Coastal Zone Management Plan

Land-use planning in coastal zones, such as using setbacks or allocating low-lying vulnerable lands to low-value uses (e.g., parks rather than housing), will help to reduce the vulnerability to sea-level rise. Additional risk-reduction measures can be created and encouraged through appropriate financial mechanisms. Each of these policies reduces the risks from current climatic variability and protects against potential sea-level rise. When these policies are grouped to form a program, it will constitute a Coastal Zone Management Plan.

References

Leatherman, S.P., R.J. Nicholls, and K.C. Dennis, 1995, "Aerial Videotape-Assisted Vulnerability Analysis: A Cost-Effective Approach to Assess Sea Level Rise Impacts," *Journal of Coastal Research,* Special Issue No. 14.

Nicholls, R.J., S.P. Leatherman, and C.R. Volonte, 1995, "Impacts and Responses to Sea Level Rise: Qualitative and Quantitative Assessments," *Journal of Coastal Research,* Special Issue No. 14.

Quelence, R.E., 1988, *Identification of Coastal Erosion Problems in The Gambia,* United Nations Environmental Programme, Washington, D.C., USA.

Climate Change Adaptation Strategies: A Challenge for Small Island Nations

J.C.K. Huang
U.S. Country Studies Program
1000 Independence Avenue, S.W., PO-63
Washington, DC, 20585, USA

Abstract

Small Pacific island nations, which have relatively small areas of arable land surrounded by boundless seas, have common environmental characteristics and concerns regarding climate change. Many such nations with strong indigenous cultures have based coastal management decisions on traditional land and marine tenure systems. This basically top-down, traditional decision-making process is not totally compatible with current socioeconomic development. An integrated coastal management program may be the best strategy an island nation could develop for adapting to climate change. This strategy would take advantage of recent advances in marine and coastal science, environmental assessment tools, and adaptive measures, yet incorporate a community-participation approach that would consider the island's unique circumstances.

Introduction

Recent scientific studies have led to a concern that human activities have adversely affected the earth's climate (Intergovernmental Panel on Climate Change [IPCC] 1990, 1992). Much attention has focused on global warming, an accelerated greenhouse effect in which the earth's atmosphere retains more heat because human activities have increased the amount of heat-absorbing trace gases, such as carbon dioxide (CO_2), nitrous oxide (N_2O), chlorofluorocarbons (CFC), and methane (CH_4). Scientists believe that global warming, if it occurs, will have serious effects on sea level, water resources, agriculture, and forestry. In fact, the international scientific community has reached a broad consensus regarding global climate change (IPCC 1991, 1992; IPCC Coastal Zone Management Subgroup [CZMS] 1990).

Direct results of global-warming-related climate change include severe and frequent cyclones, storms, floods, droughts, and accelerated sea-level rise (ASLR). According to recent estimates, ASLR will be 0.3-1.1 m over the next century (IPCC CZMS 1994). ASLR could lead to more disastrous storm surges, flooding, erosion of beaches, loss of woodlands, saltwater intrusion, destruction of coastal ecosystems, and property damage if no adaptive measures are taken. It could adversely affect coastal regions, where most of the world's population lives. Moreover, significant population growth will take place in heavily populated coastal regions, particularly in developing countries. Hence, human-induced climate change and the resulting ASLR are expected to significantly increase the stresses on coastal resources.

Scientists have made much progress in understanding coastal processes. Science offers high-technology data and tools such as computer models that can be used for examining the potential for ASLR, changes in precipitation and storm patterns, and other events that may occur within the next few generations. Such information calls upon island nations to develop adaptation strategies now to minimize the impacts of climate change and growth in the population and economy.

The IPCC CZMS urgently recommended that all nations develop and implement coastal zone management plans that use the latest knowledge and technologies to reduce vulnerability and take adaptive measures for anticipated impacts (IPCC CZMS 1991, 1992). The 1993 World Coast Conference (IPCC 1994) reaffirmed that "integrated Coastal Zone Management has been identified as the most appropriate process to address current and long-term coastal management issues, including habitat loss, degradation of water quality, changes in hydrological cycles, depletion of coastal resources, and adaptation to ASLR and other impacts of global climate change." Many island nations have declared that they will undertake integrated coastal management (ICM) within forums such as the 1992 United Nations Conference on the Environment and Development, United Nations Framework Convention on Climate Change, and others dealing with national environmental management strategies.

The ICM concept, however, is largely based on the experience of developed countries, where the balance of power between state and local authorities is well established. Such experience may not apply to small island nations, where traditional culture predominates. For most small island nations, the "coastal zone" is essentially the whole island. Furthermore, a range of political, cultural, and legal arrangements affect the use and management of resources. Traditional land tenure systems may continue to serve island populations well, even though some societies are rapidly changing. An island nation's ICM program must reflect its particular circumstances and the direction of its development initiatives; that is, small island problems require small island solutions. An approach is needed that is largely process-oriented and merges modern assessment and management tools with traditional systems. Otherwise, the possibility of obtaining the interest, involvement, and commitment of traditional communities will be limited, and these communities are at the center of the coastal management equation (Smith 1994).

This paper synthesizes work related to ICM for Pacific island nations from a workshop in Honolulu, Hawaii, in September 1994. The workshop brought together more than 40 persons from the United States, Australia, and 12 Pacific island nations, including scientists and managers from state and federal agencies, academic experts, and representatives of international organizations. The purpose of the workshop was to explore the unique problems facing Pacific island nations and suggest ways these nations might respond to global-warming-related climate change. Participants learned about the positive aspects of traditional practices in the Pacific islands (for example, the high regard placed on family values) as well as some deficiencies (for example, the inability to adopt modern management tools and skills to respond to crises). They also learned about experiments in community-based management of natural resources in the Pacific islands. They recommended that island nations absorb advanced scientific knowledge and adopt improved application tools and skills, including high-technology information and modeling systems, to develop, with community participation, coherent ICM programs based on their respective traditional practices.

Traditional Management Practice: Cultural Perspective

Traditional management practices and systems have been integral parts of Pacific island societies. These systems usually incorporate changing beliefs and perceptions. Some of these beliefs may be based on nonscientific myths that ensure proper respect for sacred ground and certain animals or marine life. Custom and tradition have been sufficient to support and enforce management practices. In the Solomon Islands, for example, it is believed that people who abuse fishing rights will be eaten by guardian sharks (Hviding and Baine 1994). In Fiji, people make offerings when visiting an area and have various totems that restrict the consumption of certain foods. In many cases, a considerable body of ecological knowledge has also accumulated during the centuries-old development of such practices.

Under traditional systems, direct relationships exist between people and the land. Kinship networks provide for community-based resource sharing and management. Resources are distributed according to local custom and a person's or family's place within a hierarchy. For example, the Matai system in American Samoa (Templet 1986) serves as a forum to mediate disputes over land and marine resources. Most Samoans consider land an inheritance from God and therefore sacred. Consequently, land is managed with great diligence. In Fiji, the native land trust board advises landowners on fair compensation and gives final approval of land-use contracts. In Yap, outsiders must secure permission from the owners to use land or fishing grounds (Iou and Smith 1994).

Traditional management systems may include various mechanisms to control use of resources. Many island communities use seasonal hunting and fishing bans, island harvest and crop rotation, and seasonal closure of reef areas to manage resources. Multicropping, terracing, and irrigation preserve soil resources and biodiversity. Traditional fishing techniques naturally limit the catch of individual fishers, and more sophisticated fishing methods are reserved for chiefs or for times of scarcity. Fish ponds and community gatherings also help to manage fish harvests. In some island nations, such as Yap, traditional management systems prescribe population migration patterns and establish family planning policies to manage population growth (Iou and Smith 1994).

Traditional management practices have weaknesses. Some customs and traditions, although of good intent and purpose, are based on nonscientific perceptions. Management based on a kinship network may not be able to cover the ever-increasing population in a rapidly developing coastal community. Educated people who accept advanced scientific knowledge and modern management tools may not follow traditional leaders. Above all, traditional systems are based on historical experiences under a natural climatic environment. They do not have the built-in flexibility to respond to irregular environmental and social impacts, such as those caused by climate change. As a result, island nations need coastal development and resource management plans.

Community-Based Management: Bottom-Up Approach

Community-based management is a successful experiment for managing an island's coastal resources. It redirects management from a top-down approach to a bottom-up one and incorporates local knowledge and experience. The community, in its capacity as resource user, becomes involved in developing and implementing the program.

Various methods exist for incorporating community-based management into management programs. In participatory rural appraisal, a multidisciplinary team of experts collects qualitative information in cooperation with members of the community (Peau 1994). It is an open process whereby not only the target audience but all villages can participate in all phases of the exercise. Communities recognize their needs, abilities, and resources as they collect data and attend idea-exchange meetings. After the communities identify their social, economic, and environmental concerns, they develop action plans to organize their responses. They also organize institutional structures to manage common resources. Resource users develop guidelines for the sustainable use of resources.

In American Samoa, for example, community-based management has been developed in the American Samoa Coastal Management Program (Aitaoto

1994). Under the Coastal Hazard and Mitigation Program, new regulations for construction in hazardous areas and a system of village-based hazard mitigation were established through participatory planning. In addition, village ordinances and regulations were developed for various woodland areas in the Community-Based Woodlands Management Project. These programs were developed through a process that included both traditional Samoan systems and Western-style regulatory enforcement components (Peau 1994). In the Federated States of Micronesia, a forestry conservation program was developed and implemented through participatory rural appraisal exercises, after the community rejected standard top-down approaches (Dahl 1994).

Adaptation Strategies and Options

Integrated coastal management is not a methodology but a set of principles offering a coherent, balanced approach for managing coastal resources and maintaining sustainable development. A typical ICM program, as recommended by IPCC CZMS (1991, 1992), is a dynamic process in which decisions on the use, development, and protection of coastal areas and resources are made to achieve goals established in cooperation with national and local authorities and with people who use those resources.

Basic Principles for ICM

An ICM program should incorporate the following:

- A continuing process for collecting and disseminating scientific data needed for assessing impacts on resources, coastal issues, functional uses and development, and the needs and desires of the population.
- A public participation process for formulating national policies and developing a coastal governance system that integrates and applies those policies.
- A process for developing, acquiring, and strengthening legal, institutional, technical, financial, and human resources for the program.

Figure 1 presents a generic model for an ICM program. It outlines a process for decision makers to use when integrating the many sectoral needs within coastal zones. These basic principles are derived mainly from the experience of continental or large island ICM programs. However, some fundamental differences exist between continental and large island states and small island nations in their general approaches to coastal management. Continental and large island states recognize their coastal zones as distinct regions with resources that require special attention, which results in a well-established sectoral approach. Although some flexibility and volunteerism are built into a continental ICM, the

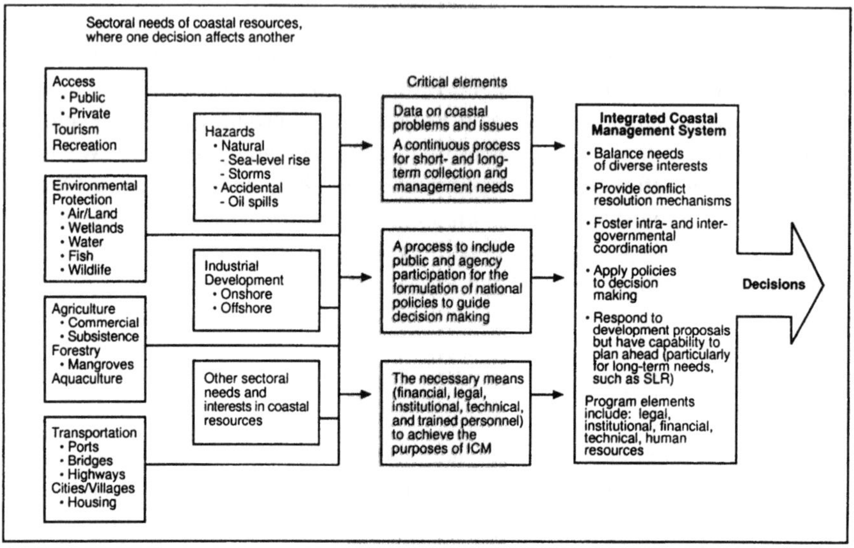

Figure 1. Integrated Coastal Zone Management Program: A Generic Model

program relies on modern knowledge and high-technology tools, with rigid legislative constraints and heavy enforcement components for execution and implementation, clearly a top-down approach.

ICM for Island Nations

The coastal areas of Pacific islands support the majority of the population, subsistence and commercial agricultural and fishery activities, and most economic development. Coastal areas are also subject to the damaging effects of natural hazards, such as cyclones (typhoons, hurricanes), storm-induced waves, and abnormally high tides. Islands at low elevations and low-lying coastal areas of the larger, higher Pacific islands are particularly at risk. Global-warming-related climate change threatens to exacerbate these hazards by accelerating the rise in sea level, increasing the frequency and intensity of cycles and storms, and causing other changes in atmospheric and oceanographic conditions. All these coastal problems highlight the need for ICM.

In small developing island nations, ICM has been based on models of sectoral management practiced by developed, continental countries. Difficulties arise in several areas: political structures in small islands are based on former colonial systems; coastal management techniques and methods for larger developed countries may not be appropriate for smaller island nations; and technical development assistance for coastal management is driven by the experiences of developed countries.

For Pacific islands with dominant traditional cultures, a different conceptual approach to ICM development should be encouraged. The customary systems, structures, and processes should form the basis of ICM; they should not merely be included in it. Such an approach requires ICM to be based on the scale of the "most effective management unit," be it the village, district, or whole island. The level at which ICM should be developed must be identified by each nation individually. National programs may be the most appropriate choice for some countries, but for others, especially those with strong traditional systems, the community level may be more appropriate. No matter what the most appropriate entry level is, however, it will be beneficial to have official sanction or legislation from the highest political levels. This approach may take longer in the short term and be more difficult to develop, but it should far more effective in the long term, especially when it encompasses larger areas (Smith 1994a).

Coastal management within a larger concept of sustainable development requires active community participation and commitment. Pacific islands generally enjoy a high level of community involvement in the management of resources, partly because of the small size of island coastal zones, but also because families and communities are organized to manage everything from fisheries to agricultural and water resources. Dependence on local resources translates into a vital interest in and commitment to sustainable development and conservation management. Pacific islands with strong indigenous cultures have traditional decision-making and management mechanisms for natural resource management. Although some ICM concepts may be new to local authorities, many cultures have mechanisms that could be adapted to new circumstances. This capability gives them a distinct advantage over Western-style organizations and institutions, which have a less intimate connection with the immediate environment. Many island cultures and communities are more closely attuned to the concepts of family and community, a condition that generally facilitates an easier understanding of the importance of allowing for the needs of future generations.

Pacific islands are living examples of traditional cultures continually adapting to modern circumstances (Smith 1994b). Although modernization threatens traditional management systems, most Pacific island coastal management programs have characteristics of both traditional and contemporary systems. In some areas outside urban regions or main islands, traditional systems have been retained, whereas in urban areas, more contemporary systems have been implemented. This dichotomy has created management problems. For effective ICM in the Pacific islands, the methodologies and approaches must be appropriate. What is deemed appropriate will vary considerably among islands. The process of implementing ICM will be critical. For example, the balance of power between the federal government and the local traditional authority must be appropriately maintained as an ICM program is implemented. In the Pacific islands, decision-making processes involve a considerable number of meetings, both as a means to provide information and as the prime method for arriving at a consensus. This is the "Pacific Way," and flexibility, not rigidity, will be required to implement ICM in the Pacific islands (Smith 1994b).

Regional ICM Efforts

The unique characteristics of the Pacific region provide significant opportunities for developing ICM. Despite its large area and diverse cultures and political systems, similar coastal management problems exist throughout the region. In addition, certain critical management information and tools, such as climate data and coastal models, are difficult, if not impossible, for individual states to implement and maintain alone. The region's common interests, combined with a history of regional cooperation, provide a foundation on which to develop a cost-effective, regional ICM project.

The South Pacific Region Environmental Program (SPREP) was set up to "assist Pacific island countries and territories to protect and improve their shared environment, and to manage their resources to enhance the quality of life for present and future generations" (Smith 1994b). Created under the unique geographic, social, cultural, and economic environment of the Pacific islands, SPREP is the institution best suited to take the lead in developing ICM programs for the region. The major objective of its ICM efforts is to answer the combined concerns of most islands by providing technical assistance and coordination efforts and by fostering cooperation among all island nations. The formulators of the SPREP and all the individual island ICM programs, together with a technical working group, have strongly emphasized that the unique characteristics of the Pacific region be taken into account by:

- Identifying and developing appropriate approaches, such as those that incorporate traditional management systems and decision-making processes;
- Translating the high degree of subsistence use of and dependence on coastal resources into a vital interest in and commitment to sustainable development;
- Taking into consideration the generally high level of community involvement in coastal zone resource management;
- Building on the importance of family and community to foster an understanding of the need to consider future generations; and
- Focusing on the relatively small areas in which the human impact on coastal resources is concentrated in most islands (Smith 1994b).

Workshop Recommendations Concerning ICM

Participants at the Honolulu workshop unanimously endorsed the IPCC CZMS recommendation that each island nation should develop its own ICM program

(IPCC CZMS 1992). They identified ways traditional practices and community-based management could be incorporated into ICM programs. Plans for small islands must consider the entire island, not just the coastal or marine areas. In addition, island governments must reinforce, honor, and include community and traditional management practices and decision-making systems in ICM programs. Villages and their leaders should be allowed to prioritize management concerns and establish the pace and direction of management programs through community-based processes. Individual islands and regions must be allowed to establish their own combinations of traditional and contemporary management systems, including new technologies and management tools, in their respective ICM programs.

Workshop participants recommended the various actions regarding ICM. These actions are discussed in the following paragraphs.

Traditional management practices can be used as the basis for ICM programs, including those that address the problems of climate change. Such problems should consider the knowledge and experience of traditional management systems. New ICM programs should be decentralized and flexible enough to allow individual regions to establish their own mix of traditional and contemporary practices. This capability would allow areas to develop site-specific programs. Traditional management systems and practices should also be reinforced by legislation and community education programs.

The kinship and community networks that form the basis of traditional systems can be used to address the intricate sociocultural problems that would arise from climate change. The management of environmental refugees and subsequent land-use and resource access problems associated with climate change could be addressed through the social structures built into traditional management systems. Customary practices are more suited than contemporary programs to deal with social problems associated with access to resources.

Individual villages should be encouraged to develop resource management plans that incorporate their own specific traditional management systems. Existing community institutions for resource management should be used in developing ICM plans; however, modern legal policies and regulations should also be considered. Traditional agroforestry methods could be used when agricultural lands are lost as a result of ASLR. Seasonal and locational bans could be used to limit access to resources made scarce by climate change. In addition, "disaster food" could be planted in preparation for possible famine or weather-related disasters.

Traditional and community-based management practices should be integrated with modern management technologies to form site-specific ICM programs to address climate change. The first step should be consultation between the government and local communities to identify site-specific problems, such as shoreline erosion, inundation, saltwater intrusion, social disruption, coral reef destruction, and deforestation. Community resources, traditional management practices, and alternative solutions should be inventoried in the early phases. On the basis of these findings, a local management program that incorporates the community's specific traditional management practices and modern management technologies should be initiated. Moreover, regulations for resource use that reinforce community-based management and enforcement practices should be developed. A community team should lead the planning and monitoring of management, in coordination with government personnel. Costs of the management program should be evaluated and divided among the community. Resource sharing should be encouraged to avoid duplication of effort and promote sharing of responsibility. Community education and workshops should be reinforced by government policy and regulations. The existing social systems should be used to disseminate information on ICM and the impacts of climate change. The pace of the development process should be dictated by local conditions and customs.

Conclusions

The probable adverse effects caused by global-warming-related climate change and ASLR have alarmed island nations and encouraged them to incorporate early adaptive measures. The IPCC CZMS (1994) recommended that all coastal and island nations develop ICMs for sustainable development. The Honolulu workshop explored the unique problems facing Pacific island nations and suggested ways they can follow up on the IPCC's recommendation to develop their ICMs. All essential systems concerning island resources management, from traditional land tenure systems, to community-based management systems, to Western continental-style coastal management systems, to ICM programs built on traditional and community systems, were discussed. The workshop generated recommendations on ways to facilitate and accelerate the development and implementation of ICMs for island nations and for the region. An ICM program based on a traditional land tenure system and integrating traditional systems with centralized, legislative management systems will result in more effective decisions on land and resource use, help retain cultural traditions, and sustain essential coastal resources.

Acknowledgments

The author wishes to thank all participants of the 1994 Regional Workshop for Pacific Island Nations sponsored by the U.S. Country Studies Program for their excellent contributions. The materials synthesized here represent the synergy generated by all the participants. Special appreciation goes to B. Mieremet for his dedication in assisting the island nations and his contributions to the workshop. Special thanks also go to the reviewers for their constructive comments.

References

Aitaoto, A., 1994, "Traditional Management Methods Utilized in the Pacific Island American Samoa," in *Climate Change and Adaptation Strategies for the Indo-Pacific Island Nations Workshop: Workshop Proceedings* (P. Rappa, A. Tomlinson, S. Zeigler, eds.) University of Hawaii, Honolulu, Hawaii, USA, p 46-47.

Dahl, C., 1994, "Participatory Rural Appraisal and Community-Based Planning on Pohnpei, FSM," in *Climate Change and Adaptation Strategies for the Indo-Pacific Island Nations Workshop*, (P. Rappa, A. Tomlinson, S. Zeigler, eds.) University of Hawaii, Honolulu, Hawaii, USA, p 49-52.

Hviding, E., and Baine, G.B.K., 1994, "Community-Based Fisheries Management, Tradition and the Challenges of Development of Mavovo, Solomon Islands," *Development and Change* 25:13-39.

Iou, J., and Smith, A., 1994, "Example of Traditional Methods in Yap States, FSM," in *Climate Change and Adaptation Strategies for the Indo-Pacific Island Nations Workshop*, (P. Rappa, A. Tomlinson, S. Zeigler, eds.) University of Hawaii, Honolulu, Hawaii, USA, p 44-46.

IPCC, 1990, *Climate Change: The IPCC Scientific Assessment*, J.T. Houghton et al. (eds.), Intergovernmental Panel on Climate Change, Working Group I, Cambridge University Press, Cambridge, United Kingdom.

IPCC, 1992, *Climate Change 1992: The Supplementary Report to the IPCC Scientific Assessment*, J.T. Houghton et al. (eds.), Intergovernmental Panel on Climate Change, Working Group I, Cambridge University Press, Cambridge, United Kingdom.

IPCC, 1994, *Preparing to Meet the Coastal Challenges of the 21st Century*, Conference Report, World Coastal Conference 1993, Noordwijk, the Netherlands.

IPCC CZMS, 1990, *Strategies for Adaptation to Sea Level Rise*, R. Misdorp et al. (eds.), Intergovernmental Panel on Climate Change, Response Strategies Working Group, Coastal Zone Management Subgroup, Ministry of Transport and Public Works, The Hague, the Netherlands.

IPCC CZMS, 1991, *Assessment of the Vulnerability of Coastal Areas to Sea Level Rise: A Common Methodology*, Rev. 1, Intergovernmental Panel on Climate Change, Response Strategies Working Group, Coastal Zone Management Subgroup, Ministry of Transport and Public Works, The Hague, the Netherlands.

IPCC CZMS, 1992, *Global Climate Change and the Rising Challenge of the Sea*, L. Bijlsma et al. (eds.), Intergovernmental Panel on Climate Change, Response Strategies Working Group, Coastal Zone Management Subgroup, Ministry of Transport, Public Works, and Water Management, The Hague, the Netherlands.

Peau, L.M., 1994, "Community-Based Resource Management in American Samoa: Successes and Lessons Learned," in *Climate Change and Adaptation Strategies for the Indo-Pacific Island Nations Workshop*, (P. Rappa, A. Tomlinson, S. Zeigler, eds.) University of Hawaii, Honolulu, Hawaii, USA.

Smith, A., 1994a, "Response Strategies to Climate Change in the Pacific Insular Region," in *Climate Change and Adaptation Strategies for the Indo-Pacific Island Nations Workshop*, (P. Rappa, A. Tomlinson, S. Zeigler, eds.) University of Hawaii, Honolulu, Hawaii, USA, p 37-39.

Smith, A., 1994b, "The ICZM Program of the South Pacific Environmental Program," in *Climate Change and Adaptation Strategies for the Indo-Pacific Island Nations Workshop*, (P. Rappa, A. Tomlinson, S. Zeigler, eds.) University of Hawaii, Honolulu, Hawaii, USA, p 54-58.

Templet, P.H., 1986, "American Samoa: Establishing a Coastal Area Management Model for Developing Countries," *Coastal Zone Management Journal* 113(3/4):241-264.

Vulnerability and Adaptation Assessments of Climate Change and Sea-Level Rise in the Coastal Zone: Perspectives from the Netherlands and Bangladesh

Frank R. Rijsberman and Andre van Velzen
Resource Analysis, Zuiderstraat 110, 2611 SJ Delft, the Netherlands

Abstract

Analysis and consensus building are two stages associated with preparing strategies for adapting to climate change. The analysis stage involves vulnerability and adaptation assessments, which have been undertaken in case studies of sea-level rise and other impacts of climate change on the Netherlands and Bangladesh. In both case studies, the Common Methodology for assessing vulnerability to climate change, developed by the Intergovernmental Panel on Climate Change, was used as a starting point. In the consensus-building stage, computer-based decision-support tools may be useful. These tools can be used to communicate the results of adaptation assessments to decision makers. The tools may also aid stakeholders in negotiating alternative strategies to address climate change and thus potentially increase the social acceptability of implementing adaptive strategies.

Methodology and Approach to Climate Change Adaptation

Preparing strategies to respond to climate change involves two stages: analysis and consensus building. During the analysis stage, a country's vulnerability to climate change is assessed through a process called vulnerability assessment, and adaptive responses that could decrease vulnerability are identified through a process called adaptation assessment. Once vulnerability and response options have been analyzed, the results of the analyses have to be communicated to decision makers. Because both the impacts of climate change and the costs of adaptive strategies will be unevenly distributed throughout economic and social groups, achieving socially acceptable adaptive response strategies will require negotiation among stakeholders. Communication and negotiation are referred to in this paper as the consensus-building stage.

Analysis Stage

The Intergovernmental Panel on Climate Change (IPCC) has developed a methodology for analyzing the preparation of adaptive strategies for sea-level rise (IPCC 1992). This methodology, the Common Methodology for assessing vulnerability to sea-level rise, has been used, wholly or in part, to carry out several vulnerability assessments for sea-level rise, but relatively few of these studies have addressed adaptive response strategies. This paper reviews the experience with this methodology for two case studies that addressed the implementation of adaptive strategies to climate change: the Netherlands and Bangladesh.

Consensus-Building Stage

International methods for building a consensus regarding adaptive responses to climate change in coastal zones are lacking. Therefore, in the declaration of the World Coast Conference ([WCC] 1994) and in the Agenda 21 adopted at the United Nations Conference on Economic Development in 1992, the international community agreed that countries with low-lying coastal zones need to introduce and implement Integrated Coastal Zone Management (ICZM) programs to prepare for, and adapt to, the impacts of climate change and sea-level rise. In such an integrated approach, issues related to climate change should be placed in the context of short-term urgent decisions about and long-term solutions to problems in coastal zones. In an integrated approach, different interests can also be represented, thereby allowing various stakeholder groups to take on an enhanced role, which will also help in consensus building.

In addition, decision makers are gaining experience in using computer-based decision-support tools to raise public awareness of strategies for addressing sea-level rise in an ICZM context. This paper describes the experience with such tools and proposes their use for the consensus-building stage to prepare for climate change.

Vulnerability and Adaptation Assessments for the Netherlands

The IPCC (1992) Common Methodology was the basis for a case study in which vulnerability and adaptation assessments were applied for the Netherlands (Peerbolte et al. 1991; Baarse et al. 1992). The Netherlands case study aimed to (1) assess the vulnerability of the Netherlands to sea-level rise under different scenarios and adaptive strategies and (2) contribute to further development of the Common Methodology for the execution of similar case studies elsewhere.

Vulnerability Assessment

For the Netherlands case study, a simulation model was developed to carry out the vulnerability assessment. This model describes the main physical mechanisms and their socioeconomic and ecological impacts, as well as the costs and effects of selected strategies. The central part of the model assesses losses and flood risks in land areas. In addition, the model considers (1) the effects of beach and dune erosion, (2) changes in the salinity of (fresh) surface and groundwater bodies, and (3) changes in water management with respect to drainage and water discharge.

In line with the Common Methodology, the following types of impacts were considered:

- Values at loss (because of losses in area and direct exposure to sea conditions),
- Values at risk (because of changes of flood risk and/or values subject to flood risk), and
- Values at change (because of changes in physical conditions of land and water systems).

The values considered included people, capital, and ecology. In the vulnerability assessment, a number of cases were considered. Such cases consisted of a combination of the following components (Baarse et al. 1992):

- *A sea-level rise scenario.* In accordance with the IPCC (1992) guidelines at the time, two distinct situations were considered: a rise of 0.3 m per 100 years and a rise of 1.0 m per 100 years.
- *A socioeconomic development scenario.*
- *An adaptive response strategy.* Strategies considered included a zero-option (no measures) and three types of protection options.

According to the results of the zero-option strategy, the Netherlands is physically very vulnerable to the effects of sea-level rise. Thus, if no measures are taken, the most important physical effects are very substantial losses of land and related capital values, a large increase in people at risk, severe drainage problems, and a limited increase in salinity problems.

On the basis of an assessment of a protective strategy, regardless of the assumed rise in sea level and socioeconomic changes, measures can be taken to alleviate the physical effects listed above (except for salinity problems) at costs of no more than 0.5% of the yearly gross national product (GNP).

Adaptation Assessment

The IPCC (1992) considers three adaptive strategies: retreat, accommodate, and protect. In the *retreat* strategy, inhabitants of the coastal zone are resettled, and structures in currently developed areas are abandoned. Any new development is relocated at specific distances from the shore, as appropriate. In the *accommodate* strategy, residents continue to occupy vulnerable areas. A greater degree of flooding is accepted, and the risks are controlled by such accommodation measures as early warning systems. Development and land-use strategies are adjusted (e.g., farms are converted to fish ponds). In the *protect* strategy, vulnerable areas — especially population centers, economic activity centers, and natural resources — are protected by constructing dikes, embankments, and seawalls. Given the enormously high investments in coastal protective infrastructure that have already been made in the Netherlands, the country is virtually locked in to one protection strategy. Of the three defined adaptive strategies, only accommodation was considered to be a supplement to the selected protection measures.

In the Common Methodology, several aspects related to implementation of adaptive response strategies were suggested. These aspects are structured according to available resources, social and political acceptability, and institutional arrangements.

Resources can be subdivided into financial and technical resources. Problems with financial resources are not expected in the Netherlands because the calculated cost of the maximum annual response strategy is less than 0.5% of the yearly GNP. Table 1 summarizes estimated costs associated with adaptive responses. In IPCC (1992), costs of less than 1% of the GNP are classified as "not problematic." Technical problems are not expected, either, because the technical requirements are not higher than those for measures already implemented.

A potential problem in the social and political category identified in the case study is that a long period without major flooding (the last storm-related disaster in the Netherlands occurred in 1953) has reduced the public's awareness of flood risks and its willingness to pay for measures to reduce the risk. In the beginning of 1995, however, the main rivers posed a major flooding threat. More than 200,000 people were evacuated. This recent threat has, at least temporarily, raised the public's awareness of flood risks.

Institutional structures related to the management of water resources and coastal zones have been very well developed throughout the history of the Netherlands. The government has a tradition and considerable experience in long-term spatial planning and land-use management. In its recently published report on the environmental performance of the Netherlands, the Office of Economic Cooperation and Development (OECD) concluded that environmentally related planning and consultation has developed in line with the country's tradition in planning and consensus building (OECD 1995).

Table 1. Estimated Costs of Adaptive Measures to Maintain Safety against Flooding and Erosion for the Netherlands (in billions of Dutch Florins)

	Estimated Cost by Climate Scenario	
Adaptive Measure or Setting	0.6-m Rise in Sea Level in 100 Years	1.0-m Rise in Sea Level in 100 Years
Dike raising	7.2	11.5
Discharge sluices, pumping stations	1.0	2.0
Dune maintenance, beach nourishment	0.8	1.9
Flood-prone areas not protected by dikes (harbors, industry, population areas)	2.0	3.5
Total cost	11.0	18.9

Source: Peerbolte et al. (1991).

With respect to the impacts of sea-level rise and implementing ICZM programs, institutional structures and long-term planning mechanisms are important. At the World Coast Conference (WCC 1994), the development of institutional arrangements was identified as a key element in implementing ICZM programs. Furthermore, well-developed planning mechanisms are important to tune short-term concerns and measures in the coastal zone to a long-term policy on sea-level rise. Consultation and emphasis on consensus building are important because of the conflicts and opportunities that arise regarding the use of coastal resources.

The well-developed institutional structures in the Netherlands, however, also cause a potential problem because the current procedures for implementing technical works are very time-consuming. The problem is best illustrated by the flooding threat in early 1995. One cause of the flooding problems is related to continuous procedures developed and used over the last two decades for maintaining and raising dikes along the rivers. These procedures integrate flood-prevention measures in the present cultural-historical landscape along the rivers, although few measures have been implemented.

Time-consuming procedures can also delay the implementation of adaptive responses and ICZM programs. Although climate change and sea-level rise are long-term processes, the IPCC (1992) concludes that coastal nations should start implementing adaptive strategies now, not because there is an impending catastrophe, but because there are opportunities to avoid adverse impacts. These opportunities may be lost if the process is delayed as a result of time-consuming procedures.

The Netherlands case study demonstrates that strengthening institutional structures is important for implementing adaptive strategies, particularly in relation to the need for ICZM. Applying integrated approaches, however, should not decrease the decisiveness of the government. In general, taking into account the ability of the Netherlands to implement adaptive strategies, the vulnerability of the nation to sea-level rise is low.

Vulnerability and Adaptation Assessments in Bangladesh

The case study of Bangladesh was funded by the Directorate General for International Cooperation of the Netherlands Ministry of Foreign Affairs (Bangladesh Centre for Advanced Studies [BCAS] 1994). The recently completed study was conducted by BCAS and Resource Analysis from the Netherlands. (See Al-Farouq and Huq [1995] for a detailed presentation of the study.)

The basis of the Bangladesh study was the general IPCC Common Methodology approach for assessing vulnerability to sea-level rise. This approach was applied in accordance with the specific conditions in and concerns about Bangladesh.

Vulnerability Assessment

Climate change in Bangladesh is expressed in terms of variables, called "agents of change." The following agents have been considered because of their effects on the natural systems of Bangladesh: sea-level rise, temperature/evaporation, precipitation, river discharges, and cyclone intensity.

The upper and lower values of these agents of change were estimated on the basis of IPCC scenarios and studies of regional climate changes in East Asia. The upper and lower sets of values were considered severe and moderate climate change scenarios, respectively. Agents of change first affect the natural system. Primary physical effects occur in Bangladesh in the form of inundations, low flow, salinity intrusion, flash floods, drought, and storm surges. These primary effects affect the country's ecological and human systems.

As in the Netherlands, a number of cases have been considered to examine the likely impacts of climate change on Bangladesh. For Bangladesh, such cases consisted of a combination of three scenarios:

- *A sea-level-rise scenario.* Severe and moderate climate change variables were defined according to the IPCC guidelines.
- *A socioeconomic development scenario.* Two variables were considered: (1) a business-as-usual variable, where demographic, social, and

economic developments continue as in the past, and (2) a high-development variable, where the growth rate of the population is reduced, education and health care improves, and economic growth accelerates from previous levels.

- *A regional development scenario.* Two variables were considered: (1) a water-sharing variable in which it is postulated that there will be a regional agreement on water sharing between all the riparian countries and (2) a water-nonsharing variable in which it is assumed that no regional water-sharing agreement is reached and that the upper riparian countries continue to abstract or divert water from the major rivers upstream of Bangladesh.

On the basis of the analysis, the coastal zone of Bangladesh is very vulnerable to sea-level rise (Table 2). In a 1-m rise in sea level, Bangladesh would face a catastrophic situation, including permanent inundation of about 15-18% of its low-lying coastal area, loss of the Sunderbans (the largest area of mangrove forest in the world), displacement of more than 10 million people, loss of valuable agricultural lands, and intrusion of salinity.

Further conclusions regarding the vulnerability of Bangladesh were as follows:

- Climate change and sea-level rise in Bangladesh will affect the entire country, not only the coastal areas.
- Bangladesh is highly vulnerable to climate change and sea-level rise, regardless of future socioeconomic developments and regional water-sharing or nonsharing conditions.
- Developments in Bangladesh and in the upstream watersheds of the main rivers bring about internal changes that are similar to those induced by severe changes in climate.
- Because development in Bangladesh is rapid, the impacts of climate change and sea-level rise are may be less severe. For example, grain production is increasing in Bangladesh. Improved management practices are expected to boost total annual production from 32 million to about 40 million Mg. Reduced production due to drought is expected to be similar in magnitude.

Adaptation Assessment

In the Bangladesh study, the following steps were taken: (1) assess the appropriate response strategy and (2) assess the capability of Bangladesh to implement the adaptive strategy. As in the Netherlands study, selecting one of the three IPCC response strategies for Bangladesh was relatively straightforward. Retreat is not a practical strategy because of the country's high population density and very low resource-to-person ratio. Implementing protective measures would require substantial external financial and technical support and is therefore

Table 2. Physical Vulnerability and Implementation Aspects of Adaptive Strategies in The Netherlands and Bangladesh

Item	The Netherlands	Bangladesh
Physical Vulnerability		
Flooding	High	High
Erosion	High	High
Salinity intrusion	High	High
Drought	Low	High
Implementation Aspects of Adaptive Strategies		
Resources available		
Financial	+	–
Technical capabilities	+	+
Acceptability		
Social	–/+	–/+
Political	–/+	+

problematic. Accordingly, accommodation seems to be the most feasible strategy. To be effective, an accommodation strategy should include such measures as creating services to improve or change agricultural practices, negotiating regional water-sharing arrangements, implementing efficient mechanisms for disaster management, developing desalination techniques, and planting mangrove protection belts.

The strategy should be formulated on the basis of traditional knowledge, and the requirements of the people should be identified through participatory planning. Implementation of a response strategy can be structured according to the same categories as those in the case study for the Netherlands: resources available, social and political acceptability, and institutional arrangements. Table 2 lists the capabilities of both the Netherlands and Bangladesh to implement response strategies.

The resources available in Bangladesh for implementing adaptive strategies are limited. Bangladesh heavily depends on donor funding. A potential problem with donor funding is that internationally funded development strategies do not always reflect the needs and priorities of the people of the recipient country. Technical capabilities are not expected to be a problem because the basis for implementing the accommodation measures will be traditional knowledge. There is political support for implementing response strategies in Bangladesh. Recently, an Interministerial Government Review Committee on Climate

Change, chaired by the Minister of Environment, was established. This high-level committee could play an important integrating role for short- and long-term development of coastal areas. Also, social organizations are aware that Bangladesh is vulnerable to climate change and sea-level rise. Public awareness and involvement can, however, be improved. Participation by local people at all stages of planning and development in the coastal zone is essential for developing appropriate and socially acceptable strategies.

With respect to institutional arrangements, Bangladesh's system of planning is centralized. Bangladesh has yet to identify and sustain local institutional arrangements for spearheading and implementing development and policy decisions. Furthermore, long-term planning mechanisms and land-use management are to be improved for the development and implementation of adaptive strategies.

Another institutional problem is that planning agencies have not been created by law, but by administrative decisions. Thus, the implication is that planning in Bangladesh is not mandatory. Planning agencies should therefore be strengthened by providing an appropriate legal mandate. Bureaucratic bottlenecks hamper the decisiveness of the institutional system in Bangladesh, and implementing development schemes is therefore problematic. This situation also affects coastal zone management projects because of their geographic distance from the urban centers of planning and monitoring.

As outlined above, several obstacles prevent Bangladesh from being able to implement the institutional arrangements needed to adapt to climate change and sea-level rise. To strengthen Bangladesh's institutional arrangements in this area, it has been recommended to establish the Government Review Committee on Climate Change (which was established for the case study project) as a permanent body.

Decision-Support Tools for Consensus Building

It is not an easy task to convince stakeholders or their representatives (whether they are ministries, private-sector organizations representing fishermen, or environmental nongovernment organizations [NGOs]) that resources should be allocated to climate change adaptation when incomes are low, development options are beckoning, and mangrove and coral reef systems are already threatened. Moreover, scientific and technical reports on vulnerability and adaptation analyses are inadequate persuasive tools. In part, this issue is addressed by incorporating adaptive response strategies in the larger framework of ICZM strategies. Another issue is the initiation of an active process to communicate and negotiate the results of the vulnerability and adaptation analyses to stakeholders to increase their social acceptability. Computer-based decision-support tools can be used to aid communication and negotiation. The purpose of these tools would be to present alternative strategies, together with their simulated or forecasted impacts associated with climate change, in terms of multiple criteria reflecting stakeholder interests, under various climate change scenarios.

Such decision-support tools should have at least the following characteristics:

- Decision makers (stakeholders or their representatives) should be able to use the decision-support tools directly, without having to rely on scientists or analysts.
- The tools should be user-friendly, attractive, and interactive. They should allow decision makers to (1) understand the structure and characteristics of the alternative policies and (2) input their preferences or judgments and see the consequences thereof.
- The tools can be thought of as a postprocessor or integrator of analyses (and modeling efforts) used to carry out vulnerability and adaptation assessments. The resulting set of adaptive response policies and associated impacts would be input for the decision-support tool proposed here.
- Impacts or scores of the response policies for criteria that cannot be easily quantified could be based on expert (qualitative) judgment or directly input by the decision maker.
- The relative weights of the criteria are another major input; the weights will vary, depending on the perspectives (interests) of the stakeholders. To this end, the tool should accommodate one or more multiple-criteria analysis supports or techniques (ranging from simple colored scorecards to formal analysis techniques).
- The tool should be structured to invite and enable discussion among decision makers using it and to stimulate the exchange of opinions. The tool could also incorporate formal negotiation procedures, such those as proposed by Ridgley and Rijsberman (1994).

A number of models have been developed with the objective of integrating the climate system, global biogeochemical cycles, and human/societal components. Such integrated models have been used in the international work of IPCC and INC for climate change issues (e.g., the IMAGE model; see Rotmans 1991 and Alcamo 1994) and in the preparation of the Convention on Long Range Transboundary Air Pollution for acid rain issues (i.e., the RAINS model; see Hordijk 1991), as discussed by Swart (1994, Chapter 3). These models are research tools built to be operated and interpreted by scientists rather than by decision makers. Although their results may well be used to support decision making, or even negotiation processes, they are not decision-support tools decision makers can use directly. Swart (1994) concludes, for instance, that gaining acceptability of the IMAGE model required the development of a user-friendly FORTRAN-based version because the original model had few and unattractive graphics and was written in a less accessible simulation language (ACSL).

User-friendly, interactive decision-support tools that satisfy the conditions listed above have been developed for climate-related issues (i.e., the analysis of response policies to sea-level rise). For the World Coast Conference, hosted by the Netherlands Government in Noordwijk in November 1993, a group of about 300 high-level decision makers with various backgrounds (as well as representatives of NGOs) successfully used decision-support tools developed for the occasion. One of these tools, COSMO, focuses on analyzing policies related to adapting to sea-level rise in the context of ICZM (Resource Analysis 1994). A second model, CORONA, focuses on the institutional requirements of implementing the resulting ICZM plans through computer-based role-playing (Rijsberman et al. 1995).

COSMO and CORONA were developed by Resource Analysis in cooperation with the CZM Centre of the National Institute for Coastal and Marine Management in the Netherlands. They were designed as awareness-raising tools and as a way to improve communication among stakeholders and are based on a fictitious area. Since their successful application at the World Coast Conference, the models have been used as training tools at workshops and seminars about ICZM and climate change/sea-level rise. In addition, similar tools have been developed for realistic areas in the Netherlands, Thailand, and West Africa that demonstrate the potential of such tools to help stakeholders develop socially acceptable adaptive strategies.

Conclusions

To adapt to climate change, it is useful to distinguish between an analysis stage and a consensus-building stage. The analysis stage involves vulnerability and adaptation assessments. The consensus-building stage involves communication and negotiation among stakeholders or their representatives. Consensus building is necessary because it helps to establish the social acceptability required for implementing response strategies.

Analysis Stage

The Common Methodology (IPCC 1992) provided a useful starting point in both case studies, although a number of limitations were identified.

First, the most relevant development factors are scenario variables in the Common Methodology. Regional development factors are not included explicitly. The Bangladesh case study, however, demonstrated that the impacts due to developments in the upstream watersheds of the main rivers (regional sharing or nonsharing of available water) are similar to those caused by climate change. In vulnerability and adaptation assessments, relevant regional development factors should therefore be taken into account as scenario variables.

Second, the methodology focuses on flooding and flood risks as the main impact category of climate change and sea-level rise. For Bangladesh, drought is a main impact category as well because climate change will significantly influence the water balance in the dry period. At the same time, more irrigation water will be required to compensate for the expected changes in evaporation and precipitation, and water will be less available because of lower river flows in the dry period and increased saltwater intrusion.

Third, the methodology concentrates on the impacts for a country as a whole or for specified regions. An ability to differentiate between different stakeholders or social impact categories was especially important in the Bangladesh case study. The Common Methodology should be extended so that vulnerability and adaptation assessments can be undertaken for different social groups.

Fourth, both countries have little choice among the three IPCC response strategies (retreat, accommodate, and protect). In the Netherlands, an implicit choice is made for the protect strategy; in Bangladesh, only the accommodate strategy seems feasible. The selection of the different strategies for both countries is predetermined on the basis of historical developments, currently available resources, and institutional arrangements. Adaptation assessment, therefore, should focus not so much on selecting a feasible strategy, but more on implementing key aspects of the selected strategy.

Fifth, both countries are physically very vulnerable to the effects of sea-level rise. However, the two countries strongly differ in their capabilities for implementing adaptive response strategies. In Bangladesh, the main bottlenecks to adaptation are the lack of available resources and institutional arrangements, although the social and political climate is aware of the need for implementing adaptive strategies. In the Netherlands, in general, the available resources and institutional arrangements are adequate for adaptation. However, one potential problem is the very time-consuming procedures for implementing technical works. These procedures tend to decrease the decisiveness of the government. A second potential problem is that in the absence of major flood disasters, public awareness of the flood risks and the social and political willingness to make sacrifices tend to decrease.

Finally, only decisive government institutions with an adequate mandate can implement a long-term policy on adapting to climate change and implementing an ICZM program. In the case studies presented here, the following important elements for strengthening the government have been identified: (1) establishing adequate institutional arrangements to address climate change issues, (2) providing access to adequate policy and planning tools, and (3) educating and training key staff members.

Consensus-Building Stage

Incorporating response strategies in the framework of ICZM is expected to help to integrate long- and short-term concerns and enhance the involvement of stakeholders. In addition, computer-based decision-support tools can aid the communication and negotiation process to increase the social acceptability of implementing response strategies.

References

Alcamo, J. (ed.), 1994, "IMAGE 2.0: Integrated Modeling of Global Climate Change," Kluwer Academic Publishers, Dordrecht, the Netherlands.

Al-Farouq, A., and S. Huq, 1995, "Adaptation to Climate Change in the Coastal Resources Sector of Bangladesh: Some Issues and Problems," in J.B. Smith et al. (eds.), *Adapting to Climate Change:An International Perspective*, Springer-Verlag, New York City, N.Y., USA , p. 335-342.

Baarse, G., et al., 1992, *Analysis of Vulnerability to the Impacts of Sea Level Rise: A Case Study for the Netherlands,* Ministry of Transport, Public Works and Water Management, Tidal Waters Division, The Hague, the Netherlands.

Bangladesh Centre for Advanced Studies (BCAS), 1994, *Vulnerability of Bangladesh to Climate Change and Sea Level Rise — Concept and Tools for Calculating Risk in Integrated Coastal Zone Management,* Resource Analysis and Approtech Consultants Ltd., Dhaka, Bangladesh.

Hordijk, L., 1991, *An Integrated Assessment Model for Acid Deposition in Europe,* Dissertation, Free University of Amsterdam, the Netherlands.

IPCC, 1992, *Global Climate Change and the Rising Challenge of the Sea,* Intergovernmental Panel on Climate Change, Ministry of Transport, Public Works and Water Management, The Hague, the Netherlands.

Organization for Economic Cooperation and Development (OECD), 1995, *Environmental Performance Reviews, The Netherlands,* OECD, Paris, France.

Peerbolte, E.B., J.O. de Ronde, L.P.M. de Vrees, M. Mann, and G. Baarse, 1991, *Impact of Sea Level Rise on Society — A Case Study for The Netherlands,* Ministry of Transport, Public Works and Water Management, The Hague, the Netherlands.

Resource Analysis, 1994, *Coastal Zone Simulation Model (COSMO) Manual,* Resource Analysis, Delft, the Netherlands, and CZM Centre, National Institute for Coastal and Marine Management, The Hague, the Netherlands.

Ridgley, M.A., and F. R. Rijsberman, 1994, *Setting Targets for Climate Policy: An Approach to Develop Equity Targets,* report prepared by Resource Analysis for the Netherlands Ministry of Housing, Physical Planning and the Environment (VROM), Delft, the Netherlands.

Rijsberman, F.R., R.S. Westmacott, and D. Waardenburg, 1995, *CORONA: Coastal Resources Management Roleplay: Trainers' Manual,* Resource Analysis, Delft, the Netherlands.

Rotmans, J., 1991, *An Integrated Model to Assess the Greenhouse Effect,* Kluwer Academic Publishers, Dordrecht, the Netherlands.

Swart, R., 1994, *Climate Change: Managing the Risks,* Dissertation, Free University of Amsterdam, the Netherlands.

World Coast Conference '93 (WCC), 1994, *Preparing to Meet the Coastal Challenges of the 21st Century,* National Institute for Coastal and Marine Management, The Hague, the Netherlands.

Adaptation to Climate Change in the Coastal Resources Sector of Bangladesh: Some Issues and Problems

A. Al-Farouq
Department of Environment, Government of Bangladesh
House 2, Road 16, Dhanmondi
S. Huq
Bangladesh Center for Advanced Studies
Dhaka, Bangladesh

Abstract

Climate change and sea-level rise are already adversely affecting economic activities in Bangladesh, one of the most densely populated, low-lying coastal zones in the world. Preventive and adaptive measures are being implemented through research, development, and dissemination of knowledge, particularly in the sectors of disaster mitigation, agriculture, aquaculture, fisheries, and tourism. Specific examples of adaptation strategies already in place and under consideration for future application are described.

Introduction

Bangladesh has one of the most densely populated, low-lying coastal zones in the world, with 20-25 million people living within a 1-m elevation from the high tide level. The country has a multitude of coastal types, including river deltas, islands, beaches, and other geomorphological features. Major economic activities in Bangladesh include shipping, ports, navigation, fisheries, aquaculture, agriculture, and tourism.

The geological history of Bangladesh is one of repeated deposition by marine, near-shore, coastal, tidal, and fluvial processes. The land mass is blessed with the largest delta, largest mangrove forest, deepest sedimentary basin, and largest deep-sea fan in the world. An argument could be made that the majority of the country is part of the world coastal ecosystem. Yet, for administrative and socioeconomic development purposes, only southern Bangladesh has been characterized as the coastal region. The general physiography of this region consists of coastal plains, coastal islands and sand bars, and large floodplains of a network of rivers and canals (Figure 1). The region is endowed with enormous, diverse, natural resources, including land, water, forests, and fish.

The Bangladesh coastline extends about 735 km, of which 125 km is covered by the Sunderbans (the major natural mangrove forest) and 85 km by Cox's Bazaar beach. The coast can be broadly divided into an east-west-oriented deltaic coastline and a north-south-oriented nondeltaic coastline. The deltaic coastline comprises the Ganges tidal plain, which represents abandoned deltas of the Ganges River, and the dynamic Meghna deltaic plain, which is the present conduit of major rivers. The nondeltaic coastline comprises the Chittagong coastal plain, dominated by mud in the north and by sand in the south.

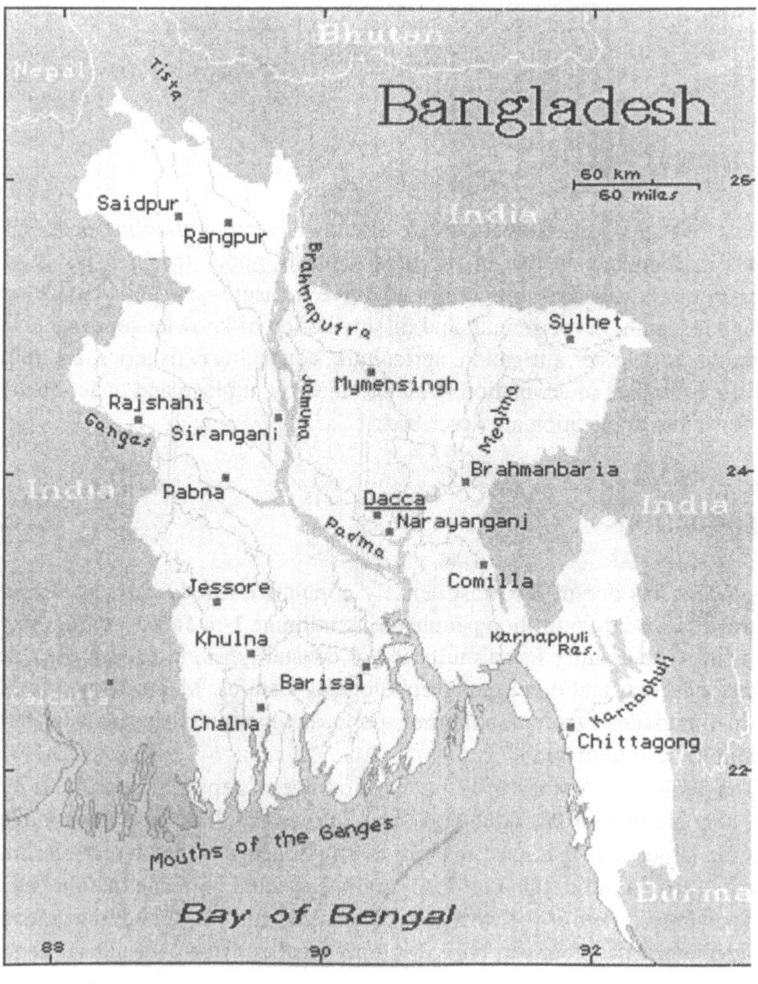

Figure 1. Map of Bangladesh

Historically, the coastline has been undergoing erosion and accretion in a dynamic equilibrium. The effect of climate change and sea-level rise may manifest in an increase in erosion and a decrease in accretion in the deltaic part of the coast. An increase in erosion over the last two decades in the Meghna estuary was observed during an assessment of changes in erosion and accretion in the area. This assessment analyzed remote sensing data obtained with geographic information systems.

Climate change and sea-level rise are expected to change the flooding and drought patterns in the Bangladesh coastal zone and affect the potential for agriculture. Climate change and sea-level rise combined may result in an increase in the surface average temperature and medium-term to permanent inundation of low-lying areas by saltwater. The increased intrusion of saltwater will probably have a strong effect on the availability of usable water and soil for all kinds of human activity. Upstream developments depend on activities in watersheds of rivers before the rivers cross the Bangladesh border. Any significant deforestation, erosion, water retention, deviation, consumption, or discharges of contaminating substances in these rivers might seriously limit the availability of potable fresh water in the entire country.

Several assessments have been conducted in Bangladesh in recent years, including *Socio-Economic Implication of Climate Change for Bangladesh*, conducted in 1993 through funding from New Zealand and the United Kingdom; *Assessment of the Vulnerability of Coastal Areas to Climate Change and Sea-Level Rise: A Pilot Study of Bangladesh*, conducted in 1993, and *Vulnerability of Bangladesh to Climate Change and Sea-Level Rise: Concepts and Tools for Calculating Risk in Integrated Coastal Zone Management*, conducted in 1994, both under assistance from the Netherlands.

Sectors of Bangladesh coastal resources identified as most vulnerable to climate change and sea-level rise are agriculture, aquaculture, fisheries, forestry, and tourism. Possible effects of climate change and sea-level rise are inundation, low flow conditions, saltwater intrusion, flash floods, droughts, storm surges, and changes in geomorphological formations. Each of these effects is discussed briefly below.

Inundations in the main flood plains of Bangladesh rivers could extend the areas vulnerable to flooding and increase the depth of flooding in unprotected areas. Such conditions are not only affected by climate change, but also by regional developments in watersheds and complex development factors within the country.

Low flow conditions are defined as water levels that fall below certain critical values and minimums. They are mainly a function of upstream developments and the subsequent cross-boundary waterflows.

Saltwater intrusion along major rivers is anticipated as a result of sea-level rise and the reduction of fresh water discharge in the rivers during dry periods. The groundwater table would also become more saline, thus affecting agriculture and drinking water.

Flash floods could occur because of changes in monsoonal precipitation and drainage congestion in the river systems, damaging agriculture and, to a lesser degree, infrastructure.

Droughts will increase in intensity under both moderate and severe climate change scenarios. During the Rabi season (October to May), the area severely affected by drought could triple, from 4,000 to about 12,000 km^2, under a severe climate change scenario. During the Kharif season (June to October), the area subject to severe and moderate to very severe drought, could quadruple, affecting both irrigated and rainfed crops.

Storm surges caused by cyclones may change in intensity, with resultant increase in the size of the affected areas and with extensive damage to agriculture, livestock, humankind, and property.

Geomorphological change is also anticipated in the form of an increase in erosion and a decrease in accretion, offsetting the prevailing dynamic equilibrium in the coastal delta system.

Until recently, the natural mangrove forest has functioned as the major deterrent to the adverse effects of calamities caused by the climate change effects described above. Greenbelts have been created along the coastal area through planned mangrove afforestation. Coastal afforestation programs have been undertaken in different districts of Bangladesh. An area of more than 112,000 ha of newly accreted land has already been planted. The mangroves also act as windbreak shelter belts in the coastal districts (Government of Bangladesh 1992).

Adaptation Options

An Integrated Coastal Zone Management (ICZM) program is considered the best mechanism to deal with possible impacts of climate change and sea-level rise. It not only provides the tools to identify short- and long-term actions in implementing national strategies for achieving sustainable development, but it also provides the framework under which stakeholders from the various strata of society act together on appropriate response strategies. The Government of Bangladesh has formed an interministerial committee, coordinated by the Ministry of Environment and Forests, to develop an advisory ICZM plan. In the future, coastal areas will be chosen to implement the plan on a pilot basis. International expertise and assistance are anticipated from the Global Environment Facility.

Possible response strategies to impacts of climate change and sea-level rise are few — limiting greenhouse gas emissions and adapting to changes. Limiting carbon emissions to 1990 levels, as recommended by the United Nations Framework Convention on Climate Change (UNFCCC), not only depends on the Government of Bangladesh but also on the cooperation of other governments. However, implementing adaptive response strategies depends largely on Bangladesh's own financial, technological, and human resources. Three adaptive measures identified by the Intergovernmental Panel on Climate Change (IPCC) are:

- *Retreat.* Resettle the inhabitants, abandon structures in currently developed areas, and require that any new development be set back at specific distances from the shore, as appropriate.
- *Accommodate.* Continue to occupy vulnerable areas, but accept the greater degree of flooding (e.g., convert farms to fish ponds). This measure includes accommodation.
- *Protect.* Defend vulnerable areas, especially population centers, economic activities, and natural resources. This measure includes building dikes, embankments, seawalls, and so on.

Retreat is not a practical solution in Bangladesh because of high population densities and very low resource-to-person ratios. Land tenure and leasing practices are also constraints. Relocation of people within the country is not a reasonable alternative because of overpopulating an already low resource base. Migration to other countries, especially to those with high resource potential, could only be possible with the consent of the international community.

Accommodation to climate change and sea-level rise has been the traditional adaptive measure taken by the people of Bangladesh, particularly those in the most hazard prone areas along the coast and rivers. Through experience in living with various adverse conditions, the people have devised many ways to adapt to hazardous living conditions. They have provided insight to planners for use in formulating appropriate adaptive development strategies. Some examples of these strategies are cited in the next section.

For some parts of the country, such as the Sundarbans mangrove forest, the best coastal protective measure against accelerated sea-level rise has been to preserve the protecting mangrove forests as much as possible and to plan for future inland migration. In the south-central and southeast areas of the country, it is possible to build new embankments or to increase the height of existing embankments to prevent inundations. But it will not be possible to prevent intrusion of saltwater into the waterways and, through capillary action, into the groundwater. Fresh water wetlands could be used as barriers to saltwater intrusion, as has been done in other parts of the world. The costs of protective measures (where feasible) are likely to be prohibitively high — too high for a poor country like Bangladesh to manage with its own resources.

Existing Adaptation Practices

The climatic conditions of the coastal areas, cyclones, floods, erosion, and accretion, are natural phenomena that have occurred for thousands of years. The people of this region, through their experience over the centuries, have devised a number of adaptive measures to cope with naturally occurring climatic conditions: growing salt-tolerant varieties of rice in the coastal areas, planting mangrove species of trees as wind and storm defenses, building high mounds as protection from cyclones, pursuing aquaculture and fishing practices in harmony with ecology and seasonality, and many others.

One example of adaptation is the highly productive aquaculture-horticulture industry developed by the local people of Swarupkati in the Barisal district, which is a tidally influenced river delta area. The local people deepened the natural creeks, enabling water to flow easily through the creek bed and to stay longer in the creeks during the dry season. In the creeks, prawns and other aquatic species are grown in "farms" formed by putting nets across the creek to prevent movement upstream or downstream. In nurseries on the creek banks, a variety of spices and fruit crops are intensively grown. The whole area has become a market center for saplings of fruit-bearing and other trees and spices. People come from all over the country to buy these products.

Nijhum Dweep, the southernmost island in the Bay of Bengal, is another area subject to erosion (at its northern end) and accretion (at its southern end). The people of this area have evolved a method of enhancing the rate of accretion. A significantly salt tolerant species of grass is grown, on which the cattle graze; as they graze, the cattle trample the grass into the soil, enhancing the soil texture and nutritional content. A highly salt tolerant species of wild rice is then grown in the soil, which is used for both fodder and soil compaction. Once the soil is ready, it can be used for a crop of rice for human consumption.

Future Adaptations

The adaptive options of retreat and protection are relatively problematic for Bangladesh because of its high population and limited resources. Bangladesh must consider other possible adaptive strategies, not only for climate change and sea-level rise in the long term, but also for better use of coastal resources in the short term. A number of strategies are suggested below.

Integrated Coastal Zone Management

The Government of Bangladesh has initiated a process of ICZM as part of its commitments under the UNFCCC, as noted earlier in this paper. Initially, the process will be on a pilot scale in a number of coastal ecosystems. These pilot-scale actions will lead to a national ICZM in the next few years.

Agriculture

Agriculture is the foundation of the economy of Bangladesh, and the coastal region is a major crop-growing area. It will be necessary to take a strong research, development, and extension approach for adapting agriculture to higher saline conditions. This approach will entail research into high-saline-tolerant crop species and extension of more salt-tolerant crops.

Aquaculture

Shrimp farming is a common practice throughout the coastal zone of Bangladesh, particularly in the south around Satkhira and the southeast around Cox's Bazaar. However, traditional farming practices produce very low yields, and the farms tend to spread horizontally over increasing areas. Adaptation involves better planning and management of the shrimp farms to enable production to increase without inundating more coastal areas.

Fisheries

Coastal fisheries in both marine and estuarine waters are very important economic activities in Bangladesh. Although these fisheries may not be adversely affected by climate change, much can be done to change the fishing practices to enable the fishers to improve their catch and livelihoods, while protecting them from calamities caused by climate change and sea-level rise. Specific measures include improved storm warnings that will allow fishing vessels to take shelter, as well as more harbors with shelters to protect the vessels.

Cities and Infrastructure

The main cities, such as Chittagong and Mongla, and the infrastructure, such as Export Processing Zones, roads, and railways, have some degree of protection in the form of seawalls or dikes in various locations. These seawalls and dikes may have to be strengthened and raised to improve protection against sea-level rise.

Tourism

Although tourism is not a major economic activity in Bangladesh, the beach at Cox's Bazaar attracts significant numbers of tourists and has some infrastructure to cater to them. Climate change effects are likely to affect tourism adversely, and the infrastructure may need protection or beach nourishment.

Forestry

The Sundarbans is already threatened by a combination of low fresh water stream flows and sea-level rise. The transition to more saline tolerant plant and animal species in the natural mangrove forests is already taking place and must be facilitated. In the other coastal areas, particularly where land accretion is occurring, efforts to stabilize the land must continue by planting mangrove species and conducting additional research and development into methods for improving coastal afforestation.

Conclusions

The coastal areas of Bangladesh, with approximately 20 million inhabitants, is a densely populated area with a multitude of important economic activities. Agriculture, aquaculture, fisheries, ports, and tourism are already vulnerable to such climate-related phenomena as cyclonic storms and periodic flooding from rivers due to high tides. These negative effects of climate-related phenomena are likely to become more frequent and extreme as global climate change and sea-level rise occur. Therefore, research and development on adaptive strategies must be undertaken, not only to combat the long-term effects of climate change and sea-level rise, but also for short-term benefits. A specific framework for working on possible adaptation options is provided by the proposed ICZM plan being prepared by the Government of Bangladesh. This plan involves all concerned government and nongovernment agencies. Potential adaptations to climate change effects should be incorporated into the plan as a preliminary step.

References

Bangladesh Center for Advanced Studies, 1991, *Cyclone 1991: An Environmental and Perceptional Study.*

Bangladesh Center for Advanced Studies, 1994, *Wetlands of Bangladesh.*

Bangladesh Institute for Development Studies, 1993, *Regional Study of Global Environmental Issues Project*, Country Study in Bangladesh.

Bangladesh Unnayan Parishad, Center for Environment and Resource Studies (University of Waikato, Hamilton, New Zealand)/University of East Anglia, 1994, *Bangladesh: Greenhouse Effect and Climate Change.*

Coastal Zone Management, 1994, *Common Framework for Integrated Coastal Zone Management.*

Government of Bangladesh, Department of Environment, 1995, *Vulnerability of Bangladesh to Climate Change and Sea-Level Rise.*

Government of Bangladesh, 1992, *Forestry Master Plan.*

Intergovernmental Panel on Climate Change (IPCC), 1992, *Climate Change: The Supplementary Report to the IPCC Scientific Assessment.*

Intergovernmental Panel on Climate Change, *Global Climate Change and the Rising Challenge of the Sea.*

5

ECOSYSTEMS AND FORESTS

Rapporteur's Statement

Jay Malcolm
University of Florida
Gainesville, FL, 32611, USA

Ecosystems and forests can be among the most difficult sectors for which to develop adaptive measures. Among the issues identified in the following chapter are:

Suitability of Assessment Methods. Models, especially "gap" models, can be useful tools for understanding adaptive measures. Extant models are too crude to be useful as adaptation tools; they do not incorporate vital components, such as pests, fire regimes, permafrost processes, and synergisms between environmental stresses. Modeling techniques that use climatic indices to predict the distributions of vegetation zones are of limited value because they assume equilibrium conditions and do not address dynamic or transitional responses. The long generation times of trees and the slow responses of ecosystems often make it difficult to evaluate adaptive measures.

Adaptive Measures. A number of policies and measures seem best suited for adapting to climate change. Some important points to consider in evaluating these options are listed below:

- Sustainable forest management implicitly incorporates adaptation. Management techniques and adaptive measures should sustain natural ecosystem processes.
- Forests are a "geographic phenomenon" and vary greatly among regions. The Russian Federation, for example, has an extremely diverse set of stand structures. As a result, appropriate adaptation techniques must vary from region to region.
- Many forest ecosystems are stressed by human activities. Stresses of anthropogenic origin cannot be ignored in strategies to adapt to future climate change. Climate change occurs within the context of current conditions. Maintaining ecologically intact forests is a prerequisite to adapting to future climate change.
- Corridors should serve as more than conduits for movement and should incorporate a significant amount of habitat to sustain viable populations.
- Adaptation strategies should use native species and ecotypes and avoid exotic species. The erosion of genetic diversity by inappropriate

management techniques is a major threat to future management options.
- Plantation schemes should mimic natural forest processes to the extent possible. Monocultures may be particularly vulnerable to climate change; an important mitigating strategy may be to increase forest diversity.

Integration of Adaptive Measures. As carbon dioxide (CO_2) sinks, forests help to mitigate future climate change. Mitigation and adaptation are thus closely aligned in the forestry sector, and the two problems can be addressed simultaneously. Policymakers often overlook the key role of forests in the global carbon cycle, which can be used to obtain much needed financial support for forestry adaptation.

Regional Collaboration. Airborne pollution causes significant damage to European forests. This first manifestation of anthropogenic atmospheric interference can be addressed only by regionwide plans to reduce emissions.

Future Research Needs. Adaptation of plants and communities to changed climatic conditions remains a high research priority. Vulnerability of highly managed forest systems to climate change is a major concern. Measures to reduce vulnerability, such as incorporating greater diversity, need additional study.

Summary Assessment. Adaptation of ecosystems and forests to impending climate change is a difficult task because of great spatial and temporal variability in system processes. Sustainable use of forest resources to preserve natural ecosystem dynamics and biodiversity should be a key component of adaptation strategies. Feasible adaptation schemes should address current anthropogenic stresses to forest ecosystems.

A Study of Climate Change Impacts on the Forests of Venezuela

Luis Jose Mata
Instituto de Mecanica de Fluidos, IMF Environmental Research Unit
Universidad Central de Venezuela
APDO 66401
Caracas, Venezuela

Abstract

The potential impact of climate change on forests in Venezuela has been evaluated by using an integrated approach based on a Holdridge Life Zone Classification model and several General Circulation Models. The results of this assessment show that Venezuelan vegetation will suffer under drier conditions. Adaptive strategies based on these results are discussed.

Introduction

Atmospheric scientists agree that increased anthropogenic emissions of greenhouse gases will raise global temperatures several degrees Celsius above the current level by the end of the 21st Century. This increase could disrupt natural ecosystems, particularly terrestrial vegetation. It is unlikely within the next decades that adequate greenhouse gas reduction measures will be taken to eliminate the possibility of climate change; therefore, some climate change is inevitable.

Regional studies have identified the impacts of climate change on tropical terrestrial ecosystems. Because ecosystems are vulnerable to climate, adaptation policies are necessary. National programs should be implemented to study, identify, and discuss measures for adapting to the possible effects of climate change on vegetation. The effects on tropical forests should be emphasized because such areas are particularly vulnerable. Venezuela's population (20.2 million) resides on 916.445 km^2 of land mostly covered by forest. This region, located at 0-6°N, is a good example of a tropical forest zone.

Methodologies and Sources of Data

The Vulnerability and Adaptation Guidelines of the U.S. Country Studies Program ([CSP] 1994) provides an overview of the relation between vegetation patterns and climate change by means of a climate-vegetation classification scheme. The close correlation between climate and vegetation has long been recognized and has led to the use of vegetation in creating climate maps and vice versa (Thornthwhaite 1948; Holdridge 1967; Walter 1985). These climate-vegetation classification schemes specifically define limits for each climate class.

The Holdridge Life Zone Classification model relates the distribution of major ecosystem complexes to the climatic variables of temperature, mean annual precipitation, and the evapotranspiration-to-precipitation ratio (Holdridge 1967). It requires only data on annual precipitation and temperature for a grid network based on latitude and longitude. (In this study, temperature was calculated as mean temperature at a monthly resolution.) Annual precipitation and temperature are then used to classify each grid cell to determine the potential land cover, "life zone," solely on the basis of climate. The resulting database of potential life zones can then be mapped.

A map based on the Holdridge model represents the potential distribution of vegetation based on climate, and its boundaries are a regular dissection of a climatic space defined by annual precipitation and growing-season temperature. Maps were generated and compared with existing maps of vegetation in Venezuela based on Holdridge life zones obtained from Venezuelan satellite images (Petroleos de Venezuela 1993) and an atlas of Venezuela from the Ministry of Environment. These sources were used to validate the vegetation analysis of this study.

The following General Circulation Models (GCMs) were used for simulating current and future climate: the Geophysical Fluid Dynamics Laboratory (GFDL) (Manabe and Wetherald 1987), Goddard Institute for Space Studies (GISS) (Hansen et al. 1983), Oregon State University (OSU) (Schlesinger and Zhao 1989), United Kingdom Meteorological Office (UKMO) (Wilson and Mitchell 1987), and Canadian Climate Change (CCCM) models.

Climate scenarios from the GCMs were transformed into a 0.5×0.5 degree latitude and longitude grid to match the results from the International Institute for Applied System Analysis (IIASA) database of global terrestrial climate (Leemans and Cramer 1990). This database was validated by temperature and precipitation data provided by the Venezuelan meteorological database.

The steps in assessing climate change impacts on forests were as follows:

- Compile a database of current environmental conditions.
- Identify regional change scenarios from GCMs.
- Generate Holdridge classification maps under the current Venezuelan climate.

- Generate Holdridge classification maps under a climate change scenario for Venezuela.
- Compare these maps to determine areas of impact.
- Identify critical areas for future analysis.

Climate data from the Ministry of Environmental and Renewable Resources database were used: 456 meteorological stations have precipitation information, and 26 have temperature information for the 30-year period suggested by the CSP guidelines. The scarcity of temperature data is a problem, however, because data are not available for the entire country. Other sources of information include the results of the GCMs (although they are of low resolution) and the IIASA climate database, which consists of data from dense networks of meteorological stations and an advanced interpolation scheme corrected for variations in topography between stations.

Analysis of Climatological Results

The GCMs generated temperature and precipitation maps of Venezuela for current conditions. Because the GCMs have low resolution, the results are not in good agreement with precipitation and temperature maps based on observed data from the Venezuelan meteorological stations (Figure 1).

Figure 2 shows a precipitation map generated from information from the IIASA database. A comparison with a map generated from the observed data (Figure 1) is remarkable. A similar analysis was done for temperature. The

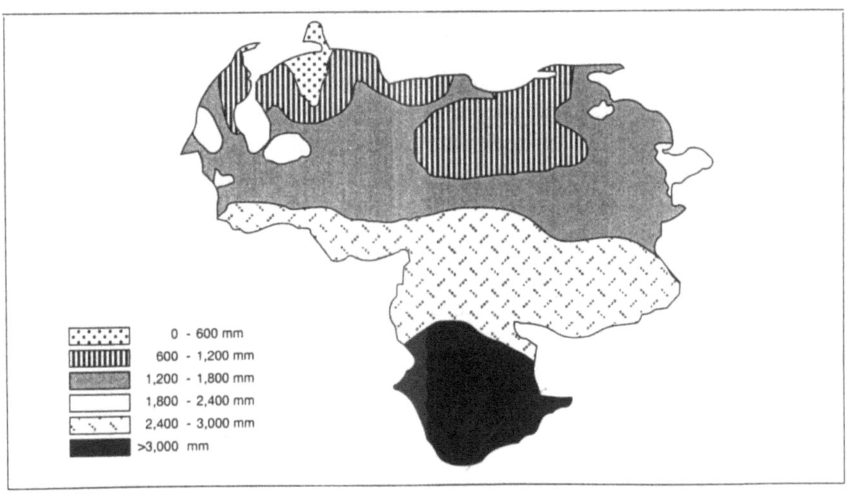

Figure 1. Precipitation Map of Venezuela Generated from Observed Data from 456 Meteorological Stations

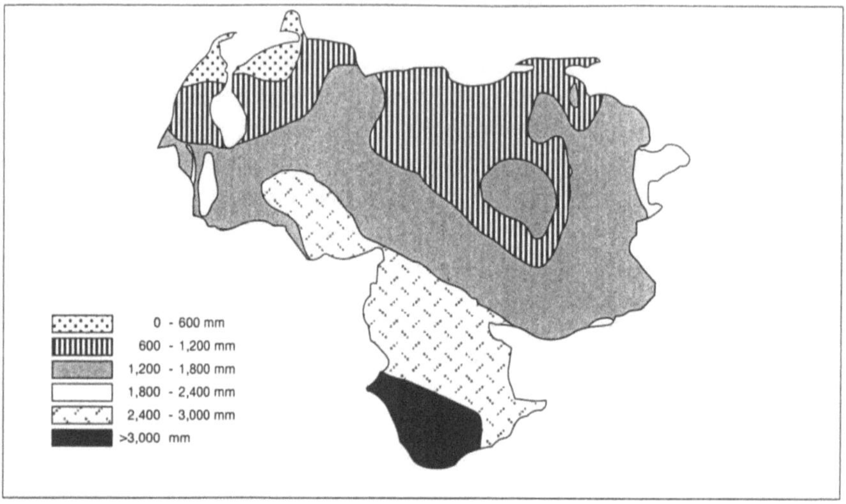

Figure 2. Precipitation Map of Venezuela Generated from Data from the IIASA Database

results are not in good agreement, probably because of the small amount of temperature data available.

Analysis of Vegetation Results

Figure 3 shows a Holdridge classification map of Venezuela based on temperature and precipitation data from the IIASA database. The results are in good agreement with the actual Venezuelan life zone map (Figure 4). Table 1 presents the forest area in square kilometers, calculated from the IIASA vegetation map.

Figures 5-8 show the results from the GISS, OSU, GFDL, and UKMO models for a doubling of atmospheric carbon dioxide concentration ($2XCO_2$) scenario. Detailed information on the areas covered by each life zone is available in IMF (1995). Each model predicts that tropical very dry forest will be the largest life zone in Venezuela for $2XCO_2$ scenarios. Table 2 presents the predicted area of this life zone.

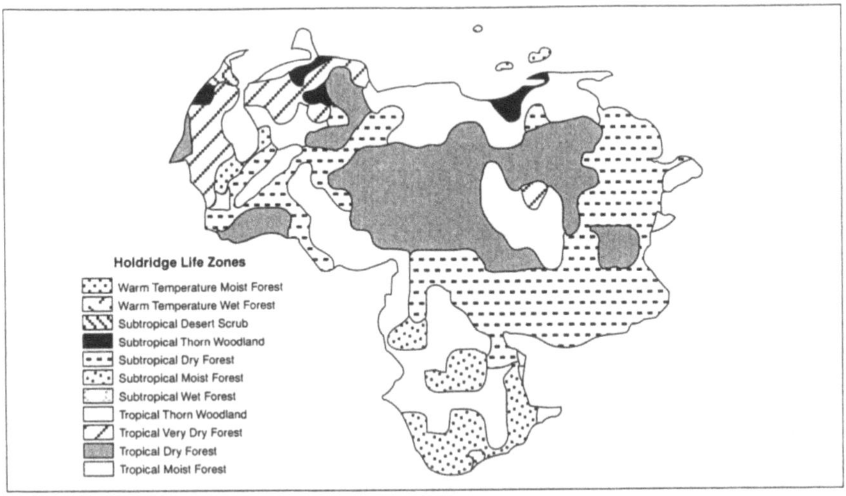

Figure 3. Holdridge Life Zones in Venezuela Based on Data from the IIASA Database

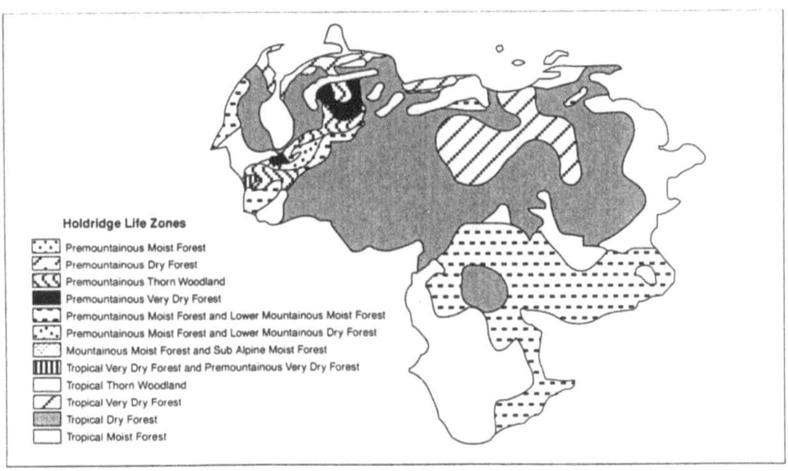

Figure 4. Actual Venezuelan Life Zone Map

Table 1. Current Forest Area in Venezuela, Classified by Life Zone

Life Zone	Area (km^2)
Subtropical moist forest	281,946.16
Tropical dry forest	241,400.98
Tropical moist forest	125,957.40
Subtropical dry forest	104,271.70
Subtropical wet forest	82,248.46
Tropical very dry forest	51,926.17
Warm temperature wet forest	12,690.12
Tropical thorn woodland	12,238.57
Subtropical thorn woodland	6,854.87

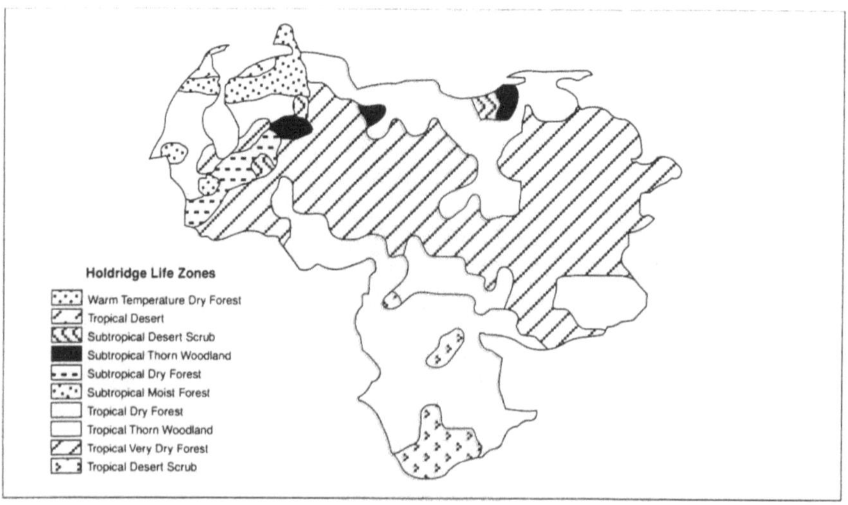

Figure 5. Holdridge Life Zones in Venezuela Predicted by the GISS Model

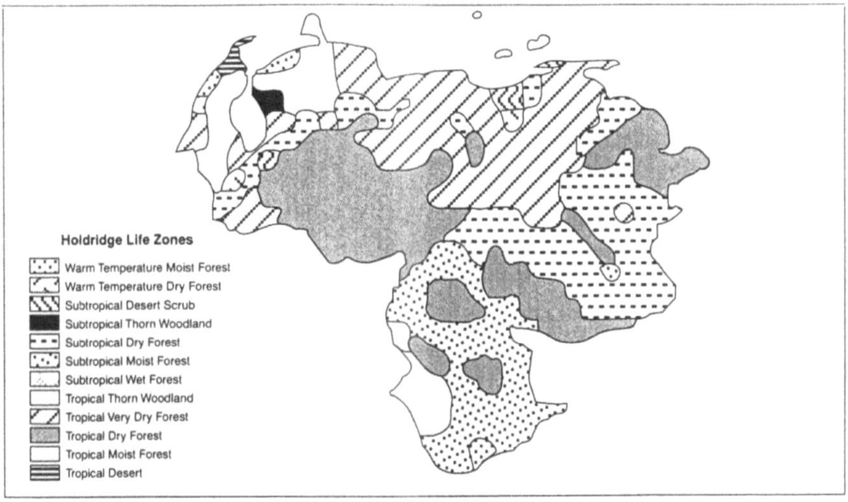

Figure 6. Holdridge Life Zones in Venezuela Predicted by the OSU Model

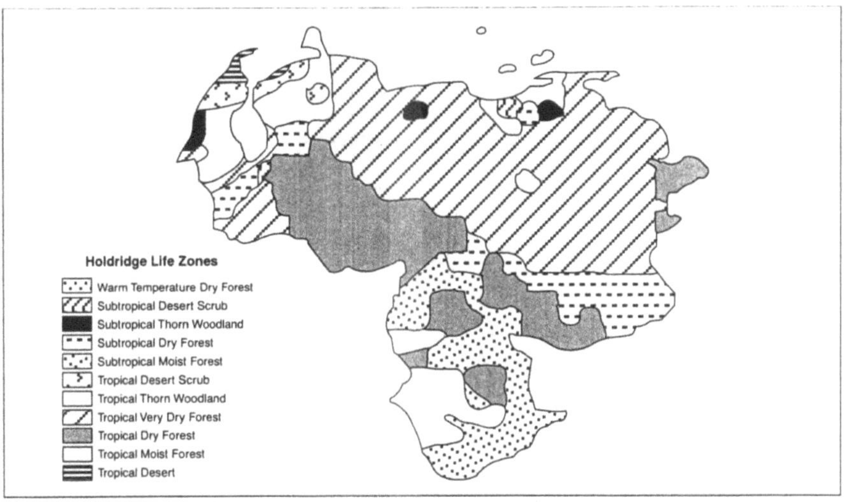

Figure 7. Holdridge Life Zones in Venezuela Predicted by the GFDL Model

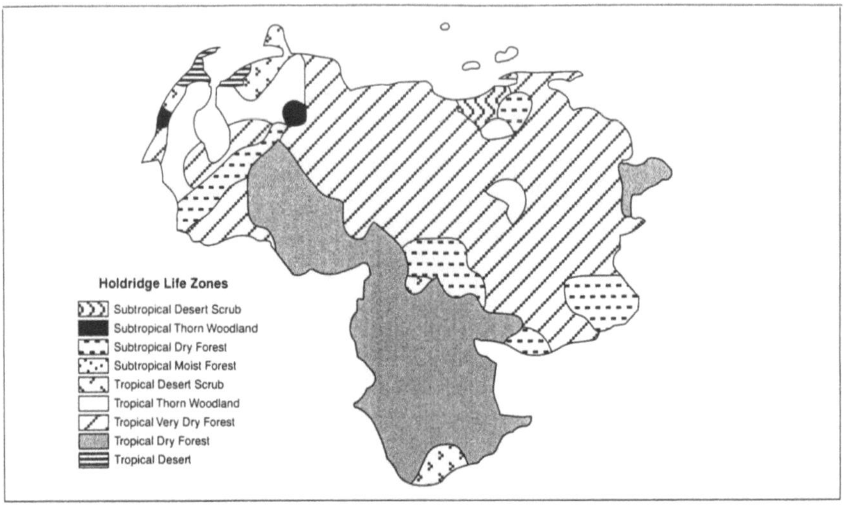

Figure 8. Holdridge Life Zones in Venezuela Predicted by the UKMO Model

Model	Area (km^2)
GFDL	383,467.72
GISS	426,907.98
OSU	223,235.58
UKMO	470,785.40

Preliminary Adaptation Assessment

A national discussion of adaptive measures should define the sensitivity of resources to future climate change and assess their potential for successful adaptation. It should also increase awareness of potential irreversible or catastrophic climate changes and help policymakers to identify and understand unfavorable trends that would make it more difficult to adapt to climate change in the future. In general, policymakers must decide whether adaptation should be a matter of responding to climate change as it occurs, or whether steps should be taken now to anticipate and mitigate the eventual effects of climate change. Policymakers must also evaluate whether the potential for natural adaptation or human intervention is low.

In the tropics, the impact of climate change may be irreversible, and valuable ecosystems could be lost permanently. Thus, choosing an adaptation policy becomes more involved because it is difficult to find adequate criteria for determining which measures should be implemented. The Venezuelan situation is complex because tropical forests are characterized by a great diversity of species. Understanding potential impacts due to a shift in climate zones, such as migration of species, reduction of inhabitable range, and altered ecosystem composition, could help to prevent current tropical dry forest from becoming very dry tropical forest (savannah) in Venezuela.

Anticipatory adaptive measures should be able to absorb surprises. Thus, society can try to anticipate rather than react to global climate change. For example, Venezuela's experience in tree planting (the *Uverito pinus caribean* project) may improve methods of enhancing tree growth rates under drought conditions. The forest industry could switch to alternative species as environmental conditions change. Another possibility could be to return the natural system to its original state of health.

Climate zones may shift hundreds of kilometers in a century, while natural rates of dispersal may be of the order of 10 during the same period. In addition, fire, hurricanes, and disease could result in a rapid dieback of existing forest. Assisting the migration of vegetation through activities such as widespread seed dispersal would be expensive and has a dubious chance of success (Smith and Tirpak 1989).

Climate change could also lead to loss of species diversity. Isolated species may find themselves in unsuitable life zones. Some local species may become extinct. The potential for human intervention to maintain species diversity in the face of climate change is limited. Migration may be possible, but barriers may have to be overcome. Corridors could be created that would allow species to migrate with climate change, but this proposition could be expensive. Selected species could be transplanted to new habitats, but this solution could be very resource intensive and would only be feasible in certain cases (Smith and Mueller-Vollmer 1993). In Venezuela, a study of *Inga alba spondias monbi* and *Noctandra grandis*, the dominant species of the Imataca Nature Forest Reserve, helped to define adaptation to a new environment.

Drought in the tropics affects soil moisture and thus plant growth. Adaptive measures must offset reductions in plant populations. Trees planted today may not survive to maturity if climate conditions change as expected. Water storage capacity should be designed to provide for the potential impacts of climate change.

Natural resource management policies are often created or changed in reaction to the most recent drought or storm. This short-term approach does not anticipate long-term changes in the future. Venezuelan regulations address harvesting, deforestation, and forest management issues, such as restrictions on areas to be preserved or reforested and a management plan based on guidelines provided by the Forest Service. However, these measures are only partial or incipient policies. Therefore, it is important to define and develop an overall adaptation plan.

Adaptation Options in Venezuela: Some General Considerations

Although adaptation options should focus mainly on reducing the recent high deforestation rates, anticipatory measures would help to maintain sustainable forest ecosystems. Given the uncertainty about forest responses to a potential climate change, however, detailed research and long-term monitoring of key variables are needed to fill important gaps in knowledge. This section discusses general considerations for adaptation strategies to reduce possible damage in both forest plantations and native forests in Venezuela.

Forest Plantations

In the near future, the Venezuelan pulp and paper industry will be based on large-scale plantations (about 500,000 ha) of Caribbean pine in Venezuela. The area is characterized by high temperatures (average annual temperature of 26°C) and low precipitation (900 mm annually), and it is currently a dry tropical forest. This study shows that this area would become a very dry tropical forest as a result of climate change.

Warming and moisture stress could become relevant issues in the development of plantations under a climate change. One adaptive measure would be to test species that might be suitable for adapting to future conditions. Soil conditions in this area are very restrictive for most forest and agricultural species; consequently, selecting viable species may be difficult. However, trials could provide an idea of their potential for adaptation. Small-scale plantations of exotic and native species (hardwood) are being established for different purposes, including soil protection. The success rates of these projects vary widely. Native species (such as mahogany) tend to have low survival rates, mainly because of pests and disease, while exotic species (such as eucalyptus) usually adapt more easily. Because planting mixed species may reduce vulnerability, efforts should promote reducing the current losses and risks of such initiatives and improving monitoring and data collection.

Because of the limited possibility of increasing tree diversity by planting mixed species in this area, researchers should be encouraged to find other alternatives for adaptive management. The plantations are located in savannah areas that are typically burned to promote pasture regrowth every year, and they face a high risk of fires. Preventive measures must be enforced to prevent losses in a warmer climate. Another option is stand thinning to determine tree growth responses to available moisture. Although thinning is a common practice for improving stand quality, it has not been used extensively at the plantations, although some stands are more than 15 years old. Thinning can also be used to favor remaining trees in drier environments.

Natural Forests

Protected areas cover almost 50% of Venezuela. These areas range from those with very restricted use, such as national parks, to those allocated specifically for sustainable timber production. If well preserved and managed, protected areas could help to conserve valuable and diverse genetic resources. However, in Venezuela, as in most tropical countries, land-use practices (such as agriculture and pastures), and other economic activities, threaten large forest areas. Northern Venezuela has suffered the most dramatic land-use change, and it would become very dry tropical forest under a climate change.

Protected areas and remaining forests, especially in the drier regions, should be maintained in their present condition to the extent possible. This option is limited by two factors: high implementation costs and high land-use pressures, originating from a complex interrelated process of socioeconomic and political variables. Another option would be to connect these areas by corridors that replicate near-natural characteristics to promote and facilitate tree migration. These sensitive areas should be given priority within a comprehensive approach to assisting natural migration.

Timber production in protected forests has been promoted through long-term concessions in large areas as a means to foster the adoption of sustainable management plans. These forest reserves contribute 50% of Venezuelan timber production and are the only alternative for sustaining the forestry industry in the long term. However, timber production in most remaining natural woodlots does not rely on sustainable management schemes; on the contrary, it is traditionally perceived as the first step of a land-clearing process. Although such lots are selectively harvested, the forest is threatened by other economic activities, and the land is ultimately cleared. Thus, the remaining forests in these areas are affected more by changes in land use than by climate change, and the design of adaptive measures should take this factor into account.

Only a few tree species are commercially harvested, which makes the forestry industry highly dependent on those species. In the north, one species, *Bombacopsis quinnata* (saquisaqui), constitutes up to 60% of the timber production from managed areas. Ensuring regeneration of this species by either establishing commercial plantations or assisting natural regeneration is a major challenge. In both cases, the outcomes have not been encouraging. In the south, no more than five species constitute 50% of the timber production, and similar problems have occurred in attempting to regenerate these species. Because the responses of these species to a changing climate are unknown, the forest industry must diversify production. Attempts have been made to introduce other tree species in the national timber market, but they have frequently failed because the technology and commercial strategies have not been consistent. Planting mixed species in suitable areas should be encouraged as a means to both preserve natural biodiversity and support a healthier forestry industry, even without a warmer climate.

Acknowledgments

This work was supported by the Venezuelan Country Study Climate Change Program under the coordination of Dra. M. Perdomo and in close collaboration with T. Smith of the University of Virginia. The author wishes to thank Ing. Yamil Bonduki for his invaluable contribution on the adaptation option for Venezuela.

References

Country Studies Program (CSP), 1994, *Guidance for Vulnerability and Adaptation Assessments*, U.S. Country Studies Management Team, Washington, D.C., USA.

Hansen, J., G. Russell, D. Rind, P. Stone, A. Lacis, S. Lebedeff, R. Ruedy, and L. Travis, 1983, "Efficient Three-Dimensional Global Models for Climate Studies, Models I and II," *Monthly Weather Review* 111:609-662.

Holdridge, L.R., 1967, *Life Zone Ecology*, Tropical Science Center, San Jose, Costa Rica.

IMF, 1995, *Venezuelan Life Zones under a Potential Climate Change*, Technical Report LJM010, FI-IMF, Caracas, Venezuela.

Leemans, R., and W. Cramer, 1990, *The IIASA Climate Database for Land Area on a Grid of 0.5 Resolution*, WP-41, International Institute for Applied System Analysis, Laxenburg, Austria.

Manabe, S., and R.T. Wetherald, 1987, "Large-Scale Changes in Soil Wetness Induced by an Increase in Carbon Dioxide," *Journal of Atmospheric Science* 142:279-288.

Petroleos de Venezuela, 1993, *Imagenes de Venezuela*, Caracas, Venezuela.

Schlesinger, M.E., and Z.C. Zhao, 1989, "Seasonal Climatic Change Introduced by Doubled CO_2 as Simulated by the OSU Atmospheric GCM Mixed Layer Ocean Model," *Journal of Climate* 2:429-495.

Smith, J.B., and J. Mueller-Vollmer, 1993, *Setting Priorities for Adapting to Climate Change*, RCG/Hagler Bailly, Inc., Boulder, Colo., USA.

Smith, J.B., and D. Tirpak (eds.), 1989, *The Potential Effects of Global Climate Change on the United States*, EPA-230-05-89-050, U.S. Environmental Protection Agency, Washington, D.C., USA.

Thornthwaite, C.W., 1948, "An Approach toward a Rational Classification of Climatic," *Geography Review* 38:55-89.

Walter, H., 1985, *Vegetation of the Earth and Ecological Systems of the Geo-Biosphere*, 3rd Ed., Springer-Verlag, Berlin, Germany.

Wilson, C.A., and J.F.B. Mitchell, 1987, "A Doubled CO_2 Climate Sensitivity Experiment with a Global Climate Model Including a Simple Ocean," *Journal of Geophysical Research* 92:315-348.

Global Climate Change Adaptation: Examples from Russian Boreal Forests

R.K. Dixon
U.S. Support for Country Studies to Address Climate Change
PO-63, 1000 Independence Ave., S.W.
Washington, DC, 20585, USA

O.N. Krankina
Department of Forest Science
Oregon State University
Corvallis, OR 97331, USA

K.I. Kobak
State Hydrological Institute, 23 Seond Line
St. Petersburg 199053, Russian Federation

Abstract

The Russian Federation contains nearly 25% of the world's timber resources and more than 50% of all boreal forests. The response and feedback of boreal forests to projected global climate change are expected to be profound. General Circulation Model scenarios suggest large shifts in the distribution (areal reduction of up to 50%) and productivity of boreal forests. Because the distribution and the productivity of future boreal forests are uncertain, the development of adaptive strategies for establishing, managing, and harvesting forests, as well as for processing wood, is complicated. Although the potential for rapid natural adaptation of long-lived, complex boreal forests is low, recent analyses suggest that Russian forest management and utilization strategies should be tested in the field to assess their effectiveness in helping boreal forest resources adapt to a changing global environment. Current infrastructure and technology can be used to help Russian boreal forests adapt to projected global environmental change. By applying this technical knowledge, policymakers can identify priorities for selecting adaptive strategies.

Introduction

In the Russian Federation, forest ecosystems occupy 884×10^6 ha, which accounts for more than 20% of the world's forest resources and about 50% of all boreal forests (Anuchin et al. 1985; Krankina and Dixon 1992). Russian forests extend from the northern tundra to the steppes in the south (East and West Siberia) and from the Baltic Sea in the west (Europe-Urals) to the Pacific Ocean (Far East) (Figure 1). This vast expanse of land is subdivided into four ecoregions that correspond to longitudinal segments of Russian territory and are significantly different from both ecological and economic perspectives (Krankina and Dixon 1994). Spruce (*Picea*), pine (*Pinus*), birch (*Betula*), and aspen (*Populus*) dominate forests in the Europe-Urals region (Figure 1), while larch (*Larix*) is the major tree genus in East and West Siberia and the Far East (Anuchin et al. 1985).

As a result of fossil-fuel combustion, deforestation, and other human activities, greenhouse gases (e.g., carbon dioxide [CO_2], methane [CH_4]) are accumulating in the atmosphere and may have already begun to change the global climate (Houghton et al. 1992; Victor and Salt 1995). The response and feedback of boreal forests to projected global change are expected to be profound (Smith and Tirpak 1989; Smith et al. 1991). Climate change scenarios

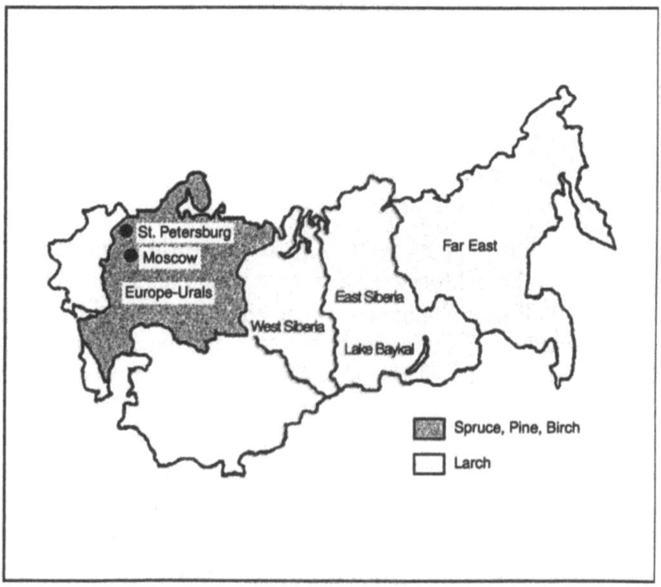

Figure 1. Four Ecoregions of the Russian Federation Showing Major Tree Species

of General Circulation Models (GCMs), such as those of the Geophysical Fluid Dynamics Laboratory (GFDL) (Manabe and Wetherald 1987) and the United Kingdom Meteorological Office (UKMO) (Mitchell 1989), suggest large shifts in the distribution and the productivity of vegetation, especially those of boreal forest systems. Tropical and temperate forests may expand by up to 20%, whereas boreal forests could decrease by up to 50%. Grasslands/shrublands may expand significantly, whereas the tundra zone may decrease by up to 50%. Because the potential redistribution of vegetation in response to global climate change is uncertain, the development of strategies to adapt to a changing global environment is complicated (King 1993). Even if the GCMs are only partially correct, the productivity of boreal forest systems will inevitably change, and the proportion of lands with marginal productivity will increase (Smith et al. 1991; Neilson et al. 1995).

Shifts in atmospheric CO_2, the timing and quantity of precipitation, wind patterns, and ambient temperature (above- and below-ground) will influence the composition, growth, health, and reproduction of Russian boreal and temperate forests (Bonan and Shugart 1989; Kokorin and Nazarov 1995). In the boreal zone, several scenarios suggest that ambient temperature will rise more in the winter than in the summer, and precipitation is likely to increase (Budyko et al. 1991). As incipient global climate change is expressed, insignificant changes in precipitation are expected, while rising ambient temperature will increase evapotranspiration. This phenomenon could increase terrestrial dryness and reduce run-off. As global climate change is manifested in the long term, increased precipitation may offset water loss due to evapotranspiration and improve conditions for plant growth in major agricultural areas (Budyko et al. 1991). These changes will affect boreal forest resources, the commodities that flow from the forest, and the forest products industry (Kokorin and Nazarov 1995; Krankina and Ethington 1995).

Adaptive responses of forests to global climate change have been the subject of some preliminary surveys and analyses (Smith 1995). These reports primarily focus on the temperate forests of North America (Dixon 1992) and Europe (Eriksson 1991). In general, the reports suggest that adaptive measures can and should be applied to manage natural resources to enhance any benefits and to mitigate harmful impacts of climate change. In a review of the impacts of global climate change on ecosystems, Budyko et al. (1991) assert that most impacts will benefit terrestrial ecosystems, especially boreal forests, and adaptation will be minimal.

The objective of this paper is to examine existing management options and technologies that can be used to help boreal forest systems adapt to global climate change in the Russian Federation. Several examples are presented.

Materials and Methods

The potential effects of climate change on global ecosystems have been estimated by using scenarios of future climate projected by GCMs and a variety of response simulation models (Smith et al. 1991; Krankina and Dixon 1994). The adaptive responses outlined in this paper are based on impacts of climate change and the Holdridge Life Zone Classification model (Holdridge 1967), which relates the major plant formations of the world to two independent climate variables: biotemperature and total annual precipitation. Climate change scenarios produced by the GFDL (Manabe and Wetherald 1987), Goddard Institute for Space Studies (GISS) (Hansen et al. 1983), Oregon State University (OSU) (Schlesinger and Zhao 1989) and UKMO (Mitchell et al. 1989) were used to drive a Holdridge Life Zone Classification model. These scenarios and model simulations project the vulnerability of boreal forest resources to global climate change, although the analyses are sometimes contradictory, and error terms are large (Smith et al. 1991). Specific regional responses to the impacts of climate change cannot currently be predicted with confidence (Dixon 1992). Given these constraints, examples of potential adaptive measures and responses were constructed by using Smith's methodology (1995).

The adaptive measures based on projected global climate change and knowledge of Russian forest dynamics presented in this paper are based on current infrastructure, technology, and understanding of potential impacts of climate change on Russian temperate and boreal forests (Krankina and Dixon 1992). The assimilation of exogenous or new technology will be important in future decades, but no attempt was made to project the impact of new technology (Scheraga et al. 1995). The case studies were not constrained by future financial or logistic limitations, although these factors are significant (Nordhaus 1993).

Results

Forest Resource Vulnerability

Terrestrial systems, particularly forest systems, are among the most vulnerable resources to be affected by global climate change (Figure 2). The rapid shift in climate zones may far exceed the ability of Russian forests to adapt or migrate (Smith et al. 1991). Climate zones may shift hundreds of kilometers in a century, while natural rates of species dispersal and colonization may be only a few kilometers in the same period (Bonan and Shugart 1989). In addition, global environmental change could increase existing forest decline and dieback (Krankina et al. 1994). Shifts in the composition and productivity of a forest ecosystem (plant and animal) are projected in circumpolar boreal forests

Figure 2. Global Areas (Black) in Which Predicted Future Vegetation is Different from Current Vegetation. (The vegetation scenario was created by using the UKMO climate scenario and Holdridge Life Form Classification System [Smith et al. 1991].)

(Budyko et al. 1991; Smith et al. 1991). The frequency and intensity of forest fire may increase, as may the incidence of exotic pathogens and insects (Dixon and Krankina 1993). All components of boreal forest ecosystems will be affected — including water resources, soil systems, and wildlife — and the combined effect may be even stronger as a result of interacting factors (Kobak and Kondrasheva 1992).

During warm periods in the geologic past (when climatic conditions were similar to those predicted for the 21st century), the northern border of the conifer forest zone was beyond 70°N latitude (Figure 2). The broadleaf forests of Russia also ranged farther south than they do today. This forest range within the Russian Federation may represent an approximation of potential forest cover under climatic conditions in the future (Kobak and Kondrasheva 1992). Some GCMs predict that boreal forest species will expand rapidly into Arctic tundra because of favorable edaphic conditions (Dixon et al. 1994). After a decade of warming in the 1930s, the tree line advanced along northern Russia's river valleys by dozens of kilometers (Bruce 1993). Receding glaciers during periods of warming altered the hydrology of watersheds, influencing water quantity and quality in rivers and lakes.

Global climate change will probably affect forest utilization and manufacturing in Russia, influencing both the industrial infrastructure and the quality and quantity of wood and nonwood products. Shifts in resource quality and supply will also influence the need for transportation, location of manufacturing facilities, and energy transmission and consumption. The chemical and physical characteristics of wood and paper products may be influenced by shifts in species composition and abundance (i.e., raw material). An enriched CO_2 environment may also influence specific gravity and the characteristics of wood fiber in some species (Dixon 1992). During periods when forests decline or expand rapidly, the abundance of nonwood products may change in transition zones (Krankina and Ethington 1995).

Forest Resource Adaptive Measures

Because forest systems are complex, relatively immobile, and long-lived, they have not been expected to adapt to certain types of changes (Smith 1995). Given the current infrastructure and technology, some adaptive measures are, however, plausible. Specific examples of potential adaptation measures are reviewed in the final sections of this paper (Tables 1, 2, and 3).

Forest Establishment

Several strategies can help to expand boreal forests or replace trees lost as a result of dieback (Table 1). For example, artificial seeding or production of tree seedlings in nurseries is one strategy to use when natural seed dispersal is not possible because either the distance from seed sources is great or yearly seed production is irregular, which is typical under the extreme northern climate (Krankina and Dixon 1992). Russia has extensive experience with artificial seeding technology, nursery production, and tree planting (Stoliarov 1990). Tree improvement methods, especially traditional selection and breeding programs, can enhance the growth rate and drought resistance of planting stock.

Another major impediment to boreal forest regeneration is widespread ground cover (mosses) that impedes the development of seedlings (Anuchin et al. 1985). For example, mosses keep soils colder and reduce the growth of seedlings. Site preparation with prescribed burning or soil scarification can substantially improve a seedling's early survival and juvenile growth. Draining certain types of mires may also improve conditions for tree growth. All of these adaptive measures can stimulate forest regeneration in zones of climate-induced dieback.

The sites and tree species for forestation at taiga-tundra borders should be selected carefully on the basis of ecological principles. River valleys, southern slopes, and light soils with moderate moss cover could provide better environmental conditions for establishing forests. New forests on these sites could then serve as bridges for forest expansion northward and toward less favorable sites. An analysis of tree planting and survival at high latitudes reveals that,

Table 1. Examples of Potential Impacts of Global Climate Change on Forest Resources of the Russian Federation and Possible Adaptive Measures or Responses

Component/System	Potential Impact of Climate Change	Adaptive Measures or Responses
Plant community	Shift in composition, productivity, and reproduction; decline, dieback, or migration of species; shifts in the ranges of insects and pathogens	Silvicultural measures, such as artificial regeneration, thinning, or fertilization; pest management
Animal community	Shift in composition; loss of habitat and diversity; species migration or invasion problems; pest problems	Habitat or species preservation; migration corridors
Water resources	Change in water quantity and quality; extreme events — loss of habitat	Water management options, such as reservoir development, flood protection, habitat protection, fish hatcheries
Soil systems	Water or wind erosion; loss of productivity; shift or loss of microorganizers, soil-moisture changes	Soil protection; fertilization or irrigation; minimize disturbance
Wetlands/peatlands	Southern peat to decompose; taiga peat expands	Peat used as bioenergy; expand agriculture in southern regions

outside the current forest zone, some planted trees survived nearly 200 years and naturally regenerated (Bruce 1993).

At the southern border of the Russian forest zone, adaptive measures could maintain the current expanse of forests or decrease the advancement of grasslands (Krankina et al. 1994). Possible strategies include forestation with drought-adapted species and shelterbelt systems. Protective forests currently occupy 11.6×10^6 ha (about 10% of the forested area) in Russia (Anuchin et al. 1985), and the demand for new shelterbelt plantations is estimated at 3.9×10^6 ha (Guiryayev 1989). These plantations may improve local climatic conditions for tree growth and agriculture by reducing evapotranspiration, enhancing water retention in soils, and decreasing wind speed.

Forest Management

Currently, the Russian Federal Forest Service manages 69.9×10^6 ha of land at the taiga-tundra border (Krankina and Dixon 1992). Of the forested area in this transition zone, 40% is occupied by closed forests, and 32% is bogs and mires.

The balance is open forests, burned and dead forests, and reindeer pastures (Anuchin et al. 1985). Clear-cutting most taiga-tundra forests is currently limited, but considering the future value of these forests as animal habitat or migration corridors, the scope and intensity of harvesting activities can be altered to accommodate impacts of a changing climate (Smith and Tirpak 1989).

Throughout the present forest zone, adaptive measures should enhance forest stability and minimize disturbance associated with rapid climate change (Budyko et al. 1991). This goal may be achieved by (1) maintaining a mixed composition of species where site conditions favor the growth of mixed stands and (2) implementing integrated fire- and pest-management approaches (Table 1). The species for plantations should be selected by considering potential climate change within the life span of the trees (King 1993). Planting trees at the southern limit of their natural range may have to be reconsidered in a rapidly changing climate (Smith et al. 1991). Perhaps the greatest potential challenge facing Russian forest managers will be managing nonindigenous pests (insects, fungi, and bacteria) that may migrate with shifting climatic conditions (Tkach 1991).

To maintain the highest degree of tree cover possible, clear-cutting should be complemented by other wood-harvesting methods (Rosencranz and Scott 1992). Harvest rotations for species well adapted to changing environmental conditions should be expanded, while forests that have achieved rotation age should be identified and harvested early (Efremov 1989). Forest health should be monitored thoroughly to identify areas at risk of decline (Krankina et al. 1994). Historically, remote-sensing techniques have been applied to monitor forest health, and this technology should be expanded to assist policymakers in the future (Krankina and Dixon 1992).

Forest Products

Again, to maintain the highest degree of tree cover possible, clear-cutting should be complemented by other wood-harvesting methods (Burdin 1991). Rotating species well adapted to changing conditions should be expanded, while forests that cannot maintain stability should be identified and harvested early (Rosencranz and Scott 1992). Silvicultural practices, alternative manufacturing technology, and development of nonwood substitutes may be viable adaptive measures in a changing climate (Krankina and Ethington 1995). Desirable wood characteristics can be fostered by altering species mix or silvicultural techniques (Efremov 1989). Nonwood product markets may expand rapidly as short-rotation plantations or agroforestry systems are established in forest transition zones (Zyabchenko et al. 1992). Alternative pulping practices and fiber mix can be adjusted in response to a shifting resource base (Table 2).

Table 2. Examples of Potential Impacts of Global Climate Change on Forest Products of the Russian Federation and Adaptive Measures or Responses

Component/System	Potential Impacts of Climate Change	Adaptive Measures or Responses
Solid-wood products	Shifts in specific gravity, fiber length, strength, and load capacity in response to CO_2 enrichment and changes in tree nutrition	Implementation of silvicultural measures to maintain desired wood characteristics; use of nonwood substitutes
Nonwood products	Shift in composition, quality, quantity; decline or dieback of nontimber species; invasion of expansion of pioneer species	Expansion of nonwood products market; Implementation of agroforestry practices in transition zones
Paper products	Possible shifts in fiber characteristics	Alter pulping practices and fiber mix

Manufacturing Infrastructure

More than 60% of the wood-harvesting and 80% of the wood-processing infrastructure is located in the Europe-Urals and West Siberia regions (Figure 1); however 75% of forest resources are east of the Ural Mountains (Krankina and Dixon 1992). Relocating manufacturing facilities northward is one adaptive measure to a shifting resource base (Table 3). Co-locating forest-harvesting operations adjacent to wood-processing facilities is also a viable option (Cardellichio et al. 1990). To accommodate the predicted future abundance of broadleaf species, wood-processing technology and products may also be altered (Efremov 1989). For example, the production of composite wood products may increase relative to the production of solid-wood products (Krankina and Ethington 1995).

The production of forest-based bioenergy may expand in a changing climate, especially if the change occurs over decades rather than centuries (Sampson et al. 1993). Because rapid climate change may preclude the long-rotation management of forests, short-rotation intensive culture in transition zones may be a viable approach (Bonan and Shugart 1989). Short-rotation culture (e.g., species, silvicultural practices) permits adaptive flexibility, and the fiber, fuel, and fodder that flow from these plantations provide multiple feedstock for bioenergy or manufactured products (Hall 1991).

Establishing new manufacturing facilities will require adaptive measures in energy production and distribution (Barr and Braden 1988). Changes in

Table 3. Examples of Potential Impacts of Global Climate Change on the Infrastructure of Forest Resources and Products and Adaptive Measures or Responses

Component/System	Potential Impact of Climate Change	Adaptive Measures or Responses
Manufacturing	Shift in resource quality and supply; changes in energy availability; emission reduction restriction	Shift in facilities and in products manufactured; altered production cycles
Energy	Changes in hydropower output; shifts in energy peak needs	Use of alternative fuels; management
Terrestrial transportation	Deterioration of roads in tundra and permafrost	Use of low-impact transportation systems
Aquatic transportation	Shifts in water supply in response to flooding or drought	Expansion of infrastructure to manage water resources

hydropower output, caused by decreases in run-off, might be addressed by increased reliance on solar and wind energy. Demand-side management may help to meet shifts in energy needs. These infrastructural modifications are complicated by the transition from a planned economy to a market economy in Russia (Krankina and Dixon 1992).

In both urban and rural areas of Russia, alternative transportation systems will be developed in response to global climate change (Anuchin et al. 1985). As permafrost recedes in Siberia and the Far East, components of low-impact transportation systems (e.g., large rubber tires vs. steel treads) can be used in forest-harvesting or transportation networks (Table 3). Reliance on aquatic transportation (e.g., river barges) may increase as a low-cost alternative to road building, if not precluded by extreme hydrologic events. Modifying the transportation infrastructure is linked to shifts in resources, forest-sector location of manufacturing facilities, and emerging consumer markets (Rosencranz and Scott 1992).

Discussion

Most infrastructural adaptations associated with forest resources, forest products, and manufacturing will be made when climate changes (Budyko et al. 1991). The forest sector will switch to alternative species as environmental conditions dictate; manufacturing facilities will be modified as the market demands; and reservoir operators will adjust pools and flux as river flow changes. These measures are reactive adaptations because they are in response to climate change

(Smith 1995). In contrast, anticipatory adaptive measures are those taken in advance of climate change (Glantz 1988). The goals of anticipatory measures are (1) to minimize the impact of climate change by reducing the vulnerability of forest resources and (2) to enable more efficient adaptation (faster or at a lower cost). The first response is robust (able to absorb surprises), while the second is resilient (able to recover from failure).

Given the uncertainty of the direction, magnitude, and rate of climate change on a regional scale (Smith et al. 1991), as well as the long time frame over which its effects may be manifested (Budyko et al. 1991), society may prefer to react to climate change rather than to try to anticipate it. International and national policymakers are reluctant to implement responses for impacts that may not occur, especially if the benefits of such anticipatory measures may not be seen for decades (Victor and Salt 1995). Furthermore, future generations may have more income and sophisticated technologies that can be used for adaptation (Nordhaus 1993). Long-term economic analyses of large-scale reactive vs. adaptive responses to environmental change are not sufficiently complete to guide policymakers.

Nevertheless, a number of potential drawbacks are associated with cautious use of adaptive measures in Russian forests:

- The impacts of climate change on forest systems may be irreversible (Budyko et al. 1991). Species extinctions and the loss of rare ecosystems are irreversible.
- The costs or impacts of climate change on the forest sector, even after adaptation, may be very high (Nordhaus 1993). Forests could experience significant reductions in range and population. Replacing manufacturing based on forest products may require intensive, long-term investments (Burdin 1991).
- Policy decisions made now may not endure to cope adequately with global climate change (Brandt 1992). For example, trees planted today with a life expectancy of decades may not survive to maturity if climate conditions change rapidly (Krankina et al. 1994).
- Rapid reaction to global climate change may imply the development of responses to extreme or catastrophic events (e.g., floods, droughts, invasive wildlife). A reactive response policy may be logistically or financially sensible, but it runs the risk of taking short-term incremental approaches and not anticipating future large-scale changes (Glantz 1988; Smith 1995).

Although anticipating climate change may be desirable in many cases, uncertainties make the design of anticipatory policies challenging, especially for countries with economies in transition (Krankina and Dixon 1992). The socioeconomic impediments of moving from a planned forest-sector economy to a market economy are well documented (Burdin 1991; Rosencranz and Scott 1992). If the impacts of climate change are concomitant with shifts to a market economy, the challenges of forest-sector adaptation could be exacerbated or perhaps consolidated. Dramatic losses in raw materials from the forest sector would negatively influence Russia's emerging market economy, since Russia increasingly depends on wood-product exports to maintain a balance of trade (Rosencranz and Scott 1992). In contrast, regional rehabilitation of manufacturing or transportation infrastructure could be implemented concurrently to cope with complementary market or climate changes, partially relieving Russia's financial burdens relative to other developed countries (OTA 1991). The challenge for Russian policymakers will be to balance adaptation associated with climate change priorities with other economic trends (Nordhaus 1993).

Uncertainties related to the impacts of future global climate change at local or regional scales may limit the development and timely deployment of specific adaptive measures (Smith 1995). However, identifying "no regrets" adaptive measures, with favorable, positive cost-benefit ratios, may help policy analysts rank future priorities (National Academy of Sciences [NAS] 1991; Victor and Salt 1995). Of the various adaptive measures discussed in this paper, such options as enhancing the health and productivity of existing plantations in Siberia and the Far East (Figure 1); expanding shelterbelt systems in southern regions; changing from clear-cutting to alternative, low-impact harvesting systems; and developing extensive systems for monitoring forest health could provide the basis for no regrets options. These measures can be applied by using existing technology and infrastructure. They are considered anticipatory responses and reduce the negative impacts associated with rapid global climate change (NAS 1991).

Adaptive strategies reviewed represent a substantial departure from traditional Russian forest resource management and utilization (Krankina and Dixon 1992). Under the conditions of climate change, the suitability of forest resources and the maintenance of an adequate manufacturing infrastructure may become a high priority in Russia (Budyko et al. 1991). Other developed countries with significant forest resources (e.g., Australia, Canada, Germany, the U.S.) have developed options for responding to natural resource policy in their National Communications to the U.N. Framework Convention on Climate Change (UNFCCC). Future development of climate change policy in Russia will have to consider carefully the fate of 20% of the world's remaining forests and 25% of the global terrestrial carbon pool resident in this vast country. Significant shifts in forest policy may be implied as Russia develops national communications with the UNFCCC.

References

Anuchin, N.P., et al., 1985, *Forest Encyclopedia*, Vols. 1 and 2, State Committee for Forestry, Moscow, USSR (former Soviet Union) (in Russian).

Barr, B.M., and K.E. Braden, 1988, *The Disappearing Russian Forest: A Dilemma in Soviet Resource Management*, Rowman and Littlefield, Totowa, N.J., USA.

Bonan, G.B., and H.H. Shugart, 1989, "Environmental Factors and Ecological Processes in Boreal Forests," *Ann. Rev. Ecol. Sys. 20:*1-28.

Brandt, R., 1992, "Soviet Environment Slips Down the Agenda," *Science* 255:22-23.

Bruce, D., 1993, "History of Tree Planting on the Aleutian Islands," in J. Alden et al. (eds.), *Forest Development in Cold Climates,* pp. 393-426, Plenum Press, New York, N.Y., USA.

Budyko, M.I., Y.A. Izrael, M.C. MacCaren, and A.D. Hecht (eds.), 1991, *Forthcoming Climate Changes,* Hydrometeorizdat, Leningrad, USSR (former Soviet Union) (in Russian).

Burdin, N.A., 1991, "Trends and Prospects for the Forest Sector of the USSR: A View from the Inside," *Unasylva* 42:43-50.

Cardellichio, P.A., C.S. Binkley, and V.K. Zausaev, 1990, "Sawlog Exports from the Soviet Far East," *Journal of Forestry* 88:12-17, 36.

Dixon, R.K., 1992, "Regional Forest Management Planning in the Southern United States," in A. Quereshi (ed.), *Forests in a Changing Climate*, pp. 378-389, Climate Institute, Washington, D.C., USA.

Dixon, R.K., and O.N. Krankina, 1993, "Forest Fires in Russia: Carbon Dioxide Contributions to the Atmosphere," *Canadian Journal of Forest Research* 23:700-705.

Dixon, R.K., S. Brown, R.A. Houghton, A.M. Solomon, M.C. Trexler, and J. Wisniewski, 1994, "Carbon Pools and Flux of Global Forest Systems," *Science* 263:185-190.

Efremov, D.F. (ed.), 1989, *Forest Resources of Far Eastern Economic Region: Their State, Use and Reproduction,* Far Eastern Forestry Research Center Publishing, Khabarovsk, USSR (former Soviet Union) (in Russian).

Eriksson, H., 1991, "Sources and Sinks of Carbon Dioxide in Sweden," *Ambio* 20:146-150.

Glantz, M.H., 1988, *Societal Responses to Regional Climate Change — Forecasting by Analogy,* Westview Press, Boulder, Colo., USA.

Guiriayev, D.M., 1989, "Healing the Land," *Forest Management* 2:25-26 (in Russian).

Hall, D.O., H.E. Mynick, and R.H. Williams, 1991, "Cooling the Greenhouse with Bioenergy," *Nature* 353:11-12.

Hansen, J., G. Russell, D. Rind, P. Stone, A. Lacis, S. Lebedeff, R. Ruedy, and L. Travis, 1983, "Efficient Three-Dimensional Global Models for Climate Studies: Models I and II," *Monthly Weather Review* 111:609-662.

Holdridge, L.R., 1967, *Life Zone Ecology,* Tropical Science Center, San Jose, Calif., USA.

Houghton, J.T., G.J. Jenkins, and J.J. Ephraums, 1992, *Climate Change: The Supplementary Report to the IPCC Scientific Assessment,* World Meteorological Organization/U.N. Environmental Programme, Intergovernmental Panel on Climate Change, Cambridge University Press, Cambridge, United Kingdom.

King, G.A., 1993, "Conceptual Approaches for Incorporating Climate Change into the Development of Forest Management Options for Sequestering Carbon," *Climate Research* 3:61-78.

Kobak, K.I., and N.Y. Kondrasheva, 1992, "Changes of Natural Zones Locations Caused by Global Warming," *Ecology* 3:9-18 (in Russian).

Kokorin, A.O., and I.M. Nazarov, 1995, "The Analysis of Growth Parameters of Russian Boreal Forests in Warming, and Its Use in Carbon Budget Model," *Ecological Modeling*.

Krankina, O.N., and R.K. Dixon, 1992, "Forest Management in Russia: Challenges and Opportunities in the Era of Perestroika," *Journal of Forestry* 90:29-34.

Krankina, O.N., and R.K. Dixon, 1994, "Forest Management Options to Conserve and Sequester Terrestrial Carbon in the Russian Federation," *World Resource Review* 6:89-101.

Krankina, O.N., R.K. Dixon, A.Z. Shvidenko, and A.V. Selikhovkin, 1994, "Forest Dieback in Russia: Cause, Distribution and Implications for the Future," *World Resource Review* 6:525-534.

Krankina, O.N., and R.L. Ethington, 1995, "Forest Resources and Wood Properties of Commercial Tree Species in the Russian Far East," *Forest Products Journal* (in press).

Manabe, S., and R.T. Wetherald, 1987, "Large-Scale Changes in Soil Wetness Induced by an Increase in Carbon Dioxide," *Journal of Atmospheric Science* 142:279-288.

Mitchell, J.F.B., C.A. Senior, and W.J. Ingram, 1989, "CO_2 and Climate: A Missing Feedback?" *Nature* 341:132-134.

National Academy of Sciences (NAS), 1991, *Policy Implication of Greenhouse Warming,* U.S. National Academy Press, Washington, D.C., USA.

Neilson, R.P., G.A. King, and J. Lenihan, 1995, "Modeling Forest Response to Climate Change: The Potential for Large Emissions of Carbon from Dying Forests," in R. La et al. (eds.), *Soil Management and Greenhouse Effect,* pp. 150-162, Lewis Publications, Boca Raton, Fla., USA.

Nordhaus, W.D., 1993, "Reflections on the Economics of Climate Change," *Journal of Economic Perspectives* 7:11-25.

Office of Technology Assessment (OTA), 1991, *Changing by Degrees: Steps to Reduce Greenhouse Gases,* OTA-482, U.S. Congress, USA.

Sampson, R.N., L.L. Wright, J.K. Winjum, J.D. Kinsman, J. Benneman, E. Kursten, and J.M.O. Scurlock, 1993, "Biomass Management and Energy," Water, *Air and Soil Pollution* 70:139-162.

Scheraga, J., et al., 1995, presented at the International Conference on Climate Change Adaptation Assessments, Coastal Resources Section, St. Petersburg, Russian Federation, May 22-25.

Schlesinger, M.E., and Z.C. Zhao, 1989, "Seasonal Climatic Change Introduced by Doubled CO_2 as Simulated by the OSU Atmospheric GCM/Mixed-Layer Ocean Model," *Journal of Climate* 2:429-495.

Smith, J.B., 1995, "Setting Priorities for Adapting to Climate Change," *The Environmental Professional* (in press).

Smith, J.B., and D.A. Tirpak (eds.), 1989, *The Potential Effects of Global Climate Change on the United States,* EPA-230-05-89-050, report to Congress by the U.S. Environmental Protection Agency, Washington, D.C., USA.

Smith, T.M., H.H. Shugart, G.B. Bonan, and J.B. Smith, 1991, "Modeling the Potential Response of Vegetation to Global Climate Change," *Advances in Ecological Research* 22:93-116.

Stoliarov, D.P., 1990, "Future of Forests: Are There Reasons for Optimism?" *Forest Management* 3:2-5 (in Russian).

Tkach, B. (ed.), 1991, *Pest Risk Assessment of the Importation of Larch from Siberia and the Soviet Far East,* U.S. Department of Agriculture Forest Service. Miscellaneous Publication 1495, Washington, D.C., USA.

Victor, D.G., and J.E. Salt, 1995, "Keeping the Climate Treaty Relevant," *Nature* 373:280-282.

Zyabchenko, S.S., T.V. Belonogove, and N.L. Zaitseva, 1992, "Non-Timber Resources of the Forest Zone, Their Use and Problems of Studying Them," *Rastit. resursi* 28:3-12.

Vulnerability and Adaptation Assessments of Forest Vegetation in Bulgaria

O. Grozev
Bulgarian Academy of Sciences
Forest Research Institute
132 Kliment Ohridski Blvd.
Sofia 1756, Bulgaria

V. Alexandrov
Bulgarian Academy of Sciences
Institute of Meteorology and Hydrology
66 Tzarigradsko chaussee Blvd.
Sofia 1784, Bulgaria

I. Raev
Bulgarian Academy of Sciences
Forest Research Institute
132 Kliment Ohridski Blvd.
Sofia 1756, Bulgaria

Abstract

The vulnerability and adaptation to climate change of forests in Bulgaria have been assessed by means of the Holdridge Life Zone Classification model. Measurements of two climate variables (mean monthly temperature and annual precipitation) from 18 meteorological stations at different elevations were used to determine the current and expected life zones. The current Holdridge life zone in Bulgaria is mainly cool temperate moist forest. If carbon dioxide concentration doubles in the next century, the life zone may become warm temperate dry forest and even subtropical dry forest at elevations less than 800 m, which includes 60.6% of the forest. Adaptive measures are being considered in new Bulgarian legislation.

Introduction

Changes in climate and weather conditions directly affect forests and agricultural production. Forests and woodland cover about one-third of the world's land surface and deserve serious consideration in assessments of the socioeconomic implications of climatic change. Foresters, consultants, extension agents, planners, and others involved in forest production face varying environmental conditions that can affect vegetation growth and development.

Recent research has focused on regional, national, and global assessments of the potential effects of climate change on forests (Intergovernmental Panel on Climate Change [IPCC] 1990). The vulnerability of Bulgarian forests to climate change is evaluated and ways to preserve them through adaptation are suggested.

Materials and Methods

This study used daily climatic data from 18 meteorological stations across Bulgaria for 1951-1990 (Figure 1). Thirty years (1951-1980) of current climate data were used in developing the baseline scenario for Bulgaria. A 30-year period is considered long enough to represent wet, dry, warm, and cool periods. A recent 30-year period is preferred, not only because it represents the current

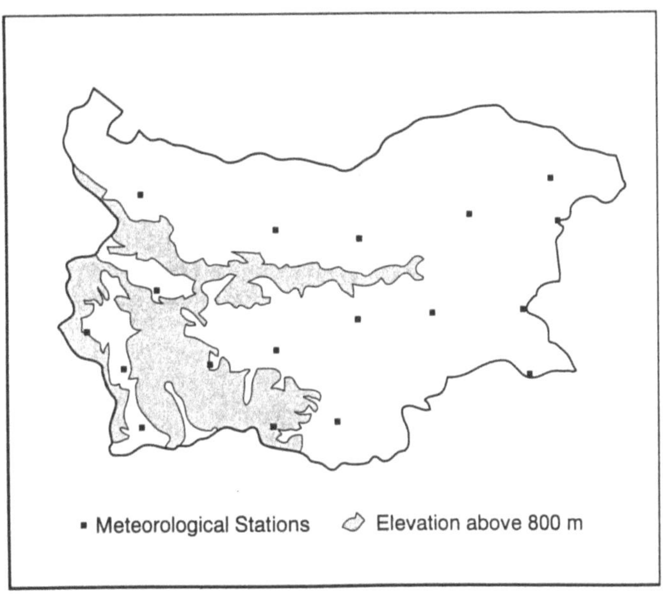

Figure 1. Locations of Meteorological Stations in Bulgaria

climate, but also because more recent data are more accurate. Bulgaria did not experience a significant warming trend during the last decade. Therefore, data from 1981 to 1990 can also be used as the current climate.

General Circulation Models (GCMs) are mathematical representations of atmosphere, ocean, and land surface processes based on the laws of physics. Such models are a credible way to estimate future climate (IPCC 1990; U.S. Country Studies Program [CSP] 1994). Climate change scenarios were created by using observed daily data from 1951 to 1980 with GCMs from the Goddard Institute for Space Studies (GISS), Geophysical Fluid Dynamics Laboratory (GFDL), and Canadian Climatic Center (CCC) to predict conditions under a doubling of atmospheric concentration of carbon dioxide ($2XCO_2$). Mean monthly changes in temperature, precipitation, and solar radiation from the appropriate GCM grid box were applied to observed current climate data to create climate change scenarios for each site.

Policymakers need information on the potential effects of climate change over the next few decades as they develop plans that will affect forests. The GCM transient scenarios can be useful in vulnerability assessments (CSP 1994). Transient scenarios are run by assuming a steady increase in greenhouse gas concentration and examining the way climate could change over time. The GFDL transient climate change scenario was used in this study.

Incremental changes in meteorological variables, such as temperature and precipitation, were also used in creating climate change scenarios (CSP 1994). The incremental scenario used to assess the vulnerability of forests to climate change combined temperature increases of 2, 4, and 6°C with 0, ±10, and ±20% changes in precipitation. The Holdridge Life Zone Classification model was used to evaluate the potential impacts of climate change on forest ecosystems (Holdridge 1967).

Results and Discussion

Climate change scenarios derived for Bulgaria evaluated potential changes in forest vegetation. The changes in temperature and precipitation corresponding to each climate change scenario were used to run the Holdridge Life Zone Classification model.

Vulnerability Assessments

Climate Change Scenarios for $2XCO_2$

Figure 2 shows the results of the CCC, GISS, and GFDL models. Five representative meteorological stations were chosen: three (Montana, Lovetch,

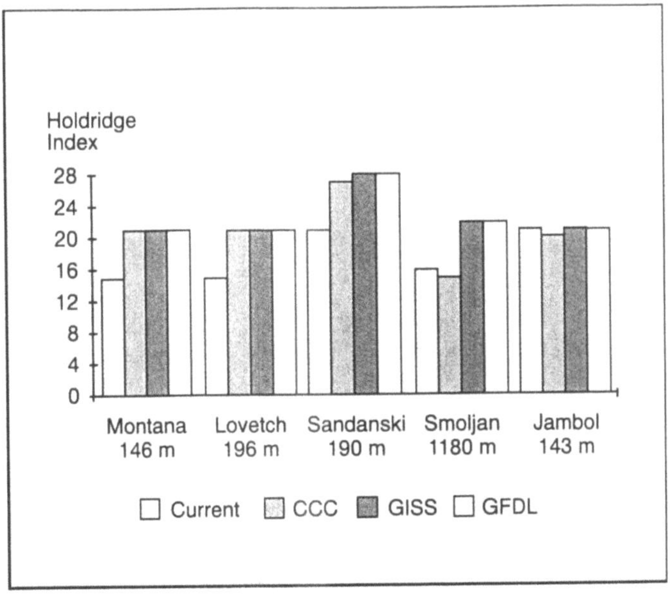

Figure 2. Changes in Holdridge Life Zones for Five Representative Meteorological Stations in Bulgaria under the $2XCO_2$ Scenario

and Jambol) for the climate characteristics of the largest part of the country (areas less than 800 m above sea level), one (Sandanski) for the warmest area, and one (Smoljan) for the mountainous area (Figure 1).

The results of the three models are similar. Cool temperate moist forest (Holdridge index 15) could become warm temperate dry forest (21) in northern Bulgaria, and warm temperate dry forest (21) could remain unchanged in southern Bulgaria. Subtropical thorn woodland (27) or subtropical dry forest (28) could be expected in the warmest region (Sandanski), which suggests drastic warming and droughts. The mountainous region (Smoljan, 1,180 m above sea level) could change from cool temperate wet forest to warm temperate moist forest. Because 60.6% of Bulgaria's forest is less than 800 m in elevation (Kostov et al. 1976), it could be vulnerable to drastic changes if the CO_2 concentration doubled in the near future.

GFDL Transient Climate Change Scenarios

The GFDL transient scenarios were used to assess the effect of expected climate change through 2006, 2036, and 2066. Figure 3 shows the expected changes in Holdridge life zones for the five representative stations: Montana, Lovetch, Sandanski, Smoljan, and Jambol.

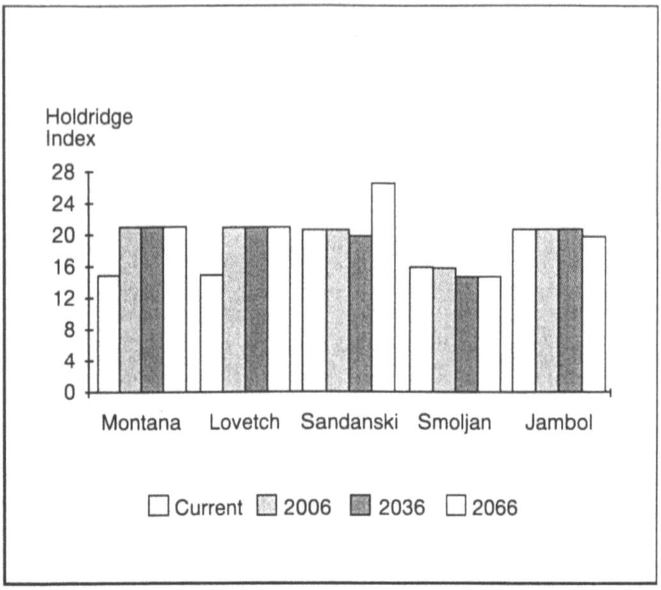

Figure 3. Changes in Holdridge Life Zones for Five Representative Meteorological Stations in Bulgaria for 2006, 2036, and 2066

In 2006, the forests in northern Bulgaria could change from cool temperate moist forest (15) to warm temperate dry forest (21). In southern Bulgaria, forests could gradually change from warm temperate dry forest (21) to warm temperate thorn steppe (20), but not until 2036 and 2066. A transition to subtropical thorn woodland (27) could be expected in the southernmost and warmest areas at the end of this period.

Incremental Climate Change Scenarios

Figure 4 illustrates the results of incremental climate change scenarios for Bulgaria. Lovetch represents the low regions of northern Bulgaria, and Jambol represents the low regions of southern Bulgaria. Temperature increases of 0, 2, 4, and 6°C and precipitation changes of 0, ±10, and ±20% were used.

For Lovetch, cool temperate moist forest (15) would predominate under a temperature increase of 2°C. With a 4°C temperature increase, for all variations of precipitation, a transition to warm temperate dry forest (21) could be expected. For a 6°C temperature increase, subtropical dry forest (28) could result, except in the case of a 20% decrease in precipitation, which could result in a transition to subtropical thorn woodland (27).

A similar situation could be expected for Jambol. However, a transition from warm temperate dry forest (21) to warm temperate thorn steppe (20) could be

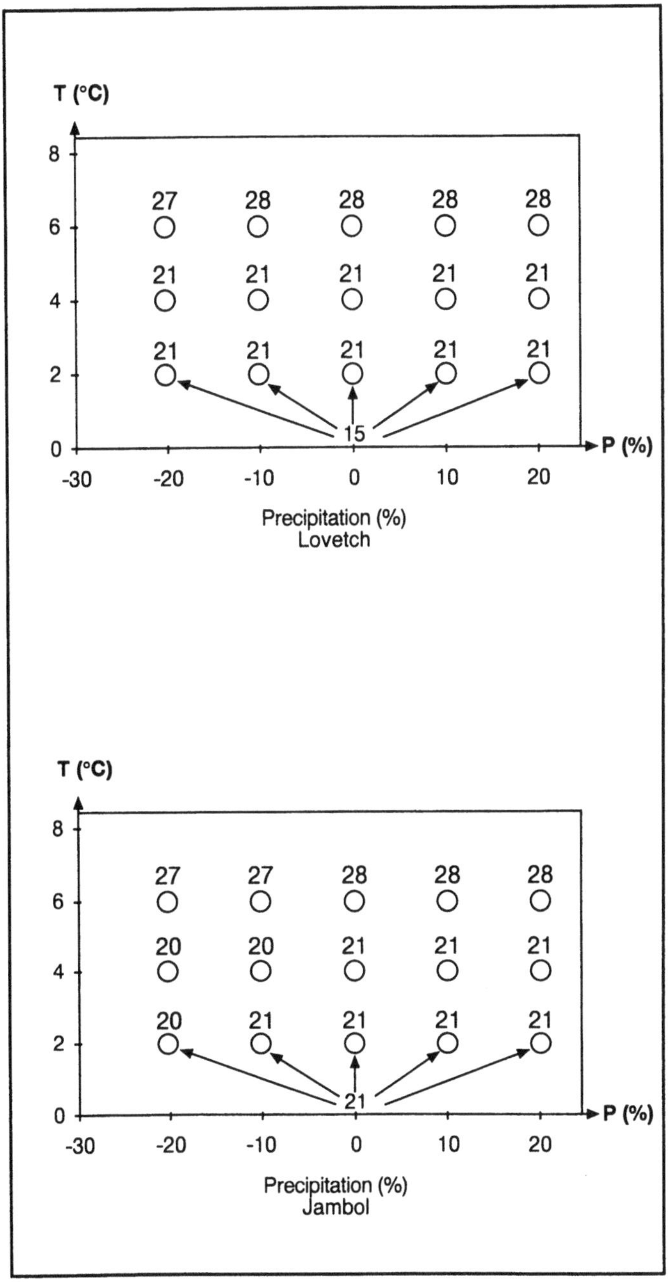

Figure 4. Changes in Holdridge Life Zones in Bulgaria According to Incremental Climate Scenarios

expected for temperature increases of 2 and 4°C. A 6°C temperature increase combined with 10 and 20% decreases in precipitation could result in subtropical thorn woodland (27).

Thus, drastic changes in Holdridge life zones could be expected in Bulgaria during the next century. A 6°C temperature increase, combined with a 10-20% decrease in precipitation, would be particularly dangerous. Warming and lengthy dryness could be expected in most of Bulgaria. Therefore, the survival of the existing forest is a concern.

Adaptation Assessments

The expected increases in temperature and changes in precipitation will significantly affect the composition of forest ecosystems, the expansion of tree-covered areas, and the number of forest tree and shrub species in Bulgaria. An overall temperature increase could move the tree line into the mountains and lead to a greater number of thermophylic trees and shrubs, especially in lower areas. It could also mean that trees naturally found in cooler mountain areas, such as the Norway spruce, would begin to disappear. Such a phenomenon occurred in the Balkan range with the European fir in the beginning of this century (Jordanov 1924). Today, natural European fir forests are rare, seen only in remote areas.

With an overall temperature increase of 1-2°C in Bulgaria, the forest would become more similar southern and lower areas, where the current climate is similar to that expected in the future. Drastic, even catastrophic, changes can be expected in the forest in the southern regions, according to the different climate change scenarios. The direction of these changes is the same for all regions of the country: toward increased warming and dryness. The Bulgarian forest, which is already subject to problems with the water balance in the low parts of the country, would be further stressed (Raev 1989). Temperature is not the decisive factor in expanding tree-covered areas; precipitation is of primary importance.

Bulgaria recently underwent a bitter afforestation experience. Large numbers of coniferous trees, particularly Scots and Austrian black pine, were used for monosylviculture plantings covering vast areas. These tree species, which grow naturally in regions higher than 600-1,000 m above sea level, were planted at lower elevations, 0-600 m above sea level (Raev 1995). About 25% of these plantings died or suffered poor growth. The wrong choice of trees species caused this afforestation attempt to fail.

This experience outlines a major adaptation policy task: the correct tree and shrub species must be chosen for afforestation. Suitable choices should be sought within the species that occur naturally in the region to be planted. More thermophilic species or genetically proven specimens could also be used. In the low areas of Bulgaria, more than 80 taxons naturally appear, most of which are deciduous. Among these are several drought-resistant species with a wide range of adaptability (Penev et al. 1969). The introduction of a new species is another possibility. The results with the Cedar genus, especially the Atlas cedar, are exceptionally encouraging (Delkov and Grozev 1995).

Recent Bulgarian forestry activities have been directed toward correcting the problems resulting from the monosylviculture afforestation attempt. The remaining viable conifer specimens are being retained, and the unsustainable conifers are being replaced with drought-resistant, native deciduous trees, predominantly from the *Fagaceae* family. A number of deciduous tree species are being introduced, and existing ones are being preserved. Mixed forests, with more than 10-12 tree species, have been created.

Mixed forests have many advantages. They are more vital and sustainable; are more resistant to fire, insects, and disease; and have greater biodiversity. However, their greatest advantages are the large reserves and their potential for survival under the expected climate changes.

Bulgaria has been successful in other adaptive measures, such as conversion from coppice forests to seeded forests. Existing vegetation is used if it produces enough seeds, but seedlings are also used, when necessary. Another common practice is creating two-storied forests. However, the success of these activities can be guaranteed only when appropriate sites and combinations of tree species are chosen. These adaptive measures could also mitigate climate changes.

Such adaptive measures are already in place in Bulgaria. Regional and national meetings have been held to prepare for acceptance of a new forest strategy. New forest legislation has been widely discussed, in which the new adaptation policy fully coincides with the ecological and sustainable development of the forest (Karamfilov 1995).

The following are possible methods for adapting Bulgarian forest to future warming and drought:

- A change in the forestry strategy is necessary for preserving forest ecosystems under the greenhouse effect.
- For afforestation, the more heat- and drought-resistant species are preferred. The more resistant species may also be genetically improved.
- The need for creating and maintaining resistant forest ecosystems in a warmer, drier climate must be considered in economic activities in forests.
- The condition of forests should be monitored.
- Forests should be maintained in good condition by such measures as fellings and treatment against pests.
- Model (demonstration) projects of the adaptive measures should be created.
- The problems of climate change and the adaptation of forest vegetation to such a change should be studied further.
- Specialists from the forestry administration (Committee of Forests), as well as from the Ministry of Environment; scientific institutions dealing with the problems of the forests (Forest Research Institute at the Bulgarian Academy of Sciences, Higher Institute of Forestry and Forest

Industries); and high state authorities should study the problems of the forests in the next century.

- International collaboration in the study of vulnerability and adaptation of the forest vegetation should continue.

Conclusions

Because of the many remaining uncertainties about possible future climate changes (particularly in regional precipitation patterns and the rate of climate change over time), specific impacts on forests cannot be forecast. Climate change scenarios can help to identify the potential direction of effects and magnitude of impacts, and the relative sensitivity of a sector to changes in different meteorological variables. Thus, it is possible to identify the types of impacts that might occur under possible climate changes and, by inference, the types of climate change that constitute a significant risk or benefit to forests.

The Holdridge life zone model shows that regions may change from cool temperate moist forest to warm temperate dry forest, subtropical thorn woodland, or subtropical dry forest because of the climate change expected in the next century. As much as 60.6% of the Bulgarian forest would be affected. This possibility gives a new meaning to the Bulgarian forestry strategy. New legislation has been directed to measures concerning environmental, ecological, and economic activities in Bulgarian forests, which would support mitigation and adaptation policies.

Acknowledgments

The authors wish to thank the U.S. Country Studies Management Team for the opportunity to participate in the U.S. Country Studies Program, for the information provided, and for the encouraging and useful suggestions concerning this study.

References

Delkov, A., and O. Grozev, 1995, "Results from the Introduction of the Atlas Cedar in Bulgaria," *Proceedings Anniversary Scientific Conference "70 Years VLTI,"* Vol. 1, pp. 221-225.

Holdridge, L., 1967, *Life Zone Ecology*, Tropical Science Center, San Jose, Costa Rica.

Intergovernmental Panel on Climate Change (IPCC), 1990, "The Potential Impact of Climate Change on Agricultural and Forestry. Likely Impacts of Climate Change," Report from Working Group II.

IPCC: See Intergovernmental Panel on Climate Change.

Jordanov, D., 1924, "On the Phytogeography of the Western Balkan Range," Vol. 1, Reprint, Sofia University, Sofia, Bulgaria.

Karamfilov, V., 1995, "To Work for the Bulgarian Forests Is Our Supreme Obligation," *Forest* 4:1-2.

Kostov, P., I. Jelev, P. Beljakov, S. Angelov, K. Kaludin, K. Tashkov, and R. Nasylevska, 1976, *The State of the Forest Resources and Consideration about Their Complex Usage in Bulgaria*, Zemizdat.

Penev, N., M. Marinov, D. Garelkov, and Z. Naumov, 1969, *Die Waldtypen in Bulgarien*, pp. 10-24, Verlag der Bulgarischen Akademie der Wissenschaften, Sofia, Bulgaria.

Raev, I., 1989, *Investigations on the Hydrological Role of the Coniferous Forest Ecosystems in Bulgaria*, Forest Research Institute, Bulgarian Academy of Sciences, Sofia, Bulgaria.

Raev, I., 1995, "The Main Reasons for the Decline of Coniferous Forest Stands in Bulgaria," *Forest* 1:25-28.

U.S. Country Studies Program (CSP), 1994, *Guidance for Vulnerability and Adaptation Assessments*, U.S. Country Studies Management Team, Washington, D.C., USA.

Biodiversity and Wildlife: Adaptation to Climate Change

Adam Markham
World Wildlife Fund
1250 24th St. N.W.
Washington, DC, 20037-1175, USA

Jay Malcolm
Center for Conservation Biology
University of Florida
Gainesville, FL, 32611, USA

Abstract

Human-induced climate change is expected to have far-reaching implications for the conservation of biological diversity; under even the least dramatic global warming scenarios, ecological impacts are anticipated. Because of the certainty of ecological change during the next decades, plus the unpredictable responses of species to change and the consequent alteration of ecosystems, adaptive strategies must be developed. Some potential consequences of climate change on biodiversity are outlined. The study then identifies the most sensitive ecosystems and reviews potential strategies to minimize negative ecological impacts. Conclusions show that maintaining ecological complexity and ecosystem resilience are the most important factors in decreasing the impacts of climate change.

Introduction

Climate change is expected to severely affect ecosystems at all latitudes (Dobson et al. 1989). Alterations in such abiotic factors as carbon dioxide concentrations, rainfall quantity and distribution, temperature, seasonality, and extreme events will force changes in ecosystem structure and function and subsequently lead to changes in the distribution of flora and fauna.

Nevertheless, a combination of uncertainty in modeling outputs and gaps in the scientific understanding of ecological responses to climate perturbations and changes makes it difficult to conduct detailed assessments of potential impacts. Current Global Circulation Models (GCMs) for developing future climate change scenarios are limited in their use for predicting possible changes in habitat and wildlife distribution because they lack the high spatial resolution, ability to reproduce reliable major climate anomalies (such as El Niño/Southern Oscillation [ENSO] and tropical storms), and sophistication to confidently predict inter- and intra-annual rainfall distribution patterns and seasonality.

For these reasons, it is difficult to predict, with any degree of certainty, the impacts of climate change on any particular species or ecosystem. Strategies for adapting to climate change must be developed in the absence of precise information about either climate change or ecological responses. The aim of adaptive strategies should be reducing vulnerability to climate change, and such strategies should be developed within the context of global, regional, and national biodiversity conservation priorities. Reducing vulnerability to climate change is likely to yield additional benefits by simultaneously reducing ecosystem vulnerability to other environmental and anthropogenic stresses. Conversely, current conservation strategies aimed at maintaining ecological complexity and biodiversity should be regarded as important tools for reducing the threat of damage from climate change.

Impacts of Climate Change on Ecosystems

A number of attempts have been made during the last decade to assess the potential worldwide implications of climate change for biological diversity (e.g., Houghton et al. 1990; Leemans and Halpin 1992; Peters and Lovejoy 1992; Markham et al. 1993). These assessments have identified certain biomes and types of ecosystem that may be at greatest risk from climate change (Table 1).

While these reviews are useful for assessing potential global risks to biological diversity and can help in determining the focus of national-level vulnerability assessments, they provide little site-specific information. Although all ecosystems may be at risk in parts of their distribution and geographic ranges, researchers need much more information to be able to determine potential threats to a particular species, habitat, or site. Of particular importance is the level of human pressure already contributing to ecosystem stress or degradation in a particular place. For example, coral reefs, worldwide, are under a great deal of stress as a result of coastal development, marine pollution, tourism, and over-fishing, and they are already threatened in many parts of their range. Climate change is expected to have a much greater negative impact on these stressed reefs than on unaffected reefs (Smith and Buddemeier 1992). Although an ecosystem type may be at risk throughout most of its range, there may be significant areas in which climate impacts will have an opposite, positive

Table 1. Examples of Ecosystem Types Sensitive to Climatic Change

Ecosystem/Biome	Key Climate Sensitivities	Authors
Mangrove forest	Sea-level rise, storms	Ellison and Stoddart (1991)
Boreal forest	Temperature, fire regime, soil moisture	Shugart et al. (1992)
Tropical forest	Drought, seasonality, fire regime, hurricanes	Hartshorn (1992) Whitmore (1993) Bawa and Markham (1995)
Alpine/montane ecosystems	Temperature, precipitation	Halpin (in press) Beniston (1994)
Arctic ecosystems	Temperature	Alexander (1990) Chapin et al. (1992)
Coastal wetlands	Sea-level rise, storms	Stevenson et al. (1986)
Coral reefs	Sea-surface temperature, storms	Smith and Buddemeier (1992) Agardy (1994)
Island ecosystems	Sea-level rise, temperature, storms	Rose and Hurst (1991)

effect. Over the long term, for example, warmer waters may enable some coral species to colonize new sea areas and form new reefs (Smith and Buddemeier 1992).

In the fields of ecosystem management and biodiversity conservation, planning for adaptation to climate change will be challenging. Current scientific knowledge about the reactions of ecosystems and species to rapid change is not yet advanced enough to allow accurate predictions. Nevertheless, several general rules have been developed in the climate impacts literature (Peters and Darling 1985; Rose and Hurst 1991; Peters and Lovejoy 1992; Markham et al. 1993; World Wildlife Fund 1994; Halpin, in press), and they can be used to guide adaptation planning.

- Ecosystems do not move in response to climate change — species do. Therefore, climate change will disrupt or eliminate existing communities and species assemblages but will allow new ones to form.
- Species and ecosystems already under stress from environmental degradation and human pressure are likely to be the most vulnerable to the added stress of climate change.
- The faster the rate of climate change, the greater will be the risk, and the harder it will be for ecosystems to adapt.
- Some species will respond more favorably to climate change than others. Invasive species, generalists, widely distributed species, rapid re-

producers, and efficient dispersers are likely to gain competitive advantages.

- Species most likely to be negatively affected by climate change will be those at the edges of their geographic ranges or with restricted ranges. Species that occupy narrow and highly specialized niches, are rare, or have genetically impoverished populations may also be expected to be at risk.

Adaptive Strategies

The U.N. Framework Convention on Climate Change, which was signed by more than 150 nations, highlights the issue of adaptation in Article 2, the overarching objective of the treaty:

> The ultimate objective of this convention and any related legal instruments that the Conference of the Parties may adopt is to achieve, in accordance with the relevant provisions of the Convention, stabilization of greenhouse gas concentrations in the atmosphere at a level that would prevent dangerous anthropogenic interference with the climate system. Such a level should be achieved within a time-frame sufficient to allow ecosystems to *adapt naturally* to climate change [authors' emphasis], to ensure that food production is not threatened and to enable economic development to proceed in a sustainable manner.

This wording presents problems of definition, because it is not clear how to apply the term "adapt naturally" to an abstract entity such as an ecosystem. Each species and hence each system will react differently. The convention objective does, however, argue for a precautionary approach: one that would emphasize mitigation strategies rather than adaptation. All climate scenarios currently used by the Intergovernmental Panel on Climate Change indicate that concentrations of greenhouse gases will continue rising for the foreseeable future. Thus, the objective of the treaty is unlikely to be met before changes in ecosystems become apparent, and efforts should be made to develop strategies for adaptation to the inevitable climatic shifts.

Clearly, the choice of adaptive strategies for wildlife and ecosystems depends on four main factors: the success of mitigation strategies, the expected rate of climatic change, the predicted sensitivity of species or ecosystems, and the values placed on the affected systems. The last issue raises the question of different motivations for conservation in different countries. Developed nations tend to pay more attention to wildlife and habitat on the basis of aesthetic, recreational, and ethical values, while developing countries typically view conservation within the context of sustainable development and natural resource management. For this reason, adaptive strategies may differ markedly among nations.

A common basis for developing adaptive strategies aimed at preserving ecosystem processes and maintaining biodiversity is an understanding that conservation management must include the two crucial elements of long-term vision and management for change. Successfully integrating these elements into conservation strategies requires longer planning horizons and wider recognition, especially among policymakers, that ecosystems are dynamic. In the past, many conservation plans assumed that habitats and species communities were stable entities that could be protected through regulation, static management plans, and the delineation of protected areas boundaries. The recognition that ecosystems (1) are highly dynamic, often representing some transitional stage in a succession (albeit on a broad range of temporal scales), and (2) can rapidly change state under certain environmental conditions has led conservation planners to develop new paradigms. Chief among these paradigms has been a shift away from a species-based approach to a fundamental emphasis on ecosystem processes and complexity (Agardy 1994). Healthy ecosystems are able to maintain a full range of undiminished ecosystem processes and biological diversity.

The primary objective of mitigation and adaptive strategies to protect wildlife and habitats must be to maintain resilience in and among ecosystems. Arrow et al. (1995) define resilience, in the biological sense, as "a measure of the magnitude of disturbances that can be absorbed before a system centered on one locally stable equilibrium flips to another." In the context of human adaptation to climate change, Riebsame (1991) differentiated between resiliency, which he defined as "an ability to absorb shocks and then return to normal," and adaptation, which he defined as "systemic change in which social systems take on quite different forms to reduce risk." For natural ecosystems, it will be necessary to maximize resilience, allow for natural adaptation, and (in some cases) intervene, or assist, adaptation. These strategies range from strict in-situ conservation approaches to various ex-situ opportunities. Soule (1991) has identified eight categories of strategy (Table 2).

The highest importance should be placed on in-situ and inter-situ conservation strategies. These strategies are much more cost-effective than ex-situ conservation methods, and while off-site techniques can be valuable aids in maintaining species or germ plasm, there can be no substitute for conserving habitat and ecosystem processes in the field. Extractive reserves based on harvesting nontimber forest products, as exemplified by the rubber and Brazil nut economy of Acre in Brazil, have been heavily promoted (de Beer and McDermott 1989; Clay 1992), but little is known about the true ecological and economic sustainability of such enterprises (Hartshorn, 1995). These types of reserves may prove to be of great value as conservation tools, but their worth is not yet fully proven from experience in the field.

Table 2. Typology of Biodiversity Conservation Strategies

Strategy	Types of Activity
In situ	Protected areas
Inter situ	Conservation outside protected areas (i.e., habitat conservation, development restrictions)
Extractive reserves	Resource harvesting on a sustainable basis
Ecological restoration	Intensive management to restore degraded habitats and landscapes
Zooparks	Maintenance of artificial mixes of species under semi-natural conditions (e.g., at game farms)
Agroecosystems and agroforestry	High-management, production-oriented systems (e.g., plantation forests, forest gardens)
Living ex situ	Zoos, botanical gardens, aquaria
Suspended ex situ	Germ plasm storage (e.g., seed banks)

Source: after Soule (1991).

Strategies, Management, and Planning of Protected Areas

Planners and managers of protected areas must begin to take into account the long-term implications of climate change for their reserves. One modeling study (Leemans and Halpin 1992) that investigated potential vegetation zone changes in more than 2,600 nature reserves of more than 1,000 ha found that for four different GCM scenarios, shifts from one major ecoclimatic zone to another could occur in 41-58% of all reserves. These types of studies are necessarily crude and do not take into account such issues as microclimate, habitat heterogeneity, or topography, but they do demonstrate that climate change may be a significant factor in future reserve environments. Vulnerability assessments can provide valuable input to the development of a national protected areas review. Even without climate change, completing such a review is an essential step toward developing a network of reserves with clear priorities, a comprehensive plan of action, research objectives, and a monitoring system (Bridgewater 1992). Protected areas reviews assess the status and management of existing reserves and identify needs for areas to be added in the future. Where sufficient data exist on biodiversity and landscape features, a gap analysis can be undertaken to identify such needs (Kavanagh and Iacobelli 1995).

Without a protected areas review, it may be difficult to assign priority to adaptive strategies for use in a changing climate. Funds for reserve management are often strictly limited, and it will be necessary to determine where often-diminishing financial resources can be most effectively used in an increasingly stressed natural environment. Halpin (1995) notes that it is important to know the objective of adaptive protected-area management in a changing climate — is it to try and maintain the existing mix of species in a protected area or to allow for adaptation and evolutionary response of the represented ecosystems? As with any adaptive strategy (Figure 1), a first step is to define the objectives and goals before assessing and then recommending adaptive strategies (Carter et al. 1994).

As a rule, adaptive strategies to climate change for protected areas should promote resilience in protected areas systems and the habitats that they protect. Many rules that hold true for increasing the effectiveness of reserve networks, such as large reserves, well-managed buffer zones, and connectivity and redundancy among reserves, if implemented, will assist adaptation to climate change. Adaptive strategies should probably emphasize redundancy. Just as poorly enforced regulation or political instability can be arguments for an increased number of reserves and, therefore, greater redundancy (Soule 1991), so can the uncertainty of climate change impacts. Many authors have noted the importance of maintaining the evolutionary potential of species within protected areas (Grumbine 1994). This concept becomes of paramount importance in a changing climate, and because of the potential for climate-induced species movements and migration, it must be reinforced by redundancy and connectivity.

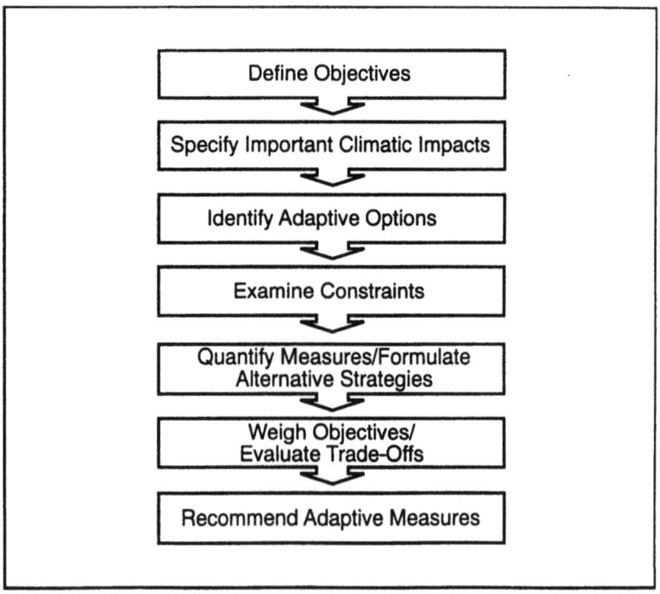

Figure 1. Adaptive Strategy (Source: Carter et al. 1994)

Reserve Planning and Choice

A number of adaptation strategies have been proposed for maintaining biological diversity and ecological integrity in reserve systems. Peters and Darling (1985) suggested that in locations where it is still possible to plan for new protected areas, it would be important to site reserves near the range limits (in the direction of the expected migration or dispersal under changed climates) of species targeted for protection. They also noted the importance of heterogeneous topographic and soil conditions within reserves, as well as the need to maximize the size and number of reserves. Increased reserve size, however, may not necessarily correspond with climate ecotones (Halpin, 1995), so it is important to balance area with landscape diversity and ecological complexity.

Altitudinal range within reserves is important because species may be able to migrate upslope to avoid the consequences of warming (Peters and Darling 1985; McNeely 1990). However, the invasion of flora and fauna from lower elevations to new, higher altitude zones may bring additional problems. In the northwestern United States, where subalpine forests have been moving upward into alpine meadow habitats (Rochefort et al. 1994), managers of protected areas may have to actively prevent meadow invasion. At Mount Rainier National Park, for example, approximately 60% of visitors come to see flower meadows (Rochefort and Peterson 1991). Altitudinal range also will not help Upper Montane species, which cannot migrate to a higher elevation, although this effect may be moderated by lateral movement around the mountain to cooler microclimates (Korner, personal communication).

For long-lived species (such as trees), elevational shifts are unlikely to be smooth progressions (Halpin, 1995). As conditions become unsuitable for trees at the edge of climatic zones, disturbance events (such as fires) will rapidly and permanently remove stands from formerly suitable habitat, and new plant communities will replaced the old. Such a scenario could hasten the loss of Montane habitat, such as that of whitebark pine in Yellowstone National Park. Whitebark pine, which provides significant food resources for such species as Clark's nutcrackers and grizzly bears, could lose 90% of its available habitat with only a 460-m upward altitudinal shift in ecoclimatic zones (Romme and Turner 1990).

Aside from the issues of size, heterogeneity of topography, and habitat and microclimate in reserves, other criteria should be examined when deciding on future reserves. Gap analysis systems used for identifying conservation needs based on representative ecosystem and enduring landscape features will have to become more dynamic (Halpin, 1995). Current gap-analysis projects generally focus on existing species and habitat distributions (Noss 1995). Reserve planning in a changing climate should also examine future needs in places where present-day conservation needs are few, but where migration and dispersal of valued species may make protection necessary or desirable. Climate change may cause new associations or assemblages of species that do not exist under current conditions, but they might qualify for protection as representative ecosystems (Markham et al. 1993). Gap analysis needs to allow for this possibility.

Flexible zoning of reserve boundaries and the development of more effective buffer-zone management have been proposed as adaptive strategies for protected areas (Peters and Darling 1985; Parsons 1991). The concept of being able to change the boundary of a park is an attractive one, but one that, as yet, has little precedent in the real world (Halpin, 1995). Buffer zones (zones that abut or surround protected areas and where land uses reduce pressure on the core reserve), however, are a common feature of nature reserve management. Buffer zones have generally failed to deliver the desired result, but if they can be made to work, they do offer some adaptive potential in a changing climate.

The biosphere reserve concept was developed by the U.N. Educational, Scientific, and Cultural Organization in 1979 to try and develop, worldwide, a network of sites in which strict management core areas would be surrounded by controlled-use buffer and transition zones. The success of this system has been limited and could be improved through (1) better planning (especially with regard to setting scientific management objectives), (2) greater input from social scientists, and (3) better integrated local community involvement (Dyer and Holland 1991).

Connectivity and Fragmentation

Connectivity between reserves in human-dominated landscapes is regarded as a key component of a well-planned protected areas network (Noss 1995). Connectivity is the opposite of fragmentation in that it joins landscapes, thereby benefiting wildlife and ecosystems. Noss and Cooperrider (1994) define the two major roles for landscape connectivity: (1) to provide dwelling habitat for plants and animals and (2) to provide a conduit for movement. They further subdivide the conduit role into three main categories: (1) daily and seasonal movements, (2) facilitation of dispersal and gene flow, and (3) allowance of long-distance range shifts.

In times of rapid climatic change, corridor systems may be of particular importance in that they allow the migration of species in response to biogeographic range changes (Peters and Darling 1985; Graham 1988; Hunter et al. 1988). Despite the clear need to increase connectivity to facilitate ecological adaptation to climate change, current scientific understanding of the utility of corridor systems is limited (Halpin, 1995). Corridor plans often target the dispersal of a particular species, and little attention is given to the overall impacts on ecosystems. For example, corridors may aid the movement of invasive and alien species or provide a channel for the spread of disease. Response to climate change will require more than just movement down corridors, because the biological range changes that will be occurring are expected be rapid, with permanent loss of previously suitable habitat. For this reason, corridors will need to function as habitat rather than mere transit lanes (Simberloff et al. 1992). Edge effects in narrow corridors may prevent their effective use by flora and fauna undergoing climatically forced range shifts.

Edge effects also have implications for the maintenance of biological diversity in fragmented ecosystems. For example, edge creation in tropical moist forests leads to pervasive habitat changes (Malcolm 1994) and, therefore, important implications for flora and fauna. Edges may also exacerbate the impacts of climate change in some cases. For example, heat-flow models suggest that fragmented landscapes are particularly susceptible to warming and drying trends (Malcolm, 1995).

Fragmentation may be the single biggest barrier to ecosystem adaptation in a changing climate. Even where paleoecological studies indicate that species may have been able to adapt to rapid climatic changes in the past, current habitat fragmentation patterns and human barriers to migration will prevent adaptation. Fragmentation may be the single most important anthropogenic factor increasing the vulnerability of tropical forests to such climate changes as increased drought and altered seasonality (Bawa and Markham 1995). Fragmentation probably affects other ecosystems similarly. Thus, reducing fragmentation rates is a critical climate-mitigation strategy, and increasing connectivity (although not necessarily by the creation of strict corridors) is a high-priority adaptation response.

Habitat Management and Intervention

Several habitat management and intervention techniques can become part of an adaptation strategy. Many of these techniques are already used in protected areas and managed reserves worldwide and these techniques can be adapted for use under a new set of climatic conditions (Table 3).

Although plans for managing and developing protected areas systems are essential elements of an adaptive strategy to climate change, many nations find this approach limiting. In most developing countries, the major cause of loss of biological diversity is (and is likely to continue to be for the foreseeable future) due to habitat destruction and degradation as a result of demographic and land-use changes. Human population growth is driving both the conversion of natural lands and the production of pollution (including greenhouse gases) (Vitousek 1994). Adaptive strategies in regions of high human pressure (such as southern Africa) will have to take into account interactions among human populations, ecosystems, and wildlife populations.

Table 3. Intervention Strategies

Strategy	Examples of Current Use	Potential Use in a Changing Climate
Ecosystem restoration	Restoration of water meadows and riparian habitat along the Rhine in the Netherlands	Restoration of degraded land to provide connectivity between existing reserves
Prescribed fire and fire exclusion	Use of fire to maintain age-suitability of jack pine habitat for Kirtland's warbler breeding in Michigan, USA	Prevent conversion of savannah to shrub-dominated communities
Species relocation	Removal of elephants from Zimbabwe to South Africa to relieve population pressure in parks	Remove species from newly unsuitable habitat and relocate them in new areas to which natural migration would be impossible
Removal of impediments to migration and colonization	Closure of logging roads in western USA	Prepare land for colonization of desired species (e.g., removal of scrub to allow forest recruitment) Remove dike or road systems preventing inland migration of coastal wetlands Remove fences or provide bridges/tunnels
Assisted migration or reintroduction	Reintroduction of wolves to Yellowstone National Park, USA	Capture and move animal species past obstacles to migration (e.g., agriculture, industrial developments) in direction of climatically forced population migration Plant seeds, move soil
Control of alien or invasive species	Eradication programs in native forest in Haleakala National Park, Hawaii, USA	Monitor for new invasive species and prevent their spread Minimize disturbance (i.e., maintain canopy opening in tropical forest) to reduce susceptibility to invasion
Control of disease	Tsetse-fly control programs in southern Africa	Monitor for changes in disease distribution and expansion of range of disease vectors Plan disease-control strategies
Irrigation or drainage	Creation of shallow, brackish coastal lakes at Minsmere Reserve (United Kingdom) to provide shorebird habitat	Use water-management technologies to reduce impacts of drought or sea-level rise Create new wetlands
Food and water provision	Provision of water in game areas during droughts in Botswana	Plan or expand programs aimed at ameliorating impacts of drought or famine

During the last century, Africa has experienced a warming trend similar in magnitude to the globally observed average. Two marked warm periods have been observed: the first peaking in the early 1940s and the second running through the 1980s (Hulme et al. 1995). The 25-year desiccation of the sub-Saharan Sahel zone represents the most severe long-term rainfall reduction in recorded history, with rainfall in some years dropping to less than 50% of levels received during the earlier decades of the century (Hulme and Kelly 1993). Shorter-term variation has been observed in recent years in southern Africa. The devastating failure of rains in southern Africa in 1991-1992 was associated with a prolonged ENSO event (Hulme et al. 1995). The overall frequency of drought has increased in countries such as Zimbabwe during the last decade (Muchena 1994), and the rainy season of 1994-1995 again exhibited a severe drop in rainfall throughout much of southern Africa. Some preliminary indications from climate modeling suggest that the duration of the dry season may increase significantly in the region (Hulme and Viner, in preparation).

The impacts of severe, long-term trends toward warming and drying in southern Africa fall most heavily on the human populations, but feedbacks that affect natural ecosystems and biodiversity are likely to be large. Crop model outputs for Zimbabwe indicate that even small increases in average temperature (2°C) are liable to decrease the productivity of maize, the primary food crop, by 8-14% (Muchena 1994). Similar changes in temperature would allow significant disease vectors, such as the tsetse fly, to spread into previously unsuitable areas (Rogers and Randolph 1993). Changes in fire regimes can also be expected (Goldammer 1992), as can alterations in river flow and lake levels (Magadza 1994). Offshore, changes could occur in coastal fisheries (Crawford et al. 1990).

All of these potential changes, and many more as yet unspecified, would strongly affect biodiversity. Increased rates of land conversion and degradation and greater human/wildlife conflict are among the chief effects. Hulme et al. (1995) propose four types of adaptation that could apply to the southern African case:

- Ad-hoc changes to accommodate the impacts of climate change.
- Reduction of vulnerability and increased resiliency.
- Purposeful adjustments to reliable predictions of change.
- Responses to crisis situations brought about by unpredictable changes or poor planning.

With regard to biological diversity, and given the current uncertainty about the predicted impacts in the region, the second option is attractive. Reducing vulnerability and increasing resilience will, however, be a substantial challenge. A first step will be to implement some of the recommendations on protected areas management and planning, but perhaps an equally important step will be to develop strategies for landscape-wide management. Any measures that can reduce current pressures on ecosystems and minimize the development and expansion of those pressures in the future will provide a conservation benefit in a changing climate. For this reason, it is useful to examine adaptive strategies that perform the dual role of benefiting both wildlife and human populations. Of particular interest is the trend toward developing multispecies animal-production systems in southern African rangelands. A recent review of the development of these systems reveals that 17% of the region is already under some type of multispecies management and suggests that 30-50% of southern Africa could eventually support this kind of rangeland production (Cumming, 1995; see Figure 2).

Game farming and other forms of wildlife use (such as tourism and trophy hunting) are attractive economic and land-use alternatives to single-species livestock systems. Wild ungulates, such as gazelle, zebra, and eland, are increasingly bred for meat in commercial ranching operations, often in combination with cattle. By adding value to wildlife previously perceived as either worthless or as a competitor to traditional livestock (such as cattle), multispecies systems can have biodiversity benefits even outside the managed lands. Cumming (in press) identifies eight multispecies systems:

- Game viewing, photographic safaris, and ecotours (nonconsumptive).
- Safari and guided trophy hunting of wild populations.
- Combinations of game viewing and guided trophy hunting.
- Recreational hunting (unguided rifle and bow hunting).

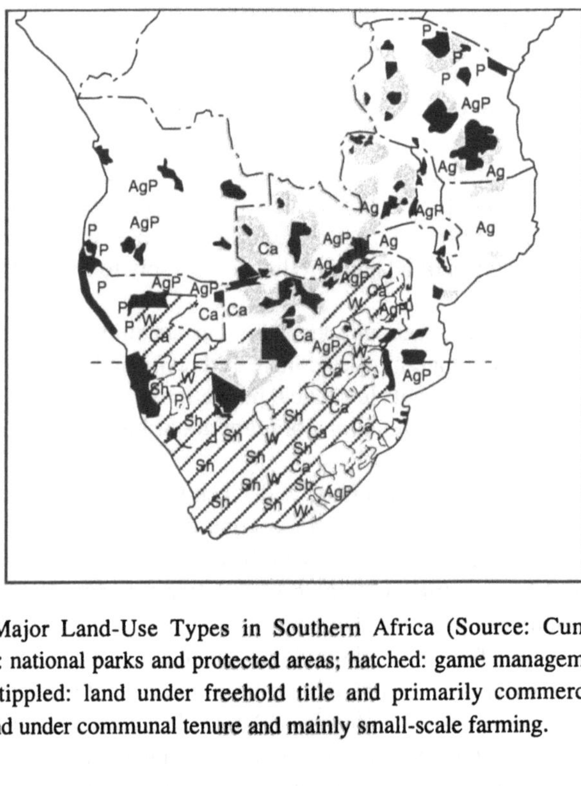

Figure 2. Major Land-Use Types in Southern Africa (Source: Cumming, 1995) Shaded areas: national parks and protected areas; hatched: game management or equivalent areas; stippled: land under freehold title and primarily commercial farmland; unshaded: land under communal tenure and mainly small-scale farming.

- Cropping wild populations for meat and other products (low- technology subsistence cropping to high-technology cropping for export).
- Cropping confined or fenced populations (game ranching).
- Combinations of sport hunting and cropping wild populations.
- Running livestock with wildlife under the above options. Livestock may be paddocked or herded.

The economics of multispecies systems is such that in commercial farming areas, they are generally able to compete against or supplement single-species systems. In communal areas, livestock provide additional benefits, including milk and manure.

Especially important is the availability of drought power from domestic livestock in the production of maize, a staple crop for many areas (Cumming, 1995). Climatic changes that increase the risk of drought and weather variability generally may lessen the benefits brought by domestic livestock. Areas prone to crop failures may depend on household purchasing power to retain food security (Cumming, 1995). Therefore, income from high-value wildlife uses could be important features of village community resilience. There is also limited evidence from the 1991-1992 drought in Zambia that many wild ungulate species, particularly browsers, were less vulnerable than domestic cattle (Markham et al. 1993; Middleton 1993).

The development of multispecies animal-production systems should be considered an important potential adaptive strategy for climate change. A trend has already started in this direction in southern Africa, but too few resources are being devoted to problems that need to be solved to increase acceptance and viability. Among these problems are disease management in mixed-animal systems, impacts of wildlife on plant-cropping systems, and the development of foreign markets for game meat. An additional issue is the distribution of income from multispecies systems (e.g., trophy hunting and ecotourism). To try and address this latter problem, a number of integrated conservation and development projects have been developed with mixed success in recent years. Under these schemes, social and economic benefits are obtained at the local level, and communities are involved in planning and control of the resource used.

Conclusions

Many approaches are available for maintaining biodiversity in a rapidly changing climate. To determine adaptive options and to choose and implement an effective strategy, conservation priorities must be determined. At a national level, strategies may be chosen through the development of a National Biodiversity Strategy (NBS). Nations that have ratified the U.N. Convention on Biological Diversity are required (under Article 6a) to "develop national strategies, plans or programs for the conservation and sustainable use of biological diversity . . . ," and for most nations, this is likely to take the form of an NBS (Glowka et al.

1994). The Convention on Biological Diversity also requires signatories to integrate the biodiversity strategy with other "relevant sectoral or cross-sectoral plans, programs and policies." Thus, priorities for climate adaptation strategies could flow from existing conservation strategies, or from a newly derived NBS.

After it has been determined what should be protected and why, the right mix of adaptive initiatives can be developed. These initiatives should also be placed in the context of other sectoral adaptive strategies (such as those for forest, water, or agricultural resources), so that the effect of the overall implementation plan is both beneficial and additive. Where biodiversity is concerned, several basic rules are proposed:

- Adaptive strategies should maintain ecological structure and processes at all levels.
- Maintenance of ecological complexity is necessary to maximize evolutionary and behavioral adaptability in species and ecosystems.
- Ecological resilience is the single most important factor influencing the ability of wildlife and natural habitats to respond to climatic change. Resiliency can be increased by conserving biological diversity, reducing fragmentation and degradation of habitat, increasing functional connectivity among habitat blocks and fragments, and reducing anthropogenic environmental stresses.
- In-situ conservation strategies are of primary importance in climate adaptation strategies, but they should be supported by a range of cost-effective, practical ex-situ methods.

By using these rules for guidance, a suite of adaptive activities can be developed. Priority should be given to planning and adaptive management of protected-areas systems, ecosystem management and environmentally sustainable development outside protected areas, and habitat restoration on degraded lands.

The potential negative impacts of climate change on biological diversity are such that although precise effects at particular sites are difficult to predict, precautionary approaches should be pursued. Climate change is an additional stress that will be added to existing threats to wildlife and habitat, and may even synergistically hasten environmental degradation and species extinctions. The most effective strategies will be those that increase resilience to all types of environmental stress and therefore maximize an ecosystem's natural ability to adapt to change. In other words, investments in conservation today may well be the best defense against climate change tomorrow.

As potential impacts become better understood, or changes begin to be manifested and identified at the ecosystem level, targeted adaptive strategies can be adopted. These strategies will most likely require increased investment in the direct management of ecosystems inside and outside of protected areas, especially to maintain biological diversity in human-dominated landscapes and to reduce human-wildlife conflicts.

References

Agardy, M.T., 1994, "Advances in Marine Conservation: The Role of Marine Protected Areas," *Trends in Ecology and Evolution* 9(7).

Alexander, V., 1990, "Impacts of Global Change on Arctic Marine Ecosystems," in J. McCullock (ed.), *Global Change: Proceedings of the Symposium on the Arctic and Global Change*, The Climate Institute, Washington, D.C., USA.

Arrow, K., B. Bolin, R. Costanza, P. Dasgupta, C. Folke, C.S. Holling, B.-O.Jansson, S. Levin, K.-G. Maler, C. Perrings, and D. Pimental, 1995, "Economic Growth, Carrying Capacity, and the Environment," *Science* 268:520-521.

Bawa, K., and A. Markham, 1995, "Climate Change and Tropical Forests," *TREE* 10(9):348-349.

de Beer, J., and M.J. McDermott, 1989, *The Economic Value of Non-Timber Forest Products in Southeast Asia*, Netherlands Committee for IUCN, Amsterdam, the Netherlands.

Beniston, M. (ed.), 1994, *Mountain Environments in Changing Climates*, Routledge, London, United Kingdom.

Bridgewater, P., 1992, *Strengthening Protected Areas*, in Global Biodiversity Strategy, World Resources Institute, World Conservation Union, and U.N. Environment Programme, Nairobi, Kenya.

Carter, T.R., M.L. Parry, H. Harasawa, and S. Nishioka, 1994, *IPCC Technical Guidelines for Assessing Climate Change Impacts and Adaptations*, University College, London, United Kingdom.

Chapin, III, F.S., R.L. Jefferies, J.F. Reynolds, G.R. Shaver, and J. Svoboda, 1992, *Arctic Ecosystems in a Changing Climate: An Ecophysiological Perspective*, Academic Press, San Diego, Calif., USA.

Clay, J., 1992, "Some General Principles and Strategies for Developing Markets in North America and Europe for Non-timber Forest Products: Lessons from Cultural Survival Enterprises, 1989-1990," *Adv. Econ. Bot.*9:101-106.

Crawford, R.J.M., W.R. Siegfried, L.V. Shannon, C.A. Villacastin-Herrero, and L.G.Underhill, 1990, "Environmental Influences on Marine Biota Off Southern Africa," *South African Journal of Science* 86:330-339.

Cumming, D.H.M. (1995), "Multispecies Systems a Viable Landuse Option for Southern African Rangelands?," in *Proceedings of International Symposium on Wild and Domestic Ruminants in Extensive Landuse Systems*, October 3-4, 1994, Berlin, Germany.

Dobson, A., A. Jolly, and D. Rubenstein, 1989, "The Greenhouse Effect and Biological Diversity," *TREE* 4(3):64-67.

Dyer, M.I., and M.M. Holland, 1991, "The Biosphere-Reserve Concept: Needs for a Network Design," *BioScience* 42(11):319-325.

Ellison, J.C., and D.R. Stoddart, 1991, "Mangrove Ecosystem Collapse during Predicted Sea-Level Rise: Holocene Analogues and Implications," *Journal of Coastal Research* 7(1):151-165.

Glowka, L., F. Burhenne-Guilmin, H. Synge, J.A. McNeely, and L. Gundling, 1994, *A Guide to the Convention on Biological Diversity*, Environmental Policy and Law Paper No. 30, IUCN, Gland, Switzerland.

Goldammer, J.G. (ed.), 1992, *Tropical Forests in Transition: Ecology of Natural and Anthropogenic Disturbance Processes*, Birkhauser, Basel, Switzerland.

Graham, R.W., 1988, "The Role of Climatic Change in the Design of Biological Reserves: The Paleoecological Perspective for Conservation Biology," *Conservation Biology* 2(4):391-394.

Grumbine, R.E., 1994, "What is Ecosystem Management?," *Conservation Biology* 8(1):27-38.

Halpin, P.N. (1995), *Global Change and Natural Area Protection: Management Responses and Research Directions.*

Hartshorn, G.S. (1995), "Ecological Basis for Sustainable Development in Tropical Forests," *Annual Review of Ecology and Systematics.*

Hartshorn, G.S., 1992, "Possible Effects of Global Warming on the Biological Diversity in Tropical Forests," in R.L. Peters and T.E. Lovejoy (eds.), *Global Warming and Biological Diversity*, pp. 137-46, Yale University Press, New Haven, Conn., USA.

Houghton, J.T., G.J. Jenkins, and J.J. Ephraums, 1990, *Climate Change: The IPCC Scientific Assessment*, Cambridge University Press, Cambridge, United Kingdom.

Hulme, M., and M. Kelly, 1993, "Exploring the Links between Desertification and Climate Change," *Environment* 35(6).

Hulme, M., D. Conway, P.M. Kelly, S. Subak, and T.E. Downing, 1995, *The Impacts of Climate Change on Africa*, Stockholm Environment Institute, Sweden.

Hunter, M.L., G.L. Jacobson, Jr., and T. Webb III, 1988, "Paleoecology and the Coarse-Filter Approach to Maintaining Biological Diversity," *Conservation Biology* 2:375-385.

Kavanagh, K., and T. Iacobelli, 1995, *Protected Areas Gap Analysis Methodology*, World Wildlife Fund, Toronto, Ontario, Canada.

Korner, personal communication to A. Markham (World Wildlife Fund, Washington, D.C., USA.

Leemans, R., and P.N. Halpin, 1992, "Biodiversity and Global Climate Change," in B. Groombridge (ed.), *Global Biodiversity: Status of the Earth's Living Resources*, Chapman & Hall, London, United Kingdom.

Magadza, C.H.D., 1994, "Climate Change: Some Likely Multiple Impacts in Southern Africa," *Food Policy* 19:165-191.

Malcolm, J.R. (in preparation), *A Model of Conductive Heat Flow in Forest Edges and Fragmented Landscapes.*

Malcolm, J.R., 1994, "Edge Effects in Central Amazonian Forest Fragments," *Ecology* 75(8):2438-2445.

Markham, A., N. Dudley, and S. Stolten, 1993, *Some Like it Hot: Climate Change, Biodiversity and the Survival of Species*, World Wildlife Fund International, Gland, Switzerland.

McNeely, J.A., 1990, "Climate Change and Biological Diversity: Policy Implications," in M.M. de Boer and R.S. de Groot (eds.), *Landscape Ecological Impacts of Climate Change*, pp. 406-428, IOS Press, Amsterdam, the Netherlands.

Middleton, T., 1993, *Drought: Wildlife v. Cattle*, Kobus, Lusaka, March.

Muchena, P., 1994, "Implications of Climate Change for Maize Yields in Zimbabwe," in C. Rosenzweig and A. Iglesias (eds.), *Implications of Climate Change for International Agriculture: Crop Modelling Study*, U.S. Environmental Protection Agency, Washington, D.C., USA.

Noss, R.F., 1995, *Maintaining Ecological Integrity in Representative Reserve Networks*, World Wildlife Fund, Toronto, Ontario, Canada.

Noss, R.F., and A.Y. Cooperrider, 1994, *Saving Nature's Legacy: Protecting and Restoring Biodiversity*, Island Press, Washington, D.C., USA.

Parsons, D.J., 1991, "Planning for Climate Change in National Parks and Other Natural Areas," *The Northwest Environmental Journal* 7:255-269.
Peters, R.L., and J.D.S. Darling, 1985, "The Greenhouse Effect and Nature Reserves," *BioScience* 35(1):707-716.
Peters, R.L., and T.E. Lovejoy (eds.), 1992, *Global Warming and Biological Diversity*, Yale University Press, New Haven, Conn., USA.
Riebsame, W.E., 1991, "Sustainability of the Great Plains in an Uncertain Climate," *Great Plains Research* 1(1):133-151.
Rochefort, R.M., and D.L. Peterson, 1991, "Tree Establishment in Subalpine Meadows of Mount Rainier National Park," N*orthwest Environmental Journal* 7(2):354-355.
Rochefort, R.M., R.L. Little, A. Woodward, and D.L. Peterson, 1994, "Changes in Sub-Alpine Tree Distribution in Western North America: A Review of Climatic and Other Causal Factors," *The Holocene* 4(1):89-100.
Rogers, D.J., and S.E. Randolph, 1993, "Distribution of Tsetse and Ticks in Africa: Past, Present and Future," *Parasitology Today* 9(7):266-271.
Romme, W.H., and M.G. Turner, 1990, "Implications of Global Climate Change for Biogeographic Patterns in the Greater Yellowstone System," *Conservation Biology* 5(3):373-386.
Rose, C., and P. Hurst, 1991, *Can Nature Survive Global Warming?*, World Wildlife Fund International, Gland, Switzerland.
Shugart, H.H., R. Leemans, and G.B. Bonan, 1992, *A Systems Analysis of the Global Boreal Forest*, Cambridge University Press, Cambridge, United Kingdom.
Simberloff, D., J.A. Farr, J. Cox, and D.W. Mehlman, 1992, "Movement Corridors: Conservation Bargains or Poor Investments?," *Conservation Biology* 6(4):493-504.
Smith, S.V., and R.W. Buddemeier, 1992, "Global Change and Coral Reef Ecosystems," *Annual Review of Ecological Systems* 23:89-118.
Soule, M., 1991, "Conservation: Tactics for a Constant Crisis," *Science* 253:744-750.
Stevenson, J.C., L.G. Ward, and M.S. Kearney, 1986, "Vertical Accretion in Marshes with Varying Rates of Sea Level Rise," in D.A. Wolfe (ed.), *Estuarine Variability*, Academic Press, New York, N.Y., USA.
Vitousek, P.M., 1994, "Beyond Global Warming: Ecology and Global Change," *Ecology* 75(7):1861-1876.
Whitmore, T.C., 1993, *Will Tropical Forests Change in a Global Greenhouse?*, unpublished presentation to U.N. University Global Environmental Forum III, Tokyo, Japan, Jan. 18.
World Wildlife Fund, 1994, *Critical Levels and Ecological Limits to Climate Change: Outputs from a WWF Workshop in Support of the IPCC Process*, white paper contributed to the IPCC Special Workshop on Article 2 of the UNFCCC, Fortaleza, Brazil, Oct. 17-21, World Wildlife Fund, Washington, D.C., USA.

6

FISHERIES

Rapporteur's Statement

Ian Burton
Atmospheric Environment Service, Environment Canada
and
Institute for Environmental Studies
University of Toronto, 33 Willcocks Street
Toronto, Ontario, Canada, M5S 3B3

The impact of climate change on fisheries is not well understood. Important issues discussed in this chapter are discussed in the following paragraphs.

Suitability of Assessment Methods for Fisheries. Forecasting by Analogy (FBA) is a case study method in which specific climate events or conditions are examined empirically to gain insight into impacts and responses to post-climate change. This method systematically captures lessons learned from experience. The main strength of this method is that it identifies adaptive responses to climate extremes actually used by society. The principal weakness of the FBA is that it does not provide quantitative estimates of how society might adapt to climate change.

Suitability of Adaptive Measures. Micronesia is a good example of how an island nation copes with the effects of climate change on fisheries. Commercial fishing within the 200-mile exclusive economic zone (EEZ) has expanded considerably under licenses issued by the Government of Micronesia. The catch is difficult to monitor, and such monitoring primarily relies on self-reporting by fishing companies. The fish stocks are under increasing pressure, constituting a major threat to Micronesia, because much of its foreign exchange comes from selling licenses. Modernization of the artisanal fishing industry also places the immediate coastal fisheries under stress.

Even without climate change, the fishing industry in Micronesia is becoming increasingly vulnerable. There is an immediate need for improved conservation, research, monitoring, and management in both pelagic and artisanal fisheries. Because of their vulnerability, even small changes in climate, when added to existing conditions, can have catastrophic effects.

The boundaries of the Micronesian EEZ are determined by some small islands and atolls on the edge of the territory. If any of these islands or atolls were to disappear as a result of sea-level rise, the legal basis of the existing boundaries of the Micronesian EEZ might be questioned. Although the existing boundaries could be frozen, the international community needs to examine this aspect as soon as possible.

The vulnerability of fisheries to the incremental effects of climate change shown in Micronesia also applies worldwide to all fisheries. Optimists say that unless the world fishing industry can be managed on a more sustainable basis, climate change could have a severe effect. Pessimists say that climate change does not matter because by the time it begins to have an effect, no commercially viable fish will be left.

Forecasting by Analogy: Local Responses to Global Climate Change

M.H. Glantz
Environmental and Societal Impacts Group
National Center for Atmospheric Research
P.O. Box 3000
Boulder, Colorado 80307-3000, USA

Abstract

Limitations associated with General Circulation Models, particularly in societal impact assessment and local and regional climate change policies, have led some researchers to seek alternative methods for assessing the potential impacts of climate change. One of these methods, forecasting by analogy (FBA), is based on the premise that past responses to extreme climate-related events provide a first approximation of how society might respond to the effects of climate change. Since its development and initial application in 1987, FBA has been used to assess climate-related impacts in various economic sectors and ecosystems. The potential value and use of FBA in impact assessment, the use of analogs in the physical sciences, and the shortcomings and potential problems associated with the approach are discussed. Some cases in which the FBA approach was applied to marine fisheries to assess the impacts of climate change and climate variability are studied.

Introduction

Over the past several years, a great deal of effort has been invested in determining how climate change might affect managed and unmanaged ecosystems and societies. During this period, General Circulation Models (GCMs) of the atmosphere have been commonly used to predict the impacts of climate changes decades into the future. Because of various limitations associated with GCMs, particularly in assessing societal impacts and local and regional climate change policies, some researchers are using alternative approaches to climate change impact assessment. For example, many economists (and other modelers who examine the potential impacts of climate change on particular sectors or economies) design a set of scenarios that may or may not be based on GCM scenarios; these provide insights regarding the sensitivity of their sectors to climate change and allow them to examine potential societal responses to climate change impacts (e.g., Cohan et al. 1994).

The focus here is on an approach called forecasting by analogy (FBA) that is designed to evaluate societal responses to recent, extreme climate-related events, identifying both the strengths and the weaknesses of the responses. The approach is based on the premise that past responses to extreme climate-related events provide a first approximation of how society might respond to the effects of climate change. Since its development and initial application in 1987, FBA has been used to assess climate-related impacts in various economic sectors and ecosystems in Vietnam (Ninh et al. 1992), Brazil (Magalhães and Glantz 1992), Russia (Zonn 1994), the U.S. (Glantz 1988), and at 15 fisheries around the globe (Glantz 1992a).

When the FBA approach was first proposed, some physical scientists challenged the idea of using analogies to predict possible or (to use a word favored by climate modelers) plausible responses by society to climate change. They argued that analogies are too qualitative or too anecdotal for use in scientific research and, therefore, in policymaking. Yet, a close scrutiny of scientific research related to the global warming issue revealed that physical scientists, too, are very dependent on analogs and analogical reasoning in their studies (Glantz 1991).

Climate change is not just something that will happen 50 or 80 years from now. Climate changes — in the form of short-term variability or decadal-scale fluctuations — are occurring now. In fact, climate is constantly changing on a variety of time scales. And whether the global climate becomes warmer or cooler or remains as it has been for the last several decades, variability will surround that trend line. The social sciences need baseline data about societal responses to climate-related change, just as physical scientists need a lengthy time series of information on which to base their analyses. Forecasting by analogy can bridge the gap between climate variability and climate-change issues; by understanding the ways in which societies deal with the impacts of today's extreme meteorological events, researchers can help them prepare to meet the challenges presented by future climate changes. Whether or not the average global temperature increases, societies will have learned about and improved upon the way they address current, extreme climate-related events — in essence a win-win situation for society. In the context of FBA, a win-win situation refers to the educational aspects of the approach.

This study is based on information obtained from a set of critical reviews[1] of the books in which the FBA approach was used (Glantz 1988; 1992a). The following sections describe the potential value and use of FBA in impact assessment, the use of analogs in the physical sciences, and the shortcomings and potential problems associated with the approach. The final section provides some case studies in which the FBA approach was applied to marine fisheries to assess the impacts of climate change and climate variability.

FBA Approach

The FBA approach is based on the premise that, to know how well society might prepare itself for the consequences of a future change in climate (the characteristics of which are still unknown), today's methods of how society copes with the extremes of climate variability and its societal and environmental impacts must be identified.

There has been considerable speculation about how the warming of the global atmosphere by several degrees Celsius will affect regional climate and human activities that depend on the climate. The basis for this speculation is the output of atmospheric GCMs, specifically sensitivity studies associated with a hypothesized doubling of atmospheric carbon dioxide (CO_2). Speculation about the ecological impacts of increased CO_2 and other greenhouse gas emissions has also been based on historical climate analogs, such as the Climate Optimum that occurred earlier in this millennium. Other climate analogs include the Altithermal of 6,000-8,000 years ago (Kellogg 1977) and epochs tens of thousands of years ago when the earth's atmosphere was up to a few degrees warmer than it is today (Flohn 1981; Budyko and Sedunov 1990).

Because GCMs can provide insights into how changes in some climate-related parameters might affect atmospheric, oceanic, and ecological processes, they are very important and useful heuristic devices. Modelers encounter limitations, however, when they attempt to use GCMs to identify

---

[1] The reviews used in this reassessment of the FBA approach came from *The Quarterly Review of Biology* (Houde 1993), *The Bloomsbury Review* (Miller 1989), *BioScience* (Ray 1993), *Environment* (Wright 1991; Muckleston 1993), *Environmental Science & Technology* (1992), *Nature* (Shepherd 1992), *State Legislatures* (Runyon 1990), *Journal of Regional Science* (1990), *Climatic Change* (Cortner 1990), *Oceanus* (Broadus 1989), *Recent Publications on Government Problems* (1989), *Bulletin of the American Meteorological Society* (Pittock 1989), *Future Surveys* (1989), *The Environmental Professional* (Templer 1989), *Population and Environment* (Abernethy 1990), and *Global Environmental Change* (Crawford 1994).

appropriate societal responses to climate-change-related impacts on the environment in general and on human activities specifically.

The output of GCMs is more consistent with some regional parameters, such as temperature changes associated with increased concentrations of greenhouse gases in the atmosphere, than with how those global temperature changes might affect regional precipitation patterns. Although regional temperature outputs from the various GCMs do not necessarily agree with one another, they do fall within a well-defined range. However, there is considerable disagreement about how global warming might translate into precipitation changes at the regional and local levels. This scientific disagreement has not hindered speculation within the scientific community about regional and local climate changes and their possible socioeconomic impacts.

While climate change scenarios generated by GCMs have considerable scientific credibility within the nonscientific community, which includes policymakers, the potential changes in global and regional climate several decades into the future that are now predicted by the scientific community are beyond the scope of experience of policymakers. Such suggested changes appear to lack reliability; this leads to a real problem for policymakers, who are faced with uncertainty regarding the local impacts of global warming, and further uncertainty within the scientific community regarding the accuracy of their predictions. As a result, national and international policymakers are put in the position of either accepting the predicted changes without question as the products of an objective scientific community that lays all its information on the table (including questions of uncertainty and perhaps even dissenting views), or disregarding them as the speculative theories of laboratory scientists who have chosen to speak out for whatever reason.

Scenarios about the world of the future derived from events that have occurred within human experience — analogies — have a degree of political and social credibility that computer-generated scenarios lack. If an extreme meteorological event occurred once, it is likely (and a statistical probability) that it could happen again. Policymakers or, more broadly speaking, decision makers (such as farmers or water resources managers) who have been directly involved in solving problems caused by climatic anomalies in the recent past use that experience as a guide to address current issues. Such experience is passed on to prospective decision makers through education, just as the experiences of the U.S. Great Plains drought in the 1930s have been passed from one North American generation to the next. More recently, the U.S. government's Federal Emergency Management Agency's (FEMA's) poor response to the devastation in south Florida caused by Hurricane Andrew in 1992 prompted FEMA managers to respond much more quickly and effectively during the disasters related to the unexpected flooding of the upper Mississippi River system in the summer of 1993 (Myers and White 1993).

When compared to computer models, experience-based scenarios seem to lack scientific credibility. They are often challenged as just site-specific "case studies" and discounted with such statements as, "the past is no guide to the future." Some researchers contend that, with a changing climate, past climatic features will not represent future changes. Those who apply the FBA approach believe that, while the climate of the future may not be like that of the present or the recent past, societal responses to extreme meteorological events in the *near future* will most likely be similar to those of the *recent past* and the present (barring revolutionary political change). The purpose of studying the recent past, then, is to determine how flexible (or rigid) societies are or have been in dealing with climate-related environmental changes.

Societies everywhere have already shown the tendency to prepare for future extreme events based on their responses to a previous extreme climate event that affected them. However, such disasters seldom recur exactly in the same place, in the same socioeconomic setting, with the same intensity, or with the same socioeconomic impacts. Nevertheless, reviews of societal responses in different times and places can provide valuable information for shaping societal behavior in response to other disasters. Analogs cannot predict the future — no forecasting system has yet been invented that can do that — but they can help to identify societal strengths and weaknesses in responding to past extreme meteorological events so that the strengths can be reinforced and the weaknesses reduced.

Forecasting societal responses to events in the near future by analogy can be a fruitful approach for improving understanding of how well a society might be able to cope with the unknown regional effects of a potential climate change some decades in the future. Decisions made today must reflect the need to maintain as much flexibility as possible in the face of future unknowns. The FBA approach can expand the range of possible responses to a change in climate; scenario-driven methods tend to narrow the range of possible responses.

Each approach, GCM, limited area modeling, and FBA, has its strengths and its shortcomings in producing climate change scenarios. It is important that all approaches be considered to maintain an objective view in discussions of future climate change. Scenarios produced by any single approach, considered alone, can be misleading. In fact, focusing on any one future scenario would most likely lead to policy decisions that are very different from those that would result from focusing on different (but equally plausible) scenarios. Researchers need to find a way to combine the strengths of all of the approaches.

The FBA approach relies on case studies. Although each case may be unique to a region or to a specific human activity, taken together they can provide general insights into human responses to changes in local and regional climate. While only a handful of such studies have been undertaken, researchers have called upon people in other regions, in other disciplines, and in various economic sectors (each with different extreme meteorological events to contend with) to undertake individual "regional" FBA assessments. FBA is a low-technology approach for glimpsing the future: to get started, all that is needed are a pencil and knowledge of some local history.

Use and Potential Value of Forecasting by Analogy

The FBA approach offers potentially valuable lessons in efforts to reduce societal vulnerability to extreme meteorological events. The basic approach is a search through historical experiences to identify appropriate lessons from societal responses to slow onset (i.e., creeping) and rapid-onset climate-related environmental change. Examples of creeping environmental problems include deforestation, desertification, ozone depletion, and global warming. Rapid-onset problems include tropical storms, tornadoes, and t sunamis.

As a method for impact assessment, FBA enables researchers to explore the impacts of recent natural and man-made environmental changes and societal responses, and use them as a guide to understanding changes that scientists predict may occur in the future, such as those associated with an increase in climate variability and the aspects of that variability that cause the greatest concern to society (i.e., extreme climate-related events).

Those who use FBA (which is only one of the methods used to assess climate impacts on society) hope that the data collected for climate change impact assessments can later be used in more formal socioeconomic modeling (whether simple or complex) efforts. An attractive aspect of FBA is that it identifies information from the data collected before any formal modeling exercise is initiated. This information may be useful for decision makers. In other words, in addition to glimpsing the future, FBA data can provide some first approximations about societal vulnerabilities and resiliencies, while the modeling and, more generally, the research communities attempt to reduce the uncertainties that pervade the science of climate change. Policymakers will probably continue to look to climate modelers to provide "answers" about future climate change. Nevertheless, FBA studies can provide decision makers with complementary information about societal responses to and perceptions about climate-related environmental change.

The results of FBA assessments can also help social science researchers to identify the learning-bias factor in society and determine whether societies can buffer their activities and well-being from the vagaries of climate variability and change. It is not difficult to show that, to date, societies around the world have pursued resource management practices that have essentially failed to avert major biological, ecological, economic, or social disruptions.

Several reviewers of FBA publications have pointed out benefits of the approach. Comments on the value and problems in the use of FBA taken from their reviews are used in this study. The following specific comments have been drawn from published reviews of the FBA approach, as discussed in two edited volumes: *Societal Responses to Regional Climate Change* (Glantz 1988) and *Climate Variability, Climate Change and Fisheries* (Glantz 1992a):

- Forecasting by analogy can provide a sobering guide to the political, social, and economic problems likely to occur during the coming decades as a result of climate variability, if not of climate change (Wright 1991).
- While FBA may offer few solutions, it can provide much information of general historical interest (Shepard 1992).
- FBA provides a reasoned foundation for anticipating the impacts of global climate change, as well as improving understanding of how society copes with climate variability today (Cortner 1990).
- Analogs . . . can help to identify society's strengths and weaknesses in dealing with climate events (Cortner 1990).
- Society is likely to look to familiar and proven methods of dealing with climatological events (Cortner 1990).
- FBA can help to identify how societies in the past have dealt with climate-related environmental change, regardless of cause (Cortner 1990).
- FBA identifies the importance of involving state and local government in the climate problem-solving process (Cortner 1990).
- FBA gives prominence to questions of intergenerational equity (Cortner 1990).
- FBA gives prominence to the importance of coalition-building for increasing awareness, as well as coping with climate-related problems (Cortner 1990).
- FBA identifies the problems with the tendency to muddle through and favor ad hoc responses over longer-term planned responses (Broadus 1989).
- Analogs identify the need for a specific catalyst to prompt policy action (Cortner 1990).
- FBA can identify potential societal responses to a possible climate change by using actual, prolonged, outlying climate events (*Sage Public Administration Abstracts 1989).*
- To know how well society might prepare itself for a future change in climate, scientists must identify how well society today deals with climate variability and its societal and environmental impacts.
- Given that scientists suggest that variability and extreme events, such as floods and droughts, will likely increase, FBA can help to identify and evaluate the effectiveness of societal responses to extreme meteorological events. It can also assess the extent to which

these actions and responses could be used to cope with extreme events in the future (Pittock 1989).

FBA as an Educational Tool

FBA should be of interest to environmental policymakers and administrators, planners, lawyers, engineers and scientists, environmental educators, multidisciplinary research groups, philosophers, water management experts, city and regional planners, and the public in industrialized and developing countries. The research findings for specific case studies of societal coping mechanisms at the regional and local levels can invoke healthy discussion on the following climate-related issues: how policy and emergency response decisions are made, the importance of climate anomalies and climate variability to human activities and social well-being, and the need to develop an alternative to the winner/loser mentality that pervades the debates surrounding various aspects of the global warming issue.

The FBA method and its findings are easy for nonexperts to understand, making the results of impact assessments more accessible to a wide range of interests related to the climate-related environmental change issue. Although the results of FBA studies may provide only a first glimpse of societal coping mechanisms, they will increase awareness of the extent to which human activities are entwined with climate variability and extremes, and with a longer-term change in average climatic conditions.

Decision Making

Societies tend to react to and prepare for the events on the basis of the most recent environmental stress, but no two extreme events hit the same way and with the same intensity. For this reason, policymakers must make decisions that are not rigid — that is, decisions that might reduce response options for future, yet-to-be-determined, climate-related hazards.

In many cases, societies know the sources of their climate-related environmental vulnerabilities, but for one reason or another, seldom deal with them directly. For example, the cause of global warming is known. It is also known who is responsible for emitting relatively large amounts of greenhouse gases into the atmosphere. A clearer picture of the deleterious effects caused by such emissions is emerging. But, society (especially policymakers) is reluctant to act, except to convene international conferences on the subject.

FBA results can provide some guidance to policymakers about how society should (or should not) prepare for the possible consequences of global warming. FBA enables policymakers to prepare for today and the not-too-distant future, which they *can* do. Many researchers believe that policymakers are incapable of developing a long-term response to a problem; they respond only on a short-term, incremental (ad hoc) basis. These researchers also believe that policymakers have only a short-term interest in any particular catastrophic, climate-related event. As time passes following the event, the political and public interest in it dissipate. Still, it is in the interest of policymakers to develop long-term contingency plans to address both societal vulnerability to climate variability and extreme events, and judicious (i.e., sustainable) use of limited and, in many cases, dwindling natural resources.

Knowledge gained from examining the success or failure of past response measures can be used to identify and assess future policy options. Policymakers must be aware of (and must counter) their collective tendency to reduce the importance of scientific research once they have made decisions to address some aspects of the climate-related scientific issue. Policymakers tend to regard an issue as resolved once some aspect of a particular local or regional climate-related problem has been addressed; but the interest must be maintained for the sake of intergenerational equity. While our generation might buy some time with ad hoc decisions, future generations will likely face the consequences of climate change if the issue is not addressed directly.

Local Impacts, Local Responses

By assessing how well societies in the past have dealt with climate-related environmental changes, regardless of cause, researchers and policymakers can obtain baseline data on climate/society interactions.

FBA provides one of the few ways of getting a glimpse of *local* responses to climate-related environmental changes. Deciding, with some degree of scientific confidence, whether the global climate has changed may take decades. However, if it has changed over that period, local communities around the globe will have already unknowingly been responding to the consequences of the slow, imperceptible (i.e., creeping) climate-related environmental changes without really knowing the causes of the changes or the impacts to their regional climate. Most climate changes will be felt at the regional and local levels, but the resolution of GCMs at these levels is poor.

An important component of societal baseline data is the set of case studies of specific weather-related impacts and responses by state and local policymakers to societal and environmental needs. Such information can provide insights into how states (or other local administrative units) can respond to global warming. At present, there is little *reliable* information on climate-related environmental change for policymakers to use at the regional or local levels, although the natural hazards literature is replete with weather-related case studies.

Involvement by state and local levels of government is extremely important in undertaking FBA impact assessments because these are the levels at which societies will most likely respond, at least initially. Local responsibility is important because local decision making and involvement in the development of response tactics and strategies are generally preferred.

Forecasting by analogy can establish a framework from which to identify the principal problems and issues facing societies in responding to extreme meteorological events, even if the climate does not change. After all, humans have to live with climate variability, whether or not that variability is induced by human activities.

FBA Case Studies

There has been an ongoing "battle" over the use of case studies in research. Some reviewers of the FBA approach have questioned the ability to generalize from a set of studies that have taken place in different locations, at different times, and by different researchers — each of whom harbors their own set of assumptions about which details to include and which to exclude. Despite this debate, case studies are commonly used to understand phenomena of interest to a researcher because they are instructive; they are rich in detail; they provide input for the scenarios; and, despite the known risk of case studies being eclectic, of variable quality, and difficult to compare, they can be used (with care) to draw some generalizations. They also provide a level of detail that allows for concrete recommendations for action.

The judicious use of such case studies can provide some insights into the problems relating to planning, especially at the local and regional levels, for the possible consequences of climate variability and climate change. Societies respond to extreme events and if, as the scientists suggest, there is a strong probability that the frequency and severity of extreme events will increase with a global warming (e.g., Emanuel 1987; Mearns 1989; Strezpek and Smith 1995), lessons gained from recent coping experiences can lead to improved decision making, as well as to reduced vulnerability. Policy responses to the impacts of climate-related environmental change are likely to come from the local and regional levels of government, and the decision makers in these institutions are the least educated on the issue.

Whatever the value of case study research, the FBA approach provides information grounded in real events and cannot easily be dismissed as mere theorizing or scare tactics. This is both its strength and its weakness — its strength because researchers can use these cases to evaluate the appropriateness and adequacy of present-day and near-future societal responses to extreme meteorological events; its weakness because societies often change some aspects of their behavior following an extreme meteorological event. So, researchers must take into account the fact that societies, too, are constantly undergoing change. It is the focus on change and societal responses to it that may be one of the most valuable outcomes of forecasting by analogy.

Shortcomings of the FBA Approach

In the published reviews of the FBA approach, reviewers have pointed out various problems with the method. Many of these have also been explicitly acknowledged by the researchers who use the FBA approach to assess climate change impacts on society. This section focuses on such problems.

A key problem with forecasting by analogy is that its utility depends on the aptness of the analogy. To various readers, the appropriateness or relevance of some of these case studies to the climate change issue might vary widely. It is important to identify and make explicit assumptions about the analogy and to note similarities and differences (e.g., the consequences for a regional water balance of global warming of the atmosphere decades in the future versus the drawdown of the Ogallala Aquifer in the agricultural heartland of the U.S. [Glantz and Ausubel 1984]). This applies to strengths and weaknesses of the various methods of forecasting climate change and the human responses to those changes. Specific predictions by any method are subject to error.

FBA is designed to provide only a first approximation of society's ability to respond to change — a glimpse of societal preparedness. It is not designed to project what either the state of the atmosphere or the state of society will be like decades into the future. Thus, users must articulate and respect the limits of the FBA approach. The method has some intrinsic traps and pitfalls (noted by Glantz 1988; Jamieson 1988; and Katz 1988). To go beyond using FBA to provide a first approximation of the level of societal preparedness to cope with future climate-related environmental changes at this time carries with it a high risk. Ultimately, climate change may manifest itself in many significantly different ways than it has in the past (e.g., as a result of anthropogenic as opposed to natural forcing of the climate system). So, it is important to pay attention to the experiences of the past, without getting lulled into preparing only for those specific experiences.

One reviewer suggested what others may feel: that the results of forecasting by analogy are common sense. Stating that lessons can be learned from examining the past is akin to announcing that motherhood is sacred or apple pie tastes good. Although some of the findings of the FBA case studies may reaffirm the obvious, the obvious course of action is not always the one pursued. Many decision makers know what the "right" actions are in a given situation, but they may not take them (or be able to take them) because of prevailing political, economic, cultural, or social pressures. It is necessary to reinforce the need for common-sense actions that are often neglected. FBA findings can reinforce the importance of those actions.

Another reviewer suggested that FBA fails to identify the powerful confluence (accumulation) of regional environmental problems. Such a criticism is not necessarily valid; while we do not seek to separate the various environmental problems that confront society at any given time in its history, we do assess societal responses to extreme events that occur in the midst of all other problems. The FBA approach considers the social as well as natural environments that face society at the time of its response to the meteorological event. A related concern is that FBA does not an attempt to project societal responses decades into the future but only into the near future. This is because, barring a cataclysmic, steplike social change (e.g., revolutionary change), social institutions and responses in the near term are likely to be similar to those taken in response to extreme events in the recent past. However, decades into the future, those institutions are likely to have changed considerably, and today's responses would likely be less relevant for planning purposes.

FBA Examples for Three Fisheries

After much discussion regarding the effects of global warming on sea level and on a variety of ecosystems, such as agriculture and human societies, concerns have been raised belatedly regarding its effect on fish populations and fisheries. Fisheries are important to countries around the globe because they provide an alternative source of protein, which is in short supply from overburdened agricultural lands, for growing human and animal populations. The economies of some countries (e.g., Iceland, Perú, Chile, Namibia, Spain, Poland) also depend heavily on fishing and related industries, such as building ships, manufacturing fishing gear, and processing fish.

This section presents three brief examples of FBA case studies of fisheries to suggest the types of extreme changes in availability or abundance of fish populations that could serve as possible analogs to climate change impacts on fisheries to which societies have had to respond. Each study is briefly described, and the lessons drawn from it are presented. More detailed descriptions of these studies, along with 12 additional case studies (Alaskan king crab, California sardine, northeast Pacific salmon, U.S. Gulf shrimp, menhaden fishery, Maine lobster, North Sea herring, Atlanto-Scandian herring, Poland's marine fisheries, the Far Eastern sardine, and the Indian Ocean tuna) are provided in Glantz (1992a).

The sea lamprey study was chosen as an example of the impact of inadvertent introduction of an exotic species into a stable ecosystem. In the event of global warming, the introduction of such species could disrupt many more ecosystems in a similar way. The Perúvian-Chilean fisheries study has gained great notoriety because of recent major El Niño events. The effect of global warming on the intensity, frequency, and magnitude of El Niño events is yet to be determined. The third case, the Anglo-Icelandic Cod Wars, was chosen as an FBA study to suggest that analogies for human responses to global warming can focus on changes in availability, as well as abundance, of living marine resources, regardless of the causes of those changes.

Great Lakes Sea Lamprey Study

Description

As a possible analog of some unpredictable phenomena that might accompany global warming, this case study focuses on the impact of the inadvertent introduction of the sea lamprey *(Petromyzon marinus)* on fisheries in the Upper Great Lakes (Huron, Michigan, and Superior) in the United States between the late 1930s and the early 1960s (Regier and Goodier 1992). Ecologists are not certain how the sea lamprey first invaded the Great Lakes. According to a popular view, it was inadvertently introduced into the lowest Great Lake, Ontario, in the early 1800s as a result of construction of the Erie Canal. Sea lamprey have been observed to attach themselves to the hulls of ships, by using their suction-cup mouths, and ride along. Sea lamprey are predators, and their prey in the Great Lakes include lake trout, lake whitefish, suckers, and introduced salmon. With the exception of suckers, sea lamprey compete with fishers for these species.

Despite fishers' concerns about the appearance of sea lamprey in the lakes, it was not until lake trout catches in Lake Michigan began to decline in about 1945 that widespread alarm was voiced. In the absence of any coordinated effort to control the eruption in the sea lamprey population, their numbers grew and their impacts spread. The early delay in responding was caused, in part, because fishers did not want regulations that might eventually threaten their catches or earnings, particularly regulations made by a distant international regulatory agency unfamiliar with local situations. In fact, the sea lamprey problem may be directly responsible for establishing the first permanent international body concerned with fishing interests in the Great Lakes, despite 60 years of previous attempts to create such an organization.

In the 25-year period under review, several natural and social stresses on fish populations have been observed in the Great Lakes in addition to the sea lamprey, primarily, overfishing and pollution. The fortunes of most Great Lakes fisheries declined in the early 1960s. In the Upper Lakes, the direct and indirect cause for the collapse of various salmonid stocks was the sea lamprey. In the shallower areas, pollution and eutrophication also played a role. The harm done by the sea lamprey has only partially been reversed during the 50 years after their initial population explosion. It appears that, as of the early 1990s, the sea lamprey is under partial control in most parts of the Great Lakes.

Lessons

The objective of controlling sea lamprey provided an appropriate focus for establishing a joint, regional, interjurisdictional action for Great Lakes fisheries.

- Fishing was not reduced to compensate for the added pressures that the sea lamprey placed on the lake trout and lake whitefish fisheries.
- Some of the larger fisheries consolidated their vertically integrated operations, diversified their fishing practices, and diversified their business interests to include nonfishing enterprises to survive.
- With the loss of a preferred fish population, there was a shift to other stocks, which were often already under great fishing pressure.
- The consequences of the sea lamprey population eruption contributed to the disintegration of small towns along the shores of the Great Lakes.
- An ecosystem can undergo serious restructuring with the intrusion of a parasitic species. Climate-related environmental changes can also lead to ecosystem restructuring.

Perúvian-Chilean Fisheries

Description

The waters in the eastern equatorial Pacific are considered to be among the most productive in the world's oceans. In this upwelling region, deep, nutrient-rich, cold water wells up to the sunlit zone. These nutrients are consumed by plankton and, eventually, by fish populations. The major commercial landings in this area have traditionally been anchoveta and sardine. This region, however, has been beset by recurrent oceanic-atmospheric oscillations referred to as El Niño-Southern Oscillation (ENSO) events. Major ENSO events in the equatorial Pacific Ocean alter the structure of marine ecosystems, noticeably depressing long-term yield and catch levels. Paleoecological evidence supports the view that ENSO events have affected the upwelling regions off the coasts of Perú and Chile for millennia. However, only since the devastating ENSO event of 1972-1973 did scientific and policymaking attention increasingly focus on this phenomenon and its regional and global impacts. This case study examines these impacts on Perúvian and Chilean fisheries (Caviedes and Fik 1992).

The fisheries of Perú and Chile have definitely been affected by the frequent climatic-oceanic variations that have characterized the last decades, but their ultimate development has been determined more by internal biological controls, the impact of economic imperatives controlled by world demand and national fiscal priorities, and resource conservation policies. Not only were the targeted fish populations of coastal fishing activities changed with the introduction of industrial fisheries and fish meal factories in Perú during the 1950s, but these alterations also upset the internal structure of coastal societies and the sources of economic power in Perú. Traditional Perúvian fishers were incorporated into the labor force as a result of the increasing number of fishing boats and processing plants, as were the more numerous *serranos* (Indians from the interior Sierra), who flocked to the coastal communities to share in this economic boom. The appeal of artisanal fishing sharply decreased, overshadowed by the relatively good wages offered by the industrial fishing establishments. Fishing entrepreneurs became the *nouveau riches* of Perú, joining or replacing the traditional agrarian and mining elites.

Initially, none of the ENSO events that occurred during the speculative growth phase of the fisheries was severe enough to trigger serious fishing control measures aimed at preventing a possible collapse of the anchoveta-based fisheries. The government (military or civilian), driven by the desire to increase profits and exports, proved more than willing to maintain an atmosphere of growth and development, abstaining from imposing restrictive resource conservation measures. With reduced numbers of workers, artisanal fisheries were not in a position to take full advantage of the unexpected increase in the abundance of tropical fish that favored warm waters.

The development of Chile's industrial fisheries followed a slightly different path from that of Perú in its early phase. By shifting the emphasis from anchoveta to sardines, from sardines to jack mackerel, and from jack mackerel to anchoveta (when they made a brief comeback in 1985), Chilean fishery industrialists have maintained relatively large captures even during the ocean-warming episodes that followed the 1982-1983 El Niño, although sardine catches are showing signs of pending collapse, i.e., one young year class dominates catches at present, and this leads to an unstable situation.

Thus, the shifting of fish in the waters of Perú and Chile appears to be the consequence of certain national policies, economic circumstances, and internal social conditions that have occurred along with global and regional climatic fluctuations. These social factors have prompted the momentous exploitation of marine resources by both countries, even during periods of environmental stress that have placed unbearable pressures on the reproductive capacity of fish stocks.

Heightened research interests have resulted in a greater understanding of ENSO and its environmental and social impacts. It is clear that ENSO has a major impact on regional fisheries (specifically, anchoveta in Perú and sardine in Chile) and that forecasts of ENSO events might enable fisheries managers to better mitigate the impacts of these potentially devastating events.

Lessons

- The combined interactions of overfishing, economic pressures, and environmental variability can lead to the collapse of an otherwise productive fishery. The search for a single cause is often futile as well as misleading.
- Diversification of targeted stocks builds resilience into the fishing industry (as well as exploited stocks) in the event of fluctuations in environmental conditions.
- Concern about overfishing and stock conservation practices demands concerted action by Perú and Chile. These two countries have been antagonistic neighbors. In the light of the regional shifts in living marine resources that accompany decadal climate variability, this study suggests that the impacts of climate change might best be addressed through improved regional cooperation.
- ENSO frequency or intensity is superimposed upon these longer-term environmental changes, limiting the utility of conventional equilibrium management concepts.
- Overfishing makes fish stocks with uncertain recovery rates highly vulnerable to ENSO events.

Anglo-Icelandic Cod Wars

Description

On four occasions since World War II, Iceland unilaterally extended its fishing jurisdiction, putting it in direct conflict with other European countries such as the United Kingdom, Germany, and Belgium. At the time of each of these conflicts (between 1952 and 1976), Iceland's fisheries supplied a large portion of its foreign exports; this percentage declined from a high of 90% in the early 1950s to about 80% in the mid-1970s. Iceland depends more than any other nation on the exploitation of fish populations, outdistancing its closest competitors by a wide margin. Needless to say, there are dangers associated with such a heavy dependence on a single resource for export. Any disruption (variability or change) in the catch-production-marketing system would have major adverse impacts on Icelandic society.

Only the United Kingdom took a militant stance against these extensions, and the ensuing conflicts have been popularly referred to as the Anglo-Icelandic Cod Wars. The analogy of the Anglo-Icelandic Cod Wars to the potential impacts of a global warming on marine resources lies in societal responses to changes in abundance or availability of critically important resources, regardless of the cause of those changes.

The Cod Wars case study provides a political/legal situation analogous to climate impacts on a fishery (Glantz 1992b). Some lessons can be drawn from this analogy for possible national responses of fishing nations to the regional impacts in the marine environment of a global warming of the Earth's atmosphere.

This analogy can be useful in providing a first approximation about how societies might deal with changes in the availability, but not necessarily abundance, of commercially exploited living marine resources. Although this is a historical example, it is important to note that Cod-War-like conflicts continue to occur, such as the recent Franco-Canadian cod wars, the conflict over walleye pollock in the Bering Sea Donut Hole, and Canada's concern over the depletion of groundfish stocks that straddle the 200-mile limit in the Grand Banks region (Anon. 1990).

Lessons

- The economic, political, and social settings at the time of a change in resource availability will be crucial to the outcome of a conflict. Depending, for example, on the state of the economy or the health of the fishing sector, different (even diametrically opposed) responses could be justified.
- One cannot rely on traditional international law in times of environmental change. For example, the International Court of Justice was not useful in affecting the outcome of these disputes and, in fact, may have exacerbated conflicts in the short term.
- Countries will resort to "tie-in" tactics, bringing in other issues to strengthen the chance for success. Iceland, for example, brought NATO into the Cod Wars conflict and exacerbated Cold War fears by trading with the former Soviet Union, Poland, and East Germany.
- When a nation's economy is perceived by its leaders to be threatened, long-standing agreements and "habits" may no longer be tolerated.
- Regional arrangements should maintain flexibility in the face of changes in environmental conditions and must not become too "bureaucratized" to act. Regional political or military organizations are not good at dealing with environment- or natural-resources-related conflicts.
- Resolution of ecological problems in one location could generate new problems for fish stocks and fishing fleets in other areas.

Conclusions

Because of the problems associated with the use of GCMs to predict changes in climate at the regional level, FBA provides another approach to scenario development and to the identification of action alternatives. While policymakers wait for scientists to improve the reliability of the output of GCMs and limited-area models, the forecasting-by-analogy method can provide policymakers with complementary information about what they and their citizens may be facing on the local level and how they might become better prepared. It can provide them with a first approximation of how various sectors of society might respond to climate-related environmental change, as well as the consequences of those responses. Perhaps the historical record could suggest how well society is prepared to deal with as-yet-unknown, climate-related environmental change.

One clear lesson from FBA studies is that both rich and poor societies remain sensitive and vulnerable to climate and societal change, despite attempts by various societies to "weatherproof" their activities for such events as the recent floods in the upper Mississippi Basin in mid-1993 or in the Netherlands in January 1995. FBA helps to identify the rigidities and flexibilities of various communities dealing with climate change, enabling decision makers to reinforce the strengths and reduce the weaknesses.

As a final comment, it is important to note the following caveats: (1) FBA does not provide forecasts of how society will respond to climate change, but identifies potential vulnerabilities and areas in which society may be able to improve its ability to respond; and (2) current social and scientific conditions (e.g., technological responses that are now feasible) may be significantly different from those of the historical periods being examined, so current vulnerabilities and societal responses may differ significantly from what occurred in the past. Those who use the FBA approach must carefully select the historical period for the analysis.

It is reasonable to assume that the measures taken by society over time, as climate-related environmental changes are observed and society learns how to adjust to the effects of the changes, will reduce the impacts of certain extreme meteorological events that otherwise would be felt but could make society more vulnerable to the impacts of other events. Yet, as experience has shown, *forecasting* these responses is as difficult as forecasting the physical climate itself.

Acknowledgment

The National Center for Atmospheric Research is sponsored by the National Science Foundation.

References

Abernethy, V., 1990, "Forecasting by Analogy," book review, *Population and Environment* Vol. 6, June.

Anon., 1990, "300-Mile Limits — or More?" *Fishing News International* 29(5):5.

Broadus, J.M., 1989, "Societal Responses to Regional Climatic Change," book review, *Oceanus,* Fall.

Budyko, M.I., and Y.S. Sedunov, 1990, "Anthropogenic Climate Change," in H.J. Karpe, D. Orten, and S.C. Trindade (ed.), *Climate and Development*, pp. 270-284, Springer-Verlag, Berlin, Germany; paper presented at The World Congress on Climate and Development, November 7-10, 1988.

Caviedes, C.N., and T.J. Fik, 1992, "The Perú-Chile Eastern Pacific Fisheries and Climatic Oscillation," in M.H. Glantz (ed.), *Climate Variability, Climate Change and Fisheries,* Cambridge University Press, Cambridge, United Kingdom, pp. 355-376.

Cohan, D., R.K. Stafford, J.D. Scheraga, and S. Herrod, 1994, *The Global Climate Policy Evaluation Framework*, Proceedings of the 1994 Air and Water Management Association Global Climate Change Conference, April 5-8, 1994, Phoenix, Ariz., USA.

Cortner, H.J., 1990, "Societal Responses to Regional Climatic Change," book review, *Climatic Change* 16:361-362.

Crawford, R., 1994, "Climate Variability, Climate Change and Fisheries," book review, *Global Environmental Change* 4(4):344-345.

Emanuel, K.A., 1987, "The Dependence of Hurricane Intensity on Climate," *Nature* (326):483-485.

Environmental Science and Technology, 1992, book review, 26(9):1718.

Flohn, H., 1981, *Life on a Warmer Earth: Possible Climatic Consequences of Man-Made Global Warming*, Executive Report No. 3, IIASA 37, Laxenburg, Austria.

Future Surveys, 1989, "Societal Responses to Regional Climatic Change," book review, Vol. 2, Feb.

Glantz, M.H. (ed.), 1988, *Societal Responses to Regional Climatic Change: Forecasting by Analogy*, Westview Press, Boulder, Colo., USA.

Glantz, M.H. (ed.), 1991, "The Use of Analogies in Forecasting Ecological and Societal Responses to Global Warming," *Environment* 33:11-15, 27-33.

Glantz, M.H. (ed.), 1992a, *Climate Variability, Climate Change and Fisheries,* Cambridge University Press, Cambridge, United Kingdom.

Glantz, M.H., 1992b, "Global Warming Impacts on Living Marine Resources: Anglo-Icelandic Cod Wars as an Analogy," in M.H. Glantz (ed.), *Climate Variability, Climate Change and Fisheries,* pp. 261-290, Cambridge University Press, Cambridge, United Kingdom.

Glantz, M.H., and J.H. Ausubel, 1984, "The Ogallala Aquifer and Carbon Dioxide: Comparison and Convergence," *Environmental Conservation* 11(2):123-131.

Houde, E.D., 1993, "Climate Variability, Climate Change, and Fisheries," book review, *The Quarterly Review of Biology* 68(4):605-606.

Jamieson, D., 1988, "Grappling for a Glimpse of the Future," in M.H. Glantz (ed.), *Societal Responses to Regional Climatic Change: Forecasting by Analogy*, pp. 73-93, Westview Press, Boulder, Colo., USA.

Journal of Regional Science, 1990, "Societal Responses to Regional Climatic Change," book review, Feb.

Katz, R.W., 1988, "Statistics of Climate Change: Implications for Scenario Development" in M.H. Glantz (ed.), *Societal Responses to Regional Climatic Change: Forecasting by Analogy*, pp. 95-112, Westview Press, Boulder, Colo., USA.

Kellogg, W.W., 1977, *Effects of Human Activities on Global Climate*, Technical Note No. 156, WMO No. 486, World Meteorological Organization, Geneva, Switzerland.

Magalhães, A.R., and M.H. Glantz (eds.), 1992, *Socioeconomic Impacts of Climate Variations and Policy Responses in Brazil*, Esquel Brasil Foundation, Brasília, Brazil.

Mearns, L.O., 1989, "Variability," in J. Smith and D. Tirpak (eds.), *Potential Effects of Global Change on the United States*, pp. 29-54.,U.S. Environmental Protection Agency, Washington, D.C., USA.

Miller J., 1989, "Global Policy," book review, *The Bloomsbury Review*, May/June.

Muckleston, K.W., 1993, "Climate Variability, Climate Change, and Fisheries," book review, *Environment* 35(2):29.

Myers, M.F., and G. White, 1993, "The Challenge of the Mississippi Flood," *Environment* 35(10):6-35.

Ninh, N.H., P. Usher, H.M. Hien, M. Glantz, and S. England, 1992, *The Potential Socio-Economic Effects of Climate Change on Vietnam*, Center for Environmental Research, Education, and Development, Hanoi, Vietnam.

Pittock, A.B., 1989, "Societal Responses to Regional Climatic Change," book review, *Bulletin of the American Meteorological Society*, Vol. 9.

Ray, G.C., 1993, "Climate Change and Fisheries," book review, *BioScience* 43(9):642.

Recent Publications on Government Problems, 1989, "Societal Responses to Regional Climatic Change," book review, March.

Regier, H.A., and J.L. Goodier, 1992, "Eruption of Sea Lamprey in the Upper Great Lakes: Analogous Events to Those That May Follow Climate Warming," in M.H. Glantz (ed.), *Climate Variability, Climate Change and Fisheries*, pp. 185-211, Cambridge University Press, Cambridge, United Kingdom.

Runyon, C., 1990, "Societal Responses to Regional Climatic Change," book review, *State Legislatures*, March, p. 5.

Sage Public Administration Abstracts, 1989, book review, "Societal Responses to Regional Climatic Change: Forecasting by Analogy," July.

Shepherd, J.G., 1992, "Being Prepared," book review, *Nature*, 358:292, July 23.

Strezpek, K.M., and J.B. Smith (eds.), 1995, *As Climate Changes*, Cambridge University Press, Cambridge, UK.

Templer, O.W., 1989, "Societal Responses to Regional Climatic Change," book review, *The Environmental Professional*, Vol. 11.

Wright, H.E., 1991, "Societal Responses to Regional Climatic Change," book review, *Environment*, April.

Zonn, I.S., 1994, *Climate Variability and Change in the Commonwealth of Independent States (CIS): Forecasting Climate-Related Impacts and Societal Responses to Them*, U.N. Environmental Programme, Nairobi, Kenya.

Anticipated Effects of Climate Change on Commercial Pelagic and Artisanal Coastal Fisheries in the Federated States of Micronesia

L. Heidi Primo
Office of Planning and Statistics
Palikir, Pohnpei
Federated States of Micronesia

Abstract

The Federated States of Micronesia have two distinct fisheries sectors. Foreign offshore commercial corporations dominate the profitable Pacific longline tuna fisheries, while the local community fishes along the inshore reefs and lagoons. Modern coastal zone and marine resource management preservation tactics do not take into account community perceptions of reef ownership and indigenous fishing rights, nor do they address traditional solutions for monitoring resources. Practical adaptive measures must consider the needs and culture of the local Micronesian population. Accelerated sea-level rise and erosion run-off from possible increased precipitation due to climate change will probably affect marine ecosystems and sustainability of fish catches. Pelagic fisheries are vulnerable to climate change because changes in sea-surface temperature and ocean circulation patterns will inevitably affect migratory patterns of straddling fish stocks. Pacific Ocean fisheries are vulnerable to climate change in many ways. Adaptive measures may be crucial in planning for economic survival.

Introduction

Islanders in the Federated States of Micronesia (FSM) rely heavily on coastal and marine resources for both local consumption and export. As a developing nation with few export commodities, the FSM has little to sell to other nations, except the relaxing beauty of the coastal environment and the right to fish within the 200-mile exclusive economic zone (EEZ). Studies of the impacts of projected global climate changes indicate that Pacific fisheries may be negatively affected. The need to develop adaptation strategies for fisheries is a crucial part of the National Climate Action Plan for the FSM.

Geographic Background

A nation of scattered islands spanning 2,500 km, the FSM lies in the western Pacific Ocean between 14°N longitude and the equator, and between 136 and 166°E latitude. The country consists of hundreds of islands and atolls, although only 40 are of significant size. Many are isolated. Approximately 32 are high islands (e.g., Pohnpei and Kosrae) with dramatic peaks. In addition, the FSM has about 30 inhabited low-lying atolls with elevations barely above sea level. These islands support a total population of approximately 103,000 (1986 census). The local community exploits hundreds of unpopulated atolls for traditional gathering of resources.

Pacific archipelagoes contain a higher coastal-zone-to-land-area ratio than do continental nations. Because the most outlying islands serve as boundaries for EEZs, the 200-mile limit for the FSM covers a sea area of almost 3 million km^2 (about 1.2 million mi^2). The total land area of FSM is only about 1,815 km^2, which means that the FSM has about 4,280 times as much water as land.

Vulnerability to Climate Change

The economies and environments of the FSM are extremely fragile. They are at risk not only from global economic fluxes, but also because of their vulnerability to natural disasters, such as flooding and typhoon surges. Pacific leaders consider global warming and sea-level rise the most serious environmental threats to the region. Social and environmental consequences are perceived as severe and long term.

One of the most frightening effects of sea-level rise is the ultimate inundation of entire atolls, which would reduce the size of the EEZ as peripheral islands become permanently submerged. International negotiators are working to establish permanent agreements on existing 200-mile EEZs surrounding islands and offshore from coastlines. Inundation of outer islands would alter the FSM's boundaries, disintegrate the integrity of the EEZ fishery zone, and cause citizens and government to lose exclusive sovereign and economic rights over those fishing areas. (The open seas beyond the boundary of each state's 12-mile limit make up the largest region for deep-water commercial fishing.)

Importance of Fisheries to Coastal Communities

The two distinct fisheries sectors critical to FSM are the low-yield artisanal fisheries and the commercial pelagic fisheries dominated by foreign-owned resident fishing fleets. The FSM has many types of fishing practices — from ancient handicrafts used to make carving hooks and knotting nets thrown from outrigger canoes to the latest remote satellite sensing technology. A variety of strategies lie in between these two extremes. People who rely on this natural resource have a vested interest in protecting it and keeping it productive. Because many villagers and atoll dwellers within the FSM still live according to traditional customs, fishing remains an integral practice for them, as well as a source of revenue for their relatives who reside in towns. Islanders prone to famines caused by climatic disasters must give priority to maintaining the ecosystems in which fish survive and breed.

The "Uncertainty" Factor

Insufficient research has been done on the relationship between ocean currents and migration routes of adult pelagic fish in the Pacific to begin to factor in climate change scenarios. It is unclear whether a new system of breeding site locations and dispersal routes to feeding areas will be developed. The starting place for adaptation strategies will be understanding the present biological processes and behavioral phenomena as completely as possible.

Because of the uncertainty in scientific predictions about alteration of ocean currents and upwellings in terms of strength and geographic location, it is difficult to predict the climatic fluctuations that could affect particular aspects of the FSM's fisheries. Various combinations of adverse impacts could occur, but the aggregate effects and linkages are uncertain.

Potential Impacts of Climate Change to FSM Fisheries

Both direct and indirect impacts of declining fish stocks on FSM would result in devastating economic consequences (e.g., loss of revenue from the sale of foreign fishing permits and licenses and from diving [tourism]). More important, however, is the loss of the fish supply, which forms an essential element of the subsistence food base. Loss of fish could lead to malnutrition and protein deficiencies, causing illness or death. Table 1 lists possible additional consequences of climate change in the FSM.

Table 1. Additional Possible Consequences of Climate Change in Micronesia

- Changes in ocean chemistry. Carbon sequestration or ultraviolet radiation from thinning of the stratospheric ozone layer could affect photosynthesis, stunting growth of phytoplankton or plants that provide food for schooling fish in the upper layer of the ocean.
- Extinction of unique biological species and habitats.
- Elimination or relocation of breeding sites or spawning grounds because of increased sea temperatures.
- Diversion of hatchling (larval) dispersal. Fish would need to be "recruited" to new feeding areas outside the migration range of the FSM EEZ.
- Relocation of commercially prized species outside of the FSM geographic limits.
- Reduction of the size of standing stocks. Overfishing, followed by insufficient replenishment, could result from climatic factors.
- Warm ocean surface pools and a shallower mixed layer above thermoclines. Certain species would seek optimal temperature conditions at lower depths. This change would require different technologies to catch fish than the ones currently being promoted or developed (e.g., different gear).
- Increased frequency in tropical typhoons and cyclones, which would limit the number of fishing days.
- Storm surge damage to reefs, fishing grounds, and possible destruction of fishing vessels.
- Coral polyps expelling algae that live in symbiosis with them (due to increased sea temperature), causing coral bleaching.
- Inability of coral reef growth to keep up to accelerated rising sea levels.
- Alteration of mangrove marsh salinity levels and ecosystems because of increased (or decreased) precipitation.
- Loss of lagoons caused by barrier reef damage to channels.
- Destruction of coastal plains from changing sea levels, which could cause flooding and beach erosion, increasing siltation onto reefs.
- Potential health hazards from increased cigautara poisoning.
- Relocation of the latitude of the intertropical convergence zone.

Impacts on Aquaculture and Mariculture

The FSM has tried backyard fish ponds with little success. The labor required to maintain them is one constraint, but they generally failed because of dietary preferences. Communities on Pohnpei rejected the use of tilapia and milk fish, the most common backyard aquaculture species for Pacific islands. Communities do not want to maintain an aquaculture system to grow fish they find distasteful, especially when reef fish are bountiful, appetizing, and free. However, because global warming contributes to many uncertainties about the future, a revival of controlled aquaculture experiments and projects may be one adaptive measure, if local taste preferences are taken into account.

The state of Kosrae maintains a giant clam aquaculture nursery that takes in seawater through pipes. These pipes have cloth filters to screen out harmful bacteria. Beds in the outer atolls are seeded with juvenile clams. Fully grown clams are distributed to several islands and marketed to gourmet restaurants. In addition, sponge farming operations are active at several locations within Pohnpei's lagoon. These facilities are extremely vulnerable to an increase in temperature and sedimentation from run-off caused by excessive precipitation. Monitoring changes will be a crucial factor for determining adaptive measures in aquaculture-based industries.

Cumulative Anthropogenic Factors Affecting Fisheries

Anthropogenic influences on the coastal zone (unrelated to climate change) stress FSM marine resources, particularly in near-shore reef environments. Such influences include, but are not limited to, fertilizer use, forest clearing, road building, earth moving, construction, and other coastal development activities on land; these activities increase siltation and pollution (Richmond 1993). In addition, on many islands, sand and coral are mined, vessels discharge oil, and inadequately treated sewage is released. High population growth has increased environmental degradation in the coastal zone, which, in turn, has increased pressure on natural resources, such as fisheries.

In addition, technological improvements in electronic fish detection and location apparatus, bait storage, hook design, on-board refrigeration, navigational aids, fishing gear, and vessel architecture have contributed to the evolution of fishing techniques (Boggs 1992; Glantz 1994). Continued technological innovations will enhance fishing efforts. These innovations will occur at the same time as climate change and could multiply negative impacts on fisheries.

Adaptation Strategies for Vulnerable Ecosystems

To address adaptation strategies for the FSM fisheries, specific vulnerable ecosystems are discussed in the following sections. Each ecosystem is evaluated to determine precise risks.

Mangrove Protection

The mangrove ecosystem produces large amounts of organic matter used by fish and other creatures. In this habitat, Micronesians collect marine species by traditional methods (i.e., using nets or gathering edible creatures by hand). Mangroves serve an important function — they act as a nursery for living marine resources (e.g., prawns, mullet) and provide a permanent home for oysters and crabs. Mangroves are also a feeding area for large fish, such as snappers, which feed on smaller fish.

Changes in precipitation as a result of climate change may affect siltation, as well as salinity of the mangrove marshes. The state of Pohnpei has a law intended to prevent cutting mangroves, but it is seldom enforced. Planners from the state of Kosrae are investigating including protection of mangroves in their Island Resource Management Plan.

The best adaptation strategy for mangroves would be to monitor the health of the mangrove forests and measure salinity at regular intervals. Water that becomes too saline could inhibit root pods from uprighting themselves to implant in the sand. Techniques such as freshwater flooding or diversion of excessive run-off could help to maintain an optimal balance of salinity. The neighboring island nation of Kiribati has a pilot plan for mangrove reforestation; if their program is successful, the FSM should also consider a mangrove reforestation program in areas that have been clear-cut.

Coral Reef and Lagoon Initiatives

Micronesia has a variety of reef formations: patch, fringing, and barrier reefs, as well as offshore atolls. These formations provide food for many species of fish, and the indigenous population concentrates their fishing activities there. Although the reef suffers from stress due to previously mentioned factors, a rise in sea temperature has been associated with coral bleaching around the Pacific. Some of the barrier reefs form sheltered lagoons, which are also exploited by local populations. Where they exist, shallow lagoons are the main source of fish for the artisanal fisheries.

No specific government policies protect the coral reefs and lagoons from adverse anthropogenic or climatic impacts. Intervention measures will be necessary; the health of the reef is the nation's best protection for sea-level rise. Geomorphologists, such as David Hopley at James Cook University in

Townsville, Australia, speculate that reef growth potentially can parallel gradual sea-level rise — *if* the reef is kept healthy.

The states of Kosrae, Yap, and Pohnpei are creating coastal zone management (CZM) plans and regulations. Yap has a plan being introduced to the community for comment. The Pohnpei CZM plan is in the draft stage. The state of Chuuk has not initiated a CZM plan at this time.

Some reefs are being monitored to evaluate their health, but such monitoring is not widespread, nor carried out on a regular basis. Placement of "artificial reefs," such as fish aggregating devices, is rare in the FSM, but would be good to initiate if natural reefs become seriously endangered.

The government of each state should consider placing permanent mooring buoys to avoid some of the anchor damage to the coral reefs caused by both fishing and scuba diving boats. All four states or the national government should ban the export of coral for jewelry, such as semiprecious black corals, if global warming makes decorative coral more precious and pilfering escalates.

Government Policy Planning Framework

At some point, the FSM Congress must examine and carefully weigh the costs and benefits of licensing distant water fleets in terms of degrading the environment and depleting resources. High-level policymakers need to discuss vulnerability to climate change and sea-level rise so that adaptation strategies can be included in each state's fisheries policies and CZM plans, as well as in the next National Five-Year Development Plan. Climate change is not currently addressed. Without dialogue and open exchange of ideas, systematic schemes cannot be formulated and prioritized.

In the past, traditional management regimes avoided overexploiting marine resources considered tabu by chiefs. They also closed fishing areas temporarily during certain seasons. For example, a person could fish for a particular species only when the breadfruit (a local staple food) was ripe, twice a year. Sometimes entire islands or atolls became off-limits. Clans were restricted from consuming the species that represented their clan totem. Harvesting or eating other species was prohibited (Goldman 1994; Itimai 1995).

One formal adaptation policy the FSM government should consider is investigating and documenting traditional management strategies and conservation practices throughout the Micronesian islands, especially where a few inhabitants are surviving at a subsistence level. Researchers need to assess the social impacts related to fisheries that could change community life. These changes could alter people's perceptions, diet, or values relating to marine resources. The government should consider combining traditional systems with modern CZM practices that apply to more populous urban townships that have ports and harbors, shoreline protection, and other vulnerable infrastructure.

The impact of the aquarium reef fish trade to Guam, the U.S., and Japan must be investigated to determine whether the demand is sustainable, given climate change predictions. If small-scale collecting can continue without upsetting the ecological balance, nondestructive methods should be used. International negotiations intended to end excessive or destructive high-seas fishing practices must continue. Newly initiated stock management regimes must continue, with vigorous enforcement controls.

Another issue to be considered in international negotiations of the Law of the Sea Treaty is whether island nations will be able to maintain the integrity of their current EEZ boundaries if their outer islands are submerged. The question as to whether it is worth the expense of raising land on outlying atolls or building protective structures, such as seawalls and revetments, must be addressed now. Funding would have to be solicited from multilateral donors or foreign countries with trained and master hydraulic engineers because the FSM does not have in-country capacity or experienced personnel. Using technology to reinforce atolls from the rising seas may avoid a substantial reduction of Micronesia's EEZ. But it is not known who would pay for this solution, if that choice were made, and who would receive the greatest benefit.

Developing integrated mathematical models for optimal management will help policymakers continue to plan adaptation strategies. Evaluating the potential of aquaculture, examining community-owned fishing cooperatives, and appraising the wisdom of building new canneries and fish packaging facilities should all consider climate change factors. Using surplus fish parts for fish meal fertilizer or pet food processing may help to diversify the fishing industry by providing another export commodity or at least an import substitute.

Crucial Statistics and Database Management

Tactics for managing the reefs and lagoons begin with baseline monitoring, where information is inconclusive. Computerized geographical information system mapping may help track the data. A satellite down-link receiving station would provide data on plankton pools, sea temperature, and other vital remotely sensed information.

Another starting point for evaluating adaptation strategies for FSM's fishing sector would be to gather statistics to determine the amount of fresh fish the local population uses compared with the amount of imported canned and frozen fish and alternative meat products they consume. Distribution patterns may show that the processed imported goods are easier to obtain or less expensive, or that they are more common on islands that have urban centers with ports or airports. If the FSM citizens generally eat canned mackerel, tuna, and sardines, while most of the fresh fish caught in their EEZ is flown daily to Japan, this aspect must be assessed. Imposing trade sanctions, tariffs, or tax schemes to equalize the price structure is one possibility that could have measurable effects on the economy.

Gathering comprehensive data on resource abundance, catches, and the rate of EEZ depletion is a step in the right direction. By determining the present levels of catch and where effort is focused, the FSM can determine sustainable yields for the future. Applying climate modification scenarios will be more than an intellectual exercise. As a precaution, future climate change should be incorporated into fisheries management programs and regimes even before direct impacts can wreak havoc on ecosystems.

Another baseline database imperative for assessment is documentation of FSM's ocean and coastal biodiversity. This database is necessary to preserve the variety of species and to research potential biotechnological exportable marine products, such as pharmaceuticals, dyes, or cosmetics. As a result, the FSM can diversify its economic base, avoiding heavy reliance on fisheries in future decades.

Community Outreach and Public Awareness

Public education and training are an essential component of an adaptation strategy. Communities can be taught to monitor the quality of coastal waters if they are provided with kits that include the necessary tools and chemicals. This information must be packaged in a way that can be understood by the lay public. Within the FSM, it would be practical and suitable to incorporate indigenous chiefs or village leaders into any government program that promotes cooperative community-based resource management.

Environmental education of local people (in their own language) about land-based sources of eutrophication, especially "nonpoint source pollution" will be crucial for preserving marine ecosystems. Workshops on coral bleaching in the 10 indigenous languages in the village meeting places, or alternatively, a pamphlet translated into local languages would apprise islanders of signs that may indicate abrupt temperature change or other stress on the reef system. Teaching communities to do their own transects and assessments of coral cover and reef health would be a suitable nationwide program to generate awareness and interest in local and regional environmental changes that could affect a community's way of life.

Periodic review and debate about climate change impacts and adaptation strategies must continue via forums. Fisheries' extension personnel should conduct workshops for fishing communities on the potential impacts of climate variabilities on the FSM fisheries. Information can be updated as conclusions from ongoing climate research and ocean monitoring are disseminated.

Conclusions

The call for island and coastal people to "cherish and nurture the sea creatures so that our grandchildren and succeeding generations also will have a bountiful ocean environment" (Van Dyke 1994) is being heard around the world. The idea that the sea is limitless is no longer true; this knowledge requires adapting former lifestyles and practices to best manage existing resources, so that the impacts of climate change will be negligible. Currently, legal and institutional systems separate people responsible for managing resources, agencies responsible for protecting the environment, and people responsible for planning and managing the economy. The interlocked nature of economic and ecological systems will become evident, however, because bioeconomics is integral to renewable resource management and an inevitable aspect of sustainable living.

References

Boggs, C.H., 1992, "Depth, Capture Time and Hooked Longevity of Longline-Caught Pelagic Fish: Timing Bites of Fish with Chips," *Fisheries Bulletin* 90(4):643-658.

Glantz, M.H., 1994, *The Impacts of Climate on Fisheries*, Environmental Library, No. 13, U.N. Environmental Programme, Nairobi, Kenya.

Goldman, B., 1994, *Yap State Marine Resources and Coastal Management Plan*, Yap State Government, Marine Resources Management Division, Federated States of Micronesia (Feb.)

Itimai, F., 1995, personal communication from Itimai (Fisheries Specialist, Government Department of Resources and Development, Federated States of Micronesia) (March).

Richmond, R.H., 1993, "Coral Reefs: Present Problems and Future Concerns Resulting from Anthropogenic Disturbance," *American Zoologist* 33:524-536.

Van Dyke, J., 1993, "Beyond National Jurisdiction: Resources under the 1982 Law of the Sea Convention," in *Toward a Pacific Island-Based Tuna Industry, Proceedings from a Workshop on Sustainable Living in the Aquatic Continent: Creating Sustainable Jobs*, pp. 29-37, Lahaina, Maui, Hawaii, United States, August.

7

HUMAN SETTLEMENTS

Rapporteur's Statement

Robert Etkins
NOAA, Office of Oceanic and Atmospheric Research
1315 East-West Highway
Silver Springe, MD, 20910, USA

Evaluation of the effects of climate change on human settlements show rather disparate consequences. Among the possibilities are the following:

- Permafrost degradation in high northern latitudes.
- The prospect of increased epidemics of infectious diseases and other adverse effects on human health.
- Decreases in energy for space heating in winter (offset to some extent by increased energy use in summer for space cooling).
- The refugee problem that would result in Pacific island countries because of the inundation of low-lying islands due to sea-level rise.

Permafrost regions occupy as much as 25% of the earth's land surface, and the melting or thawing expected from warming in high latitudes will have both positive and negative consequences (e.g.,an increase in arable land, along with a possible greater increase in methane (CH_4)-generating wetlands). The effects of climatic warming on human health due, for example, to the spread of vector-borne diseases, such as malaria, dengue fever, and cholera, are mainly negative, although adaptation could possibly occur if adequate public health measures were implemented. In Micronesia, other island nations, and countries with dense populations and low-lying coastal regions such as Bangladesh, the effect of sea-level rise would be totally negative, with little prospect for adaptation.

Interesting questions arise with the juxtaposition of these various consequences. Would the quantity and quality of new lands that become available for habitation in high latitudes offset the loss of low-lying islands and coastal lands? Could the people who become homeless (more than 20 million are at risk) be resettled in the new northern territories

Permafrost and Global Warming: Strategies of Adaptation

O.A. Anisimov
State Hydrological Institute, 23 Second Line
St. Petersburg, 199053, Russian Federation

F.E. Nelson
State University of New York, Department of Social Sciences
1400 Washington Ave.
Albany, NY, 12227 USA

Abstract

Anticipated warming will be pronounced at high latitudes and may seriously affect permafrost, which currently occupies nearly 25% of the land area worldwide. In the first half of 2000, climate changes will reduce permafrost areas, first by melting its sporadic zone and then by warming and deepening its active layer. A climate-based permafrost model used several scenarios to estimate the changes in the distribution of permafrost by mid-2000. All results showed a significant reduction in the amount of permafrost in North America and Eurasia. These changes will noticeably affect vegetation and northern agriculture, hydrology, land use, engineering, and socioeconomic areas.

Introduction

Permafrost is defined as any subsurface material that remains frozen for several consecutive years. It exists in continental areas with severe climate conditions, and currently underlays nearly 25% of the land area worldwide. Although permafrost areas, such as the Arctic and sub-Arctic regions of Eurasia and North America, are not densely populated, they may play an important economic role. Because regions such as Alaska in North America and West Siberia and Yakutia in Eurasia are rich in fossil fuel resources, the oil, coal, and gas industries have turned northward as sources of inexpensive fuel in the temperate latitudes have been depleted. This continuing trend may stimulate development in the North, leading to the construction of cities, transportation lines, and other elements of a modern infrastructure. A study of the permafrost will be useful in planning land-use and socioeconomic development in Arctic and sub-Arctic regions.

Studies suggest global warming will be greater at high latitudes (Budyko and Izrael 1987; Budyko et al. 1990, 1992) and may profoundly affect the permafrost. Higher air temperatures and precipitation may cause sporadic permafrost to thin or disappear in areas where current climate conditions produce near-zero annual surface temperatures. Widespread and continuous permafrost may warm, and its active layer may thicken. Permafrost degradation will seriously affect the environment, forestry, agriculture, hydrology, engineering, and construction in these areas.

The objectives of this study are as follows:

- Develop a strategy for permafrost mapping and regionalization by using climate-based models and criteria.
- Compare several climate change scenarios and estimate the possible effects of global warming on permafrost distribution over the Northern Hemisphere.
- Analyze the socioeconomic impacts of warming associated with permafrost degradation in Arctic and sub-Arctic regions.

Permafrost Mapping

The most common classification schemes for permafrost categorize frozen ground on the basis of its seasonal or perennial nature and its areal continuity or discontinuity. Although various authors use different criteria for delineating these zones, classification conveys the relative areal dominance of permafrost and permafrost-free conditions. A classification scheme in which permafrost is characterized as continuous, widespread discontinuous, and sporadic is generally used at small geographic scales. These categories correspond to conditions under which permafrost underlies more than 80%, 50-80%, and less than 50% of the area, respectively (Harris 1986). Each category is bound to a certain range of surface temperature. The mechanical properties of permafrost depend on the surface temperature. Thus, different permafrost types require different engineering strategies.

Small-scale permafrost mapping involves interpolating observational data taken at representative sites over larger territories. Several methods have defined permafrost continuity empirically, and their results vary substantially (Nelson and Anisimov 1992). The International Permafrost Association has given highest priority to the problem of standardizing permafrost measurements and interpreting observational data. As a result, a new, highly detailed map of permafrost distribution in the Northern Hemisphere has recently been issued (Heginbottom et al. 1993).

Although regionalizations based on observational data may be satisfactory for depicting the distribution of permafrost under current climate conditions, their reliance on empirical correlations with present-day environmental factors may restrict their predictive capabilities. A promising approach for small-scale permafrost mapping uses frost numbers introduced by Nelson and Outcalt (1987). Frost numbers are easily computable combinations of climatic parameters that reflect thermal conditions at the soil surface. The simplest formulation of the frost-number-based model requires only two climatic parameters (monthly average air temperature and precipitation) to calculate the normalized ratio of freezing and thawing degree-day sums at the soil surface.

The model has been tested with modern climate data and has shown good agreement with the permafrost map prepared by the International Permafrost Association (Anisimov and Nelson 1990, 1995; Nelson and Anisimov 1990). Although the calculated and empirical permafrost maps differed in their details, the model reproduced the boundaries of continuous, widespread discontinuous, and sporadic permafrost in both North America and Eurasia.

The frost-number-based model implicitly treats the climate-permafrost system under equilibrium conditions and thus neglects the thermal lag between the temperatures at the surface and those deep in the permafrost. While nonequilibrium effects are beyond the scope of this method, the model can be used to evaluate the geocryological impacts of climatic change, at least for an upper permafrost layer of a few meters depth. Such a layer reaches thermal equilibrium with the atmosphere shortly after climate changes occur, and changes in the uppermost permafrost horizon are of primary importance for most practical applications in engineering and ecology.

Climate Change Scenarios

Two methods are generally used for predicting future changes in climatic parameters: General Circulation Models (GCMs) and paleoclimatic reconstructions. These modeling strategies are fundamentally different in conceptualization and results (Budyko and Izrael 1987).

General Circulation Models are complex, three-dimensional computer models that account for many processes involved in the climate system. They use the atmospheric carbon dioxide (CO_2) concentration as the main driving factor and can estimate such future climate parameters as monthly average air temperature and precipitation. Although the results of different models vary in their details, they generally agree that anthropogenic warming will be amplified at high latitudes (Cess and Potter 1988).

The general idea behind the paleoreconstruction method is to restore the regional temperatures and precipitation of previous warm epochs, which are considered as analogs to future climates. The reconstruction is usually based on an analysis of paleobotanical and paleogeological data and provides seasonal resolution for the reconstructed regional air temperatures and annual resolution for precipitation. The three most frequently used warm epochs are the Holocene climatic optimum (6,200-5,300 years before present [BP]), the Eemian interglacial (125,000-122,000 years BP), and the Pliocene climatic optimum (approximately 4 million years BP). The differences between the mean annual global air temperature for these epochs and the current value are about 1.2, 2.0, and 4.0°C, respectively (Klimanov 1982; Zubakov and Borzenkova 1983; Velichko 1984, 1985; Budyko and Izrael 1987).

Projected climate changes derived from GCMs and paleoreconstructions, although in qualitative agreement, differ in details. Paleoreconstructions predict higher air temperatures in Arctic and sub-Arctic regions during the next century (Budyko et al. 1991). Many attempts have been made to harmonize the results obtained from the two methods, but it is still unknown whether the divergence results from inaccurate parameterizations of the atmospheric and hydrologic processes used in GCMs or from inherent limitations of the paleoreconstruction method.

To estimate the effects of the differences between the climate change scenarios on the potential permafrost distribution under global warming, several scenarios, including selected GCMs and paleoreconstructions, were used with the frost-number-based model of permafrost. The following models were used in this study:

- The Geophysical Fluid Dynamics Laboratory (GFDL) model (Manabe and Wetherald 1980; Wetherald and Manabe 1986),
- The Goddard Institute for Space Studies model (Hansen et al. 1988),
- The United Kingdom Meteorological Office (UKMO) model (Wilson and Mitchell 1987), and
- Empirical paleoreconstructions of the Holocene and Eemian interglacial epochs (Budyko and Izrael 1987).

General Circulation Models and paleoreconstructions use different metrics for projecting future climates. Except for transient versions, GCMs are used most frequently to calculate two equilibrium climates for normal CO_2 concentrations and a doubling of CO_2 concentrations ($2XCO_2$); paleoreconstructions are bound to a prescribed value of global air temperature increase. No definite correspondence exists between an increase in global air temperature and atmospheric CO_2 concentration; estimates of the former, known as climate sensitivity, show temperature increases varying between 1.5 and 4.5°C under $2XCO_2$ (Greco et al. 1994). Complicating the problem, GCMs use different spatial grids, while paleoreconstructions are bound to the points where the probes were taken.

To compare results from GCMs and paleoreconstructions, this study combined different climate change scenarios by scaling to a standardized value of 2°C global warming and reinterpolating the results to a common global grid. Both GCMs and paleoreconstructions calculate the climatic parameters over large geographic lattices. The resolution of GCMs is typically several degrees of latitude and longitude. Paleoreconstructions are usually available as predictive maps that show izorithms of the projected air temperature and precipitation; the accuracy of the mappings depends on the number of data points used in the reconstruction.

To improve the spatial resolution of the scenarios, in particular, those obtained from the GCMs, the following procedure was used. Differences between the projected and current temperatures and precipitation values were first computed at each node by using results from different models. These increments of the climatic parameters were then used with a detailed database of mean monthly air temperatures and precipitation for the current climate at each node of a global $0.5\text{-} \times 0.5$-degree grid (Leemans and Cramer 1991). The projected values of the climatic variables were thus calculated at each node of a fine grid as the sum of its current value and an increment taken from the nearest node of a coarse grid. This procedure not only improves the resolution of the climate change scenarios, but also minimizes errors in estimating the control climate, which could be transferred to the projected future climate.

The study calculated the distribution of permafrost in the Northern Hemisphere under 2°C global warming by using the frost-number-based model and various climate change scenarios. The permafrost retreat patterns were similar for all scenarios, indicating that all permafrost zones will be displaced several hundred kilometers to the north and northeast of their contemporary positions in both North America and Eurasia (Anisimov and Nelson 1995). Although the details of the projected permafrost configuration varied, different climate change scenarios suggest that the discontinuous permafrost would be reduced less than the continuous zone, especially when using the paleoreconstruction scenario.

These results, which show that the potential effect of climatic change on permafrost is substantial, were obtained by using 2°C warming scenarios derived from paleoreconstructions and GCM experiments on $2xCO_2$. A new class of transient GCM can estimate gradual climate changes in real time. The Intergovernmental Panel on Climate Change (IPCC) examined several models (Greco et al. 1994), and more realistic scenarios of climate change for the first half of the next century were developed.

Transient Climate Scenarios and Projected Changes in Permafrost

Three transient GCMs (GFDL-89 [Manabe et al. 1991], UKMO [Murphy and Mitchell, 1992], and the ECHAM-1A model from the Max-Planck Institute [Cubasch et al. 1992]) generated the transient climate for 2050. These models and the transient climate change experiments were well documented, and more computational details can be found in the references. Following the suggestion of the IPCC climate experts (IPCC 1990; Greco et al. 1994), the climate models were harmonized by adjusting their sensitivity to a mid-range 2.5°C temperature increase to the year 2100 under the standardized CO_2 emission scenario. The rate of the climate change was estimated by a one-dimensional climate model that produced a 1.16°C mean global air temperature increase for 2050 referenced to 1990 (Wigley and Raper 1992). More details on the procedures involved in harmonization can be found in Greco et al. (1994).

Figure 1 shows the projected permafrost distribution in the Northern Hemisphere by the middle of the 21st Century. The different panels correspond to transient climate change scenarios for 2050 derived from the GFDL-89, ECHAM1-A, and UKMO models. The solid lines show the southern limit of modern permafrost.

The results for North America, which were obtained from different models, are in better agreement than those for Eurasia, where three climate scenarios predict slightly different patterns of permafrost distribution. The UKMO scenario results in the most extreme reduction of permafrost, while both the GFDL-89 and ECHAM1-A scenarios produce only small changes in the permafrost distribution in Eurasia. The ECHAM1-A scenario even predicts some expansion of areas underlain by sporadic permafrost in Siberia (compare marked permafrost areas in Figure 1 with the southern limit of modern permafrost, shown as a solid line).

Socioeconomic Impacts of Permafrost Reduction

Many permafrost regions are known to be large sources of fossil fuel, and their economic development and infrastructure are strongly related to the gas and oil industries. The arctic slope of Alaska on the North American continent and most of West Siberia in Eurasia are examples of such regions. In these areas, permafrost is often used as a natural solid base for construction, and its retreat or warming could cause serious problems. Special measures should be taken to avoid permafrost melting around pipelines (TransAlaskan

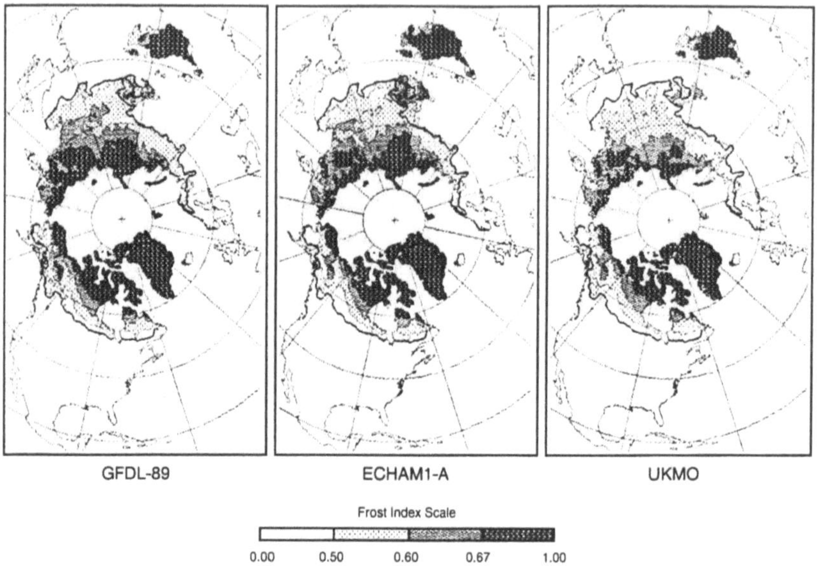

Figure 1. Comparison of Projected Permafrost Distribution in the Northern Hemisphere

in North America and TransSiberian in the Russian Federation), including artificial thermal isolation in areas where permafrost is critical for the reliability of the existing structure. Neglecting such measures could have catastrophic consequences, such as cracked pipelines, gas bursts due to uneven sedimentation, and pollution of the environment through gas and oil emissions.

Unlike the U.S. and Canada, which have practically no large cities in the permafrost terrain, the Russian Federation urbanized large territories in the sub-Arctic as its economy expanded to the north. As a result, buildings are based on permafrost in cities such as Norilsk. Warming and thawing of permafrost could become a major problem at such locations, and intensive monitoring of the subsurface thermal regime and artificial measures of permafrost preservation are necessary to avoid massive collapse of buildings in the future. Such measures involve heat-resistant covers (gravel, peat, sand), vegetation cover with selective heat conductivity (moss), reflective sun shields, and other devices that affect the surface heat balance.

The presence of ground ice is a critical factor in the response of permafrost to climate change, in particular, sedimentation. Ground ice is generally concentrated in the upper few meters of permafrost, precisely the layers that will thaw first if the permafrost degrades. The ice content tends to be highest in fine-grained surficial materials and is generally low in coarse sands, gravels, and bedrock. Wet valley floors tend to have a higher ice content than do dry ridge tops, but any given soil type or terrain unit may contain variations. Consequently, the response of the permafrost to climate change can be expected to vary greatly, even on a local scale. Accurate estimates of climate change and detailed subsurface data are necessary to predict the local response of permafrost to global warming.

In the context of the problems discussed here, it is important to develop a global system of geocryological monitoring and data management that will involve the most important information on the current state of the climate and permafrost, climatic scenarios, and models of climate-permafrost interaction. Such a system could provide a powerful tool for predicting changes in the permafrost under global warming. It would also be useful in developing strategies for adapting to climate changes in permafrost regions and solving problems in climate-permafrost interaction.

One such problem is associated with the release of methane (CH_4) from melting permafrost. Methane, having an extremely high potential for radiative forcing (about 20 times higher than that of an equal amount of CO_2), may significantly contribute to warming; the question, however, is whether this feedback to the climate system is important compared with other sources of methane.

Balancing its negative effects on engineering and construction in permafrost regions, climate change could create favorable conditions for developing agriculture, such as thickening of the active layer of soil, a longer warm period, and higher air temperatures. Warming would also allow for developing roads and transportation lines with traditional technologies. Generally, the peripheral territories of the modern cryolitozone, which are currently underlain by sporadic and widespread discontinuous permafrost, would be affected. Construction of transportation corridors in these areas could noticeably speed up their economic development.

These general conclusions were derived on the basis of current knowledge of cryological processes and natural phenomena in Arctic and sub-Arctic regions. To address these problems in more detail, further research on permafrost and its interaction with a changing climate is necessary.

References

Anisimov, O.A., and F.E. Nelson, 1990, "On the Application of Mathematical Models to Investigations of the Interaction between Climate and Permafrost," *Meteorology and Hydrology* 10:13-20 (in Russian).

Anisimov, O.A., and F.E. Nelson, 1995, "Permafrost Distribution in the Northern Hemisphere under Scenarios of Climate Change, *Global and Planetary Change* (in press).

Budyko, M.I., and Y.A. Izrael, 1987, *Anthropogenic Climatic Change,* Hydrometeoizdat, Leningrad, former Soviet Union (in Russian).

Budyko, M.I., N.A. Yefimova, and I.Y. Lokshina, 1990, "Anticipated Modifications of Global Climate," *Proceedings of the USSR Academy of Science, Geographical Service* 5:45-55 (in Russian).

Budyko, M.I., Yu. A. Izrael, M.C. McCracken, and A.D. Heckt, 1991, *Anticipated Climatic Change,* Hydrometeoizdat, Leningrad, former Soviet Union (in Russian).

Budyko, M.I., I.I. Borzenkova, G.V. Menzhulin, and K.I. Seljakov, 1992, "On Pending Changes of the Global Climate," *Proceedings of the Russian Academy of Science, Geographical Service* 4:36-52 (in Russian).

Cess, R.D., and G.L. Potter, 1988, "A Methodology for Understanding and Intercomparing Atmospheric Climate Feedback Processes within General Circulation Models," *Journal of Geophysical Research* 93:8305-8314.

Cubasch, U., K. Hasselmann, H. Hock, E. Maier-Reimer, U. Mikolajewicz, B.D. Santer, and R. Sausen, 1992, "Time-Dependent Greenhouse Warming Computations with a Coupled Ocean-Atmosphere Model," *Climate Dynamics* 8:55-69.

Greco, S., R. Moss, D. Viner, and R. Jenne (eds.), 1994, *Climate Scenarios and Socioeconomic Projections for IPCC WG II Assessment*, Consortium for International Earth Science Information Network.

IPCC, 1990, *Climate Change: The IPCC Scientific Assessment*, J.T. Houghton, G.L. Jenkins, and J.J. Ephraums (eds.), Intergovernmental Panel on Climate Change, World Meteorological Organization/U.N. Nations Environmental Programme, Cambridge University Press, Cambridge, United Kingdom.

Hansen, J., I. Fung, A. Lacis, D. Ring, S. Lebedeff, R. Ruedy, and G. Russell, 1988, "Global Climate Changes as Forecast by the Goddard Institute for Space Sciences Three Dimensional Model," *Journal of Geophysical Research* 93:9341-9363.

Harris, S.A., 1986, "Permafrost Distribution, Zonation and Stability along Eastern Ranges of the Cordillera of North America," *Arctic* 39:29-38.

Heginbottom, J.A., J. Brown, E.S. Melnikov, and O.J. Ferrians, 1993, "Circumarctic Map of Permafrost and Ground Ice Conditions," in *Proceedings of the 6th International Conference on Permafrost*, Beijing, People's Republic of China, July 5-9, 1993, Vol. 2, pp. 1132-1136, South China University of Technology Press, Wushan, People's Republic of China.

Klimanov, V.A., 1982, "Climate of Western Europe in the Holocene Climatic Optimum," in *Development of Nature in the USSR Territory in the Upper Pleistocene and Holocene,* pp. 251-258, Nauka, Moscow, USSR (in Russian).

Leemans, R., and W. Cramer, 1991, *The IIASA Database for Mean Monthly Values of Temperature, Precipitation, and Cloudiness on a Global Terrestrial Grid*, Research Report RR-91-18, International Institute for Applied Systems Analysis, Laxenburg, Austria.

Manabe, S., and R.T. Wetherald, 1980, "On the Distribution of Climate Change Resulting from an Increase in the CO_2 Content of the Atmosphere," *Journal of Atmospheric Science* 37:99-118.

Manabe, S., R.E. Stoufer, M.J. Spelman, and K. Bryan, 1991, "Transient Responses of a Coupled Ocean-Atmosphere Model to Gradual Changes of Atmospheric CO_2. Part 1: Annual Mean Response," *Journal of Climate* 4:785-818.

Murphy, J.M., and J.F.B. Mitchell, 1995, "Transient Response of the Hadley Centre Coupled Ocean-Atmosphere Model to Increasing Carbon Dioxide. Part II: Spatial and Temporal Structure of Response," *Journal of Climate.*

Nelson, F.E., and S.I. Outcalt, 1987, "A Computational Method for Prediction and Regionalization of Permafrost," *Arctic and Alpine Research* 19:279-288.

Nelson, F.E., and O.A. Anisimov, 1990, "Climate Change and Permafrost Distribution in the Soviet Arctic," in G. Weller and C. Wilson (eds.), *Abstracts of the International Conference on the Role of the Polar Regions in Global Change,* p. 159, University of Alaska Geophysical Institute, Fairbanks, Alaska, USA.

Nelson, F.E., and O.A. Anisimov, 1992, "Permafrost Zonation in Russia under Anthropogenic Climatic Change," *Permafrost and Periglasial Processes* 4:137-148.

Velichko, A.A., 1984, *Late Quaternary Environments of the Soviet Union*, University of Minnesota Press, Minneapolis, Minn., USA.

Velichko, A.A., 1985, "Empirical Paleoclimatology (Principles and Accuracy)," in *Methods of Reconstructions of Paleoclimates,* pp. 7-20, Nauka, Moscow, former Soviet Union (in Russian).

Wetherald, R.T., and S. Manabe, 1986, "An Investigation of Cloud Cover Change in Response to Thermal Forcing," *Climate Change* 8:5-25.

Wigley, T.M.L., and S.C.B. Raper, 1992, "Implications for Climate and Sea-level of Revised IPCC Emissions Scenarios," *Nature* 357:293-300.

Wilson, C.A., and J.F.B. Mitchell, 1987, "A Doubled CO_2 Climate Sensitivity Experiment with a Global Climate Model Including a Simple Ocean," *Journal of Geophysical Research* 92:13315-13348.

Zubakov, V.A., and I.I. Borzenkova, 1983, *Paleoclimates of the Upper Cenozoic*, Hydrometeoizdat, Leningrad, former Soviet Union (in Russian).

Health Adaptations to Climate Change: Need for Farsighted, Integrated Approaches

J.A. Patz
Division of Occupational and Environmental Medicine
John Hopkins School of Hygiene and Public Health
615 N. Wolfe Street, Room 7041
Baltimore, Maryland 21205-2179, USA

Abstract

The health effects attributable to climate change may include an increase in (1) heat-related mortality and morbidity, (2) infectious diseases, (3) malnutrition and dehydration, and (4) damage to the public health infrastructure. The monitoring of health-risk indicators associated with climate change must be improved, and the need to plan for adaptive measures is immediate. Adaptive options include improved and extended preventive medical care, disaster preparedness, protective technology, and appropriate professional training. Some of these options are energy-intensive (e.g., air conditioning) and will worsen greenhouse warming by creating a need for more energy, unless alternative energy sources are developed. Vector control is another response with potentially harmful side-effects. An integrated approach to environmental management is needed for realistic success. Many diseases may not respond to preventive actions; moreover, the growth of emerging epidemics in warmer environments will be difficult to anticipate. In light of these shortcomings, improved monitoring and research are needed to guide policymakers as they choose options for mitigating greenhouse gases in conjunction with implementing adaptive measures.

Introduction

According to the Intergovernmental Panel on Climate Change (IPCC), a doubling of carbon dioxide concentration ($2xCO_2$) in the atmosphere is expected to elevate the average global temperature by 1.5-4.5°C (IPCC 1992, 1994). Global warming is also anticipated to alter precipitation patterns, increase weather variability, and cause thermoexpansion of the oceans (IPCC 1992). Subsequent health effects may include an increase in (1) heat-related mortality and morbidity (Grant 1990; Kalkstein and Smoyer 1993); (2) infectious diseases, in particular those that are arthropod-borne (Longstreth and Wiseman 1989; Shope 1991; Dobson and Carper 1993; Rogers and Packer 1993; Reeves et al. 1994); (3) malnutrition and dehydration attributable to a threatened food and water supply (World Health Organization [WHO] 1990; Parry and Rosenzweig 1993); and (4) damage to the public health infrastructure as a result of weather disasters and a rise in sea level, which will be aggravated by climate-related forced human migration (Leaf 1989; Haines and Fuchs 1991).

Because of the scope of potential health consequences associated with climate change, mitigating greenhouse warming is a logical precaution. Implementing adaptive measures, however, is also in order, because the world may already be undergoing global warming, regardless of aggressive mitigation strategies. Adaptive measures are possible at different levels, including legislative, technological, or behavioral. Adaptive strategies must be well planned and should neither exacerbate global warming nor introduce new hazards. Ideally, efforts toward both mitigating and adapting to global warming need to be pursued in parallel.

Health Impacts Anticipated from Climate Change

Impact Assessment Methodology

To identify the health impacts requiring adaptation, an ecologically based human health assessment may be warranted because the way in which health effects will occur is complex. Except for heat-related mortality, many health effects anticipated from global climate change may not occur via the classical toxicological, metabolic, or infectious processes scientists are accustomed to studying (McMichael 1993; WHO, 1995). Figure 1 illustrates some of these more indirect pathways (Patz, 1995). The variability, duration, frequency, and timing of climatic events can contribute significantly to subsequent ecological change. Feedback loops and human and nonhuman adaptive responses are likely to affect the ultimate vulnerability of human health significantly because the time scales are so long. Assessments of vulnerability will, therefore, require interdisciplinary analysis and will involve integrated, predictive modeling approaches, in addition to empirical data (McMichael 1993). Thus, adaptive responses will need to be anticipatory.

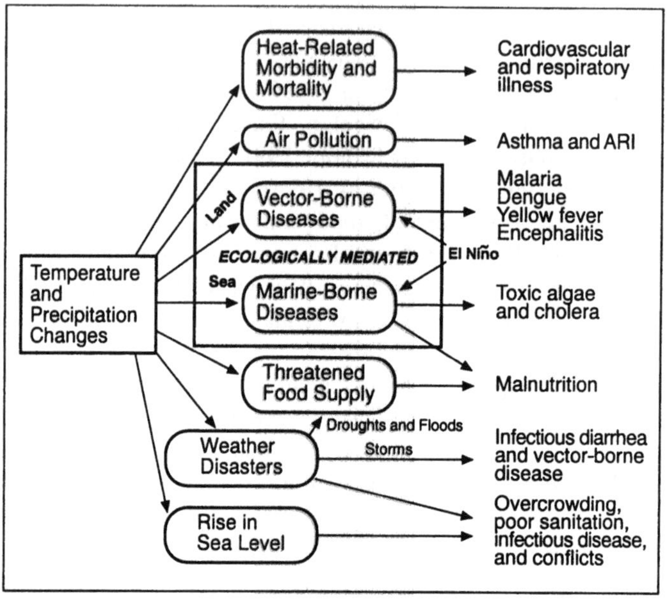

Figure 1. Causal Pathways of Impacts on Public Health from Global Climate Change

Heat-Related Morbidity and Mortality

Although humankind has a great capacity to adapt to adverse environmental conditions, there are physiological limits (Kilbourne 1989). Weather conditions exceeding threshold temperatures and persisting for several consecutive days can increase human mortality. In large urban areas, poor housing and the urban "heat island" effect further worsen conditions. Elevated nighttime temperature is the most significant meteorological variable contributing to heat-related mortality; and the greenhouse effect is predicted to especially affect minimum nighttime temperatures (Kalkstein and Smoyer 1993). Higher latitudes are expected to warm disproportionately more than tropical and subtropical zones (IPCC 1990).

Under global warming, the hazards to human health (e.g., the heat island effect and harmful air pollution) may become significant public health problems, given current trends in urbanization. Warmer temperatures combined with increased ambient ultraviolet (UV) radiation could worsen air pollution, especially over urban areas. The UV photodecomposition of nitrogen oxides in the presence of volatile organic compounds produces tropospheric (ground-level) ozone, or photochemical smog. Tropospheric ozone is a potent pulmonary irritant (Grant 1990).

Infectious Diseases

Infectious agents that cycle through cold-blooded insect vectors to complete their development are quite susceptible to subtle climate variations (Dobson and Carper 1993). Temperature influences vector infectivity by affecting pathogen replication, maturation, and the period of infectivity (Longstreth and Wiseman 1989). Also, elevated temperature and humidity intensify the biting behavior of most insects. Examples of diseases associated with pathogens, vectors, or hosts affected by climate change include malaria, dengue (or "breakbone fever"), and arboviral encephalitis.

Floods and hurricanes also significantly affect vector-borne diseases. With more aquatic breeding sites for mosquitoes, the population densities of insects are higher. Moreover, with the destruction of shelters, human populations become more vulnerable to contact with infective mosquitoes. Simultaneously, infected humans exposed to the elements become a more readily accessible reservoir of disease, thereby fueling epidemics spread by mosquitoes feeding on humans.

Globally, malaria is the most prevalent vector-borne disease and causes 1-2 million deaths annually (Institute of Medicine 1991). The control of malaria has been disappointing, and the annual number of cases is increasing. Treatment has not improved because the most virulent strain, *Plasmodium falciparum*, has become drug-resistant, and antimalarial vaccines have shown only limited efficacy. The malaria parasite cannot develop inside its mosquito host at temperatures below 16°C (Gilles 1993). Thus, unseasonably hot weather has been found to increase malaria transmission among humans in Pakistan (Bouma et al. 1994) and has been associated with outbreaks of malaria at higher altitudes in Rwanda (Loevinsohn 1994).

Over the past 15 years, dengue epidemics have increased in both number and severity, especially in tropical urban centers. Dengue hemorrhagic fever, a more serious variant of dengue usually associated with second infections of dengue virus, now ranks as one of the leading causes of hospitalization and mortality in children in southeast Asia (Institute of Medicine 1992). Dengue is on the rise in the Americas. Thus far, the range of the primary mosquito vector, *Aedes aegypti*, has been limited because freezing temperatures kill both larvae and adults. However, warming trends may shift vector and disease distribution to higher latitudes and altitudes.

The prevalence of dengue appears to be sensitive to climatic factors. In Mexico, in 1986, the median temperature during the rainy season was found to be the most important predictor of dengue transmission (Koopman et al. 1991). In the laboratory, increasing the temperature by only 2-5°C quickened virus development inside the mosquito, such that dengue transmission rates could be expected to rise by threefold (Watts et al. 1987; Koopman et al. 1991).

Marine Effects

Phytoplankton (or algae) blooms in marine and freshwater environments respond to poor erosion-control management, liberal agricultural application of fertilizers, and coastal sewage release, all of which result in effluents rich in nutrients. Warmer temperatures at the sea's surface, as would be expected with global warming, could augment this growth (Epstein et al. 1993). Harmful marine phytoplankton blooms (including red tides) cause diarrheal and paralytic diseases and amnesia attributable to shellfish poisoning. Marine zooplankton, which feed upon phytoplankton, have been found to harbor *Vibrio cholera*; thus, phytoplankton blooms could expand as an important source of cholera epidemics (Huq et al. 1990).

Food Supply

Climate change could adversely affect agriculture both by long-term changes (such as reducing soil moisture through evapotranspiration) and, more immediately, by extreme weather events (such as droughts, flooding [and erosion], and tropical storms). "CO_2 fertilization," which can enhance photosynthesis, may initially benefit plants (IPCC 1990). Nonetheless, one study found that agriculture in developing countries may suffer most; in these nations, as many as 300 million additional people will be at risk from hunger as a result of climate change attributable to 2xCO_2 (Parry and Rosenzweig 1993). Many coastal settlements depend on fish as their main food. A rapid rise in sea level caused by global warming (see next section) is expected to damage the coral reefs and estuaries that support many of today's fisheries, which are already threatened by overfishing (IPCC 1990).

Disruptive Weather and Sea-Level Rise

A rise in sea-level and increased thermal energy in the atmosphere are expected to produce more severe storms. For example, the 2XCO_2 levels could increase hurricane strength by as much as 40% (IPCC 1990). Such events would directly disrupt dwellings and infrastructures for maintaining public health, such as sanitation and storm-water-drainage systems (IPCC 1992). Vulnerable populations in low-lying coastal areas and small islands would be forced to migrate to safer locations. Infectious diseases could spread more rapidly among these environmental refugees as a result of overcrowding and poor sanitation, and native populations would be at risk from potential influxes of infected individuals (WHO 1990). Flooded water-drainage systems may further exacerbate the situation. Psychological impacts of storms must also be considered, because post-traumatic stress syndrome often occurs after major storms.

The intrusion of saline water into coastal aquifers could diminish the supply of freshwater and reduce the amount of coastal farmland. For example, a 1-meter rise in sea level would destroy 15% and 20% of agriculture in Egypt and Bangladesh, respectively (IPCC 1990).

Adaptive Strategies

Assuming that some global warming is inevitable, it would be advantageous to consider adaptive strategies to cope with anticipated climatic and subsequent ecological changes, regardless of best efforts at mitigation. Anticipatory adaptive strategies are preferred over reactive measures, given the inertia of the climate system and essential irreversibly of climate change (Table 1). Yet, even without climate change, much of the world's population already suffers from ill health because of unsatisfactory living conditions and poor access to health care. Many of the adaptive strategies described in this paper are not specific to climate change — in fact, they should not be viewed in isolation or out of context of (1) the broader problem of global environmental degradation and (2) a public health infrastructure that, in much of the developing world, has already been compromised. For example, many poor populations suffer from infectious diarrheal diseases because of a lack of sanitary practices or available technology. Similarly, population-wide malnutrition is more of a problem related to food distribution than to food availability.

Conventional preventive strategies in public health are classified as primary, secondary, or tertiary (Figure 2). Primary prevention refers to avoiding or removing a hazardous exposure or protecting individuals so that exposure to the hazard is of no consequence. Examples include removing lead paint from a home or administering childhood vaccinations. Secondary prevention, somewhat "downstream" in the disease pathway, involves early detection (or screening) of an altered human physiologic state and subsequent intervention that averts full progression to disease. Finally, tertiary prevention, which in practical terms is treatment rather than prevention, attempts to minimize the adverse effects of the already present disease. The farther "upstream" an intervention can be implemented, generally, the greater the potential health benefit to the largest number of individuals.[1]

---

[1] Global climate dynamics and global ecological change have received little consideration by environmental and public health experts. Greenhouse gas mitigation adds another level within the progression from exposure to disease outcome to which preventive steps may be applied. Mitigating the process of global warming might, therefore, be viewed as "pre-primary" public health prevention.

Table 1. Examples of Multilevel Adaptive Measures for Some Anticipated Health Outcomes of Global Climate Change

Adaptive Measure	Responses to Climate-Related Problems: Heat-Related Illness	Vector-Borne Diseases	Health and Extreme Weather Events
Administrative /Legislative	Implement weather watch/warning systems Plant trees in urban areas Implement education campaigns	Implement vaccination programs Enforce vaccination laws Monitor breeding sites Implement education campaigns to eliminate breeding sites	Create disaster preparedness programs Employ land-use planning to reduce flash floods Ban precarious residential placements Enforce strict building codes
Engineering	Insulate buildings Install high-albedo materials for roads	Install window screens Release sterile male vectors	Construct strong seawalls Fortify sanitation systems
Personal Behavior	Maintain hydration Schedule work breaks during peak daytime temperatures	Use topical insect repellents Use pyrethroid-impregnated bed nets	Heed weather advisories

Adaptive measures should neither exacerbate the primary hazardous exposure (global warming) nor introduce new hazards. In essence, the "cure" must not be worse than the "disease." Some options are energy-intensive (e.g., air conditioning) and will worsen greenhouse warming by increasing the demand for energy, unless alternative energy sources are developed (air conditioning may be financially infeasible for much of the world anyway). Vector control is another response with potentially harmful side effects. Narrowly focused policies to combat rising insect populations, either by spraying pesticides or by draining wetlands, can have serious human health and environmental ramifications; the use of pesticides (primarily for agriculture) is estimated to cost between $100 billion and $200 billion annually worldwide (Pimental 1990). Finally, it will also be important to assess in advance any risks to health from proposed technological adaptations (e.g., risk from nuclear waste or accidents).

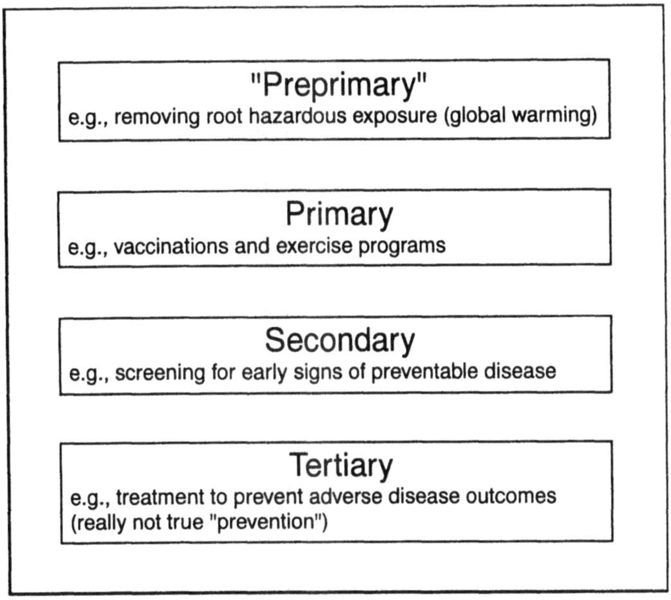

Figure 2. Levels of Health Prevention

Hierarchy of Controls and Impact-Specific Adaptive Strategies

Hierarchy of Controls

Abiding by a "hierarchy of controls" can help to develop a strategy for preventing as much human illness and suffering as possible (Figure 3). Control options range from population-wide or technology-based to steps taken by individuals. Measures fall into the sequential categories of (1) administrative or legislative, (2) engineering, and (3) personal (behavioral). As a rule, legislation can potentially affect the largest national or international populations, and directives at this level may result in either engineering or personal behavioral changes that benefit large numbers of people. Engineering advances, independent of administrative or legislative mandates, may bring substantial benefits as well. For example, advances in sanitary treatment facilities have likely prevented an enormous burden of illness globally. Of course, in many instances, legislative or administrative mandates may be required for effective dissemination and population-wide adoption of this technology to reap the potential benefits. Personal behavior change is at the bottom of the hierarchy of controls. Individual responses to health threats can be highly varied and unreliable. Also, misperceptions of relative health risks

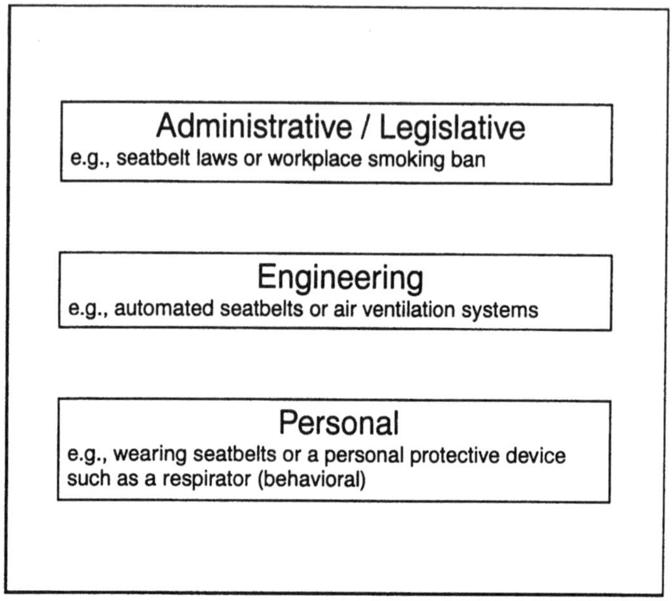

Figure 3. Hierarchy of Control Measures

are rampant in the public. As a result, only select groups within a population may take appropriate preventive steps.

Impact-Specific Adaptation Strategies

The following adaptive strategies are presented as an incomplete list of examples directed toward some of the anticipated health impacts (WHO, 1995). Cost-benefit considerations have not yet been applied. Nevertheless, as expressed in the preceding section, population-wide administrative or engineering interventions generally protect greater numbers of individuals. Ideally, adaptive strategies should apply to current health crises in addition to longer-term health challenges presented by climate change.

For direct heat-related illness:

- Insulating buildings and other design features that reduce thermal loads.
- Planting trees within cities and selecting high-albedo materials for roads, parking lots, and roofs to reduce the urban heat island effect.
- Establishing new weather watch/warning systems that focus on health-related adverse conditions, such as oppressive air masses.
- Creating public education campaigns regarding precautions to take during heat waves.

- Implementing work schedules that avoid peak daytime temperatures for outdoor laborers.

For vector-borne diseases:

- Installing window screens in areas endemic to insect-borne diseases.
- Expanding vaccination programs aimed at infectious agents likely to increase with climate change, such as yellow fever (unfortunately, no such vaccines exist for some diseases most sensitive to climate change, such as malaria and dengue).
- Increasing public education to encourage elimination of man-made breeding sites (small water containers).
- Implementing education campaigns to sensitize health-care givers in geographically vulnerable regions.
- Releasing sterilized male insects to reduce the reproductive capacity of vector populations.
- Promoting the use of pyrethroid-impregnated bed nets to deter mosquitoes.

For agricultural stresses:

- Producing climate-resistant transgenic plants (genetically engineered).
- Reducing the proportion of monocultural farming for better crop resistance to pests.
- Promoting land reforms that would favor environmentally sound land use.

For weather extremes and sea-level rise:

- Maintaining disaster preparedness programs, including tools for local public health facilities to assess health needs rapidly.
- Implementing engineering measures (such as strengthening seawalls) or requiring building contractors to follow hurricane standards in coastal areas.
- Adopting land-use planning to minimize erosion, flash-flooding, and precarious residential placements.

More explicit measures might be found in the recommendations of forthcoming documents from the IPCC. When considered in total, these adaptive measures will offer varying amounts of protection to human health, and critical evaluation of implemented strategies will need to be pursued early. Many diseases may not be amenable to preventive actions (Haines and Parry 1993). For instance, expanded distribution of vaccine may stem the spread of potential yellow fever epidemics, but vaccines are not available for trypanosomiasis or dengue. Similarly, parasite and vector resistance to therapeutic agents and pesticides, respectively, presents an ongoing challenge. Moreover, the unforeseen growth in emerging epidemics in a warmer environment will be difficult to anticipate. In light of such shortcomings, monitoring and surveillance should be emphasized to improve risk assessment, allow early intervention, and help guide policy decisions.

Surveillance and Monitoring

Improvements in risk assessment and monitoring represent some of the most important adaptive measures. If strategies for determining key biological or ecological indicators of subsequent health risk can be established, more upstream "primary" intervention becomes feasible. Monitoring the populations of vector species, for instance, may reveal early influences of climate change that would not yet have resulted in human disease, since insect and protozoan species react more quickly to changes in their environment. Monitoring these health-related bioindicators will require unprecedented multidisciplinary coordination. Yet, deficiencies in established human health surveillance systems must continue to take precedence over the monitoring of far less understood human health/ecological relationships.

One focus might be detecting changes in climate-driven infectious diseases in sensitive geographic regions that border endemic zones. Establishing sentinel diagnostic centers in such regions should be considered. Such detection will not only provide an early warning for possibly more pervasive changes in disease behavior and allow for intervention, but it will also enrich our knowledge of the climate dependency of diseases and help us create better predictive models.

Monitoring changes in habitats that influence disease-carrying insect or mammal populations and/or infectivity represents another surveillance approach that could be easy to perform and provide earlier warning of the spread of vector-borne diseases. Although such surveillance may not answer questions of vulnerability in the short term, it may provide the quantitative long-term data through which more accurate models can emerge. To the extent that early indicators or "precursors" of disease are monitored, increased opportunities for even earlier preventive measures might result.

A geographic method for organizing and storing data will be essential because much of the health risk analysis will require (1) superimposing data on disease incidence, vector populations, demographics, and climate and (2) linking this information to specific geographic locations. The use of geographic information systems (GISs) for predicting the incidence of disease is new and still under development.

Remote sensing (or remote satellite imaging) can be useful in analyzing diseases when their distribution or epidemiology depends on climate and landscape features that determine habitats. Vector-borne diseases are particularly influenced by spatial and temporal fluctuations in vector populations. Vegetation partly determines vector populations, since plant composition in a region reflects the aggregate effect of temperature, precipitation, and humidity (Washino and Wood 1994). Satellite-generated vegetative index or habitat maps have identified areas receptive to dengue fever and Lyme disease in the United States, schistosomiasis in the Philippines, and malaria in rice-growing regions (Washino and Wood 1994).

In combination, GIS and remote-sensing technology may serve as a useful analytical tool. GIS allows one to overlay maps with multiple components, some of which are easily acquired via satellite. For example, Landsat images can discern standing pools of water and locate human settlements (Epstein et al. 1993). When superimposed over maps of climatological, entomological, and epidemiological data, enhanced integrated analysis may be possible.

Mitigation and the Precautionary Principle

The precautionary principle holds that under conditions of uncertainty, if a suspected threat is substantial enough, actions should be taken in advance to avert potentially harmful future consequences. In addition to anticipatory adaptive strategies, precautionary measures reduce the root hazardous exposure (greenhouse warming) that initiates the detrimental processes leading to impacts on human health. Discussion of these mitigating steps is beyond the scope of this paper. Considering the shortcomings of many of the adaptive measures described in this paper and the uncertainty of the magnitude of potential health effects, complacency would be inappropriate, even if all the described measures were economically and technologically feasible.

With more people in the world, the demand for energy is greater and, consequently, emissions of greenhouse gases are higher. As mentioned above, the developed world, which represents only 25% of the world's population, emits 75% of the anthropogenic greenhouse gases. Therefore, conservation measures in these affluent nations must supersede any rationale for concomitant, environmentally related population policy. Population policy does become more important for long-range planning, if one assumes the inevitable industrialization of the third world. The disproportionate industrial growth expected in developing countries may accelerate stresses on the environment (World Bank 1992).

Precautions aimed at reducing greenhouse warming will have inherently beneficial near-term effects for public health. Reducing air pollution and population growth, for example, have obvious positive influences on health. Such short-term gains weigh favorably in the policy debate, which must consider potentially irrevocable major health consequences simultaneously with vast uncertainties. Yet, just as deleterious side effects of adaptive measures must be minimized, the costs of mitigation must be evaluated. These costs are particularly important in impoverished countries, where energy resources still need to be developed.

Conclusions

The health effects expected from climate change may prove difficult to define accurately, and they pose a challenge to the planning of anticipatory adaptive strategies. Impacts may result from changes in agriculture and natural resources (such as freshwater or fisheries) or via multiple species within ecosystems (such as rodents or insect vectors). Except for heat waves or weather disasters, most of these ecologically mediated threats will involve complex ecological feedbacks in nonlinear systems.

Natural, economic, and demographic factors must be integrated into a systems-based health risk assessment before any adaptive measures are initiated. Interdisciplinary links for promoting an integrated assessment need to be advocated.

Adaptive measures can occur at legislative, technological, or behavioral levels. Adaptive strategies must be well planned and should neither exacerbate global warming nor introduce new hazards. Ideally, efforts toward both mitigation and adaptation will be pursued in parallel. Finally, policymakers should be included at all levels of the analysis to help adopt appropriate adaptive steps. Such cooperation will enhance initial problem formulation and, ultimately, the risk communication necessary for making informed decisions affecting the health of the public.

Acknowledgment

This paper was funded by the U.S. Environmental Protection Agency's Office of Policy, Planning and Evaluation, Cooperative Agreement number 823143010.

References

Bouma, M.J., H.E. Sondorp, and H.J. van der Kaay, 1994, "Climate Change and Periodic Epidemic Malaria," *Lancet* 343:1440.

Dobson, A., and R. Carper, 1993, "Biodiversity," *Lancet* 342:1096-1099, Oct.

Epstein, P.R., T.E. Ford, and R.R. Colwell, 1993, "Marine Ecosystems," *Lancet* 342:1216-1219, Nov.

Epstein, P.R., D.J. Rogers, and R. Slooff, 1993, "Satellite Imaging and Vector-Borne Disease," *Lancet* 341:1404-1406.

Gilles, H.M., 1993, "Epidemiology of Malaria," in H.M. Gilles and D.A. Warrell (eds.), *Bruce-Chwatt's Essential Malariology*, Edward Arnold Division of Hodder & Stoughton, London, United Kingdom.

Grant, L.D., 1990, "Respiratory Effects Associated with Global Climate Change," in J.C. White (ed.), *Global Atmospheric Change and Public Health*, Elsevier, New York, N.Y., USA.

Haines, A., and C. Fuchs, 1991, "Potential Impacts on Health of Atmospheric Change," *Journal of Public Health Medicine* 13(2):69-80.

Haines, A., and M. Parry, 1993, "Potential Effects on Health of Global Warming," *World Resource Review* 5(4):430-48.

Huq, A., R.R. Colwell, and R. Rahman, 1990, "Detection of *Vibrio cholerae* 01 in the Aquatic Environment by Fluorescent-Monoclonal Antibody and Culture Methods," *Applications of Environmental Microbiology* 56:2370-2373.

Institute of Medicine, 1991, *Malaria: Obstacles and Opportunities*, National Academy Press, Washington, D.C., USA.

Institute of Medicine, 1992, *Emerging Infections: Microbial Threats to Health in the United States*, National Academy Press, Washington, D.C., USA.

IPCC, 1990, *Climate Change: The IPCC Impacts Assessment*, Intergovernmental Panel on Climate Change, Australian Government Publishing Service, Canberra, Australia.

IPCC, 1992, *Climate Change 1992: The Supplementary Report to the IPCC Impacts Assessment*, Intergovernmental Panel on Climate Change, Australian Government Publishing Service, Canberra, Australia.

IPCC, 1992, *Climate Change 1992: The Supplementary Report to the IPCC Scientific Assessment*, Intergovernmental Panel on Climate Change, Cambridge University Press, Cambridge., United Kingdom.

IPCC, 1994, *Radiative Forcing of Climate Change, the 1994 Report of the Scientific Assessment Working Group of IPCC: Summary for Policy Makers*, Intergovernmental Panel on Climate Change, Oxford University Press, Oxford, United Kingdom.

Kalkstein, L.S., and K.E. Smoyer, 1993, "The Impact of Climate Change on Human Health: Some International Implications," *Experiencia* 49:469-479.

Kilbourne, E.M., 1989, *Heat Waves. The Public Health Consequences of Disasters*, M.B. Gregg (ed.), pp. 51-61, U.S. Department of Health and Human Services, Public Health Service, Centers of Disease Control, Atlanta, Georgia, USA.

Koopman, J.S., et al., 1991, "Determinants and Predictors of Dengue Infection in Mexico," *American Journal of Epidemiology* 133:1168-1178.

Leaf, A., 1989, "Potential Health Effects of Global Climatic and Environmental Changes," *New England Journal of Medicine* 321:1577-1583.

Loevinsohn, M., 1994, "Climatic Warming and Increased Malaria Incidence in Rwanda," *Lancet* 343:714-718, March.

Longstreth, J., and J. Wiseman, 1989, "The Potential Impact of Climate Change on Patterns of Infectious Disease in the United States," in J.B. Smith and D.A. Tirpak (eds.), *The Potential Effects of Global Climate Change in the United States*, document 230-05-89-057, Appendix G, U.S. Environmental Protection Agency, Washington, D.C., USA.

McMichael, A.J., 1993, "Global Environmental Change and Human Population Health: Conceptual and Scientific Challenge for Epidemiology," *International Journal of Epidemiology* 22(1):1-8.

Parry, M.L., and C. Rosenzweig, 1993, "Food Supply and the Risk of Hunger," *Lancet* 342:1345-1347.

Patz, J.A. (in press), "Global Climate Change and Public Health," in G. Shahi, B. Levy, T. Kjellstrom, R. Lawrence, and A. Binger (eds.), *International Perspectives in Environment, Development and Health*, Springer Publications, New York, N.Y., USA.

Pimental, D., 1990, *Estimated Annual Worldwide Pesticide Use. Facts and Figures: International Agricultural Research,* The Rochefeller Foundation, New York, N.Y., USA.

Reeves, W.C., J.L. Hardy, W.K. Reisen, and M.M. Milby, 1994, "The Potential Effect of Global Warming on Mosquito-Borne Arboviruses," *Journal of Medical Entomology* 31(3):323-332.

Rogers, D.L., and M.J. Packer, 1993, "Vector-Borne Diseases, Models, and Global Change," *Lancet* 342:1282-1284.

Shope, R.E., 1991, "Global Climate Change and Infectious Diseases," *Environmental Health Perspectives* 96:171-174, Dec.

Washino, R.K., and B.L. Wood, 1994, "Application of Remote Sensing to Arthropod Vector Surveillance and Control," *American Journal of Tropical Medical Hygiene* 50(6):134-144.

Watts, D.M., D.S. Burke, and B.A. Harrison, 1987, "Effect of Temperature on the Vector Efficiency of *Aedes aegypti* for Dengue 2 Virus," *American Journal of Tropical Medical Hygiene* 36:143-152.

WHO, 1990, *Potential Health Effects of Climatic Change,* World Health Organization, Geneva, Switzerland.

WHO (1995), *Climate and Health in a Changing World,* World Health Organization, Geneva, Switzerland.

World Bank, 1992, *World Development Report 1992: Development and the Environment,* Oxford University Press, Oxford, United Kingdom.

Index

MIX
Papier aus verantwortungsvollen Quellen
Paper from responsible sources
FSC® C105338

If you have any concerns about our products,
you can contact us on
ProductSafety@springernature.com

In case Publisher is established outside the EU,
the EU authorized representative is:
Springer Nature Customer Service Center GmbH
Europaplatz 3, 69115 Heidelberg, Germany

Printed by Libri Plureos GmbH
in Hamburg, Germany